The Long-Range Atmospheric Transport of Natural and Contaminant Substances

NATO ASI Series

Advanced Science Institutes Series

A Series presenting the results of activities sponsored by the NATO Science Committee, which aims at the dissemination of advanced scientific and technological knowledge, with a view to strengthening links between scientific communities.

The Series is published by an international board of publishers in conjunction with the NATO Scientific Affairs Division

A Life Sciences	Plenum Publishing Corporation
B Physics	London and New York
C Mathematical	Kluwer Academic Publishers
and Physical Sciences	Dordrecht, Boston and London
D Behavioural and Social Sciences	
E Applied Sciences	
F Computer and Systems Sciences	Springer-Verlag
G Ecological Sciences	Berlin, Heidelberg, New York, London,
H Cell Biology	Paris and Tokyo

The Long-Range Atmospheric Transport of Natural and Contaminant Substances

edited by

Anthony H. Knap

Bermuda Biological Station for Research,
Inc., Ferry Reach, Bermuda

and

Mary-Scott Kaiser

Technical Editor

SPRINGER-SCIENCE+BUSINESS MEDIA, B.V.

Proceedings of the NATO Advanced Research Workshop on
The Long-Range Atmospheric Transport of Natural and Contaminant
Substances from Continent to Ocean and Continent to Continent
St. Georges, Bermuda
January 10–17, 1988

Library of Congress Cataloging in Publication Data

NATO Advanced Research Workshop on the Long-Range Atmospheric
   Transport of Natural and Contaminant Substances from Continent to
   Ocean and Continent to Continent (1988 : Saint George, Bermuda
   Islands)
     The long-range atmospheric transport of natural and contaminant
   substances : proceedings of the NATO Advanced Research Workshop on
   the Long-Range Atmospheric Transport of Natural and Contaminant
   Substances from Continent to Ocean and Continent to Continent, held
   in St. Georges, Bermuda, January 10-17, 1988 / edited by Anthony H.
   Knap ; technical editor, Mary-Scott Kaiser.
         p.   cm. -- (NATO ASI series. Series C, Mathematical and
   physical sciences ; no. 297)
     "Published in cooperation with NATO Scientific Affairs Division."

     1. Air--Pollution--Congresses. 2. Atmospheric diffusion-
   -Congresses. 3. Transboundary pollution--Congresses.   I. Knap,
   Anthony H., 1949-   . II. Kaiser, Mary-Scott. III. Title.
   IV. Series.
   TD881.N37   1988
   628.5'3--dc20   89-26706

ISBN 978-94-010-6712-6   ISBN 978-94-009-0503-0 (eBook)
DOI 10.1007/978-94-009-0503-0

Printed on acid-free paper

This book contains the proceedings of a NATO Advanced Research Workshop held within the programme of activities of the NATO Special Programme on Global Transport Mechanisms in the Geo-Sciences running from 1983 to 1988 as part of the activities of the NATO Science Committee.

Other books previously published as a result of the activities of the Special Programme are:

BUAT-MENARD, P. (Ed.) – *The Role of Air-Sea Exchange in Geochemical Cycling* (C185) 1986

CAZENAVE, A. (Ed.) – *Earth Rotation: Solved and Unsolved Problems* (C187) 1986

WILLEBRAND, J. and ANDERSON, D.L.T. (Eds.) – *Large-Scale Transport Processes in Oceans and Atmosphere* (C190) 1986

NICOLIS, C. and NICOLIS, G. (Eds.) – *Irreversible Phenomena and Dynamical Systems Analysis in Geosciences* (C192) 1986

PARSONS, I. (Ed.) – *Origins of Igneous Layering* (C196) 1987

LOPER, E. (Ed.) – *Structure and Dynamics of Partially Solidified Systems* (E125) 1987

VAUGHAN, R. A. (Ed.) – *Remote Sensing Applications in Meteorology and Climatology* (C201) 1987

BERGER, W. H. and LABEYRIE, L. D. (Eds.) – *Abrupt Climatic Change – Evidence and Implications* (C216) 1987

VISCONTI, G. and GARCIA, R. (Eds.) – *Transport Processes in the Middle Atmosphere* (C213) 1987

SIMMERS, I. (Ed.) – *Estimation of Natural Recharge of Groundwater* (C222) 1987

HELGESON, H. C. (Ed.) – *Chemical Transport in Metasomatic Processes* (C218) 1987

CUSTODIO, E., GURGUI, A. and LOBO FERREIRA, J. P. (Eds.) – *Groundwater Flow and Quality Modelling* (C224) 1987

ISAKSEN, I. S. A. (Ed.) – *Tropospheric Ozone* (C227) 1988

SCHLESINGER, M.E. (Ed.) – *Physically-Based Modelling and Simulation of Climate and Climatic Change* 2 vols. (C243) 1988

UNSWORTH, M. H. and FOWLER, D. (Eds.) – *Acid Deposition at High Elevation Sites* (C252) 1988

KISSEL, C. and LAY, C. (Eds.) – *Paleomagnetic Rotations and Continental Deformation* (C254) 1988

HART, S. R. and GULEN, L. (Eds.) – *Crust/Mantle Recycling at Subduction Zones* (C258) 1989

GREGERSEN, S. and BASHAM, P. (Eds.) – *Earthquakes at North-Atlantic Passive Margins: Neotectonics and Postglacial Rebound* (C266) 1989

MOREL-SEYTOUX, H. J. (Ed.) – *Unsaturated Flow in Hydrologic Modeling* (C275) 1989

BRIDGWATER, D. (Ed.) – *Fluid Movements – Element Transport and the Composition of the Crust* (C281) 1989

LEINEN, M. and SARNTHEIN, M. (Eds.) – *Paleoclimatology and Paleometeorology: Modern and Past Patterns of Global Atmospheric Transport* (C282) 1989

ANDERSON, D.L.T. and WILLEBRAND, J. (Eds.) – *Ocean Circulation Models: Combining Data and Dynamics* (C284) 1989

BERGER, A., SCHNEIDER, S. and DUPLESSY, J. Cl. (Eds.) – *Climate and Geo-Sciences* (C285) 1989

TABLE OF CONTENTS

Participants

Richard Arimoto
Roger Atkinson
Elliot L. Atlas
Leonard A. Barrie
Terry Bidleman
Bernard Bonsang
Patrick Buat-Ménard
Katheryn Burns
Renato A. C. Carvalho
Thomas M. Church
Antonio Cruzado
Frank Dehairs
Jan Duinker
Francois Dulac
Manfred Ehrhardt
Anton Eliassen
Bernard E. A. Fisher
James N. Galloway
Krystyna Gorzelska
Robert Harriss

Donald R. Hastie
Neils Z. Heidam
Ruprecht Jaenicke
Timothy D. Jickells
William C. Keene
Anthony H. Knap
Hiram Levy, II
Leon Mart
John Miller
Jennie L. Moody
Joseph M. Prospero
Jochen Rudolph
Alexey G. Ryaboshapko
Dennis L. Savoie
Lothar Schutz
William T. Sturges
Shinsuke Tanabe
Douglas M. Whelpdale
William H. Zoller

LIST OF FIGURES

xiv

LIST OF TABLES

We have been increasingly aware of the role of long-range atmospheric
transport in the transfer of man-made and natural substances around the
globe. The nuclear accident at Chernobyl highlighted the fact that one
nation's atmospheric emissions may profoundly affect its neighbor. Over
the past few years, we have gained insight into the processes governing
long-range atmospheric transport. Many international and interdisciplin-
ary research groups are already working to unravel the many problems in
this field and several new global research programs have just begun (such
as, the Joint Global Ocean Flux Study [JGOFS], the World Ocean Circulation Experiment [WOCE], and the International Geosphere/Biosphere Program
[IGBP]). It was, therefore, timely to organize a workshop where-in those
presently working in the field could assess the present knowledge in the
field of long-range transport, identify the gaps in our knowledge, and
make recommendations for future research.

The NATO Advanced Research Workshop, ''The Long-Range Atmospheric
Transport of Natural and Contaminant Substances from Continent to Ocean
and Continent to Continent,'' was held at the Bermuda Biological Station
for Research, Inc., in St. Georges, Bermuda, from 10-17 January 1988.
Thirty-nine scientists best known for their work in the field attended
the workshop. This volume contains the background papers, case studies,
and findings of this group of world-reknown scientists. The workshop was
organized along the lines of the successful NATO ARW entitled ''The Biogeochemical Cycling of Sulfur and Nitrogen in the Remote Atmosphere''
that was held at the Bermuda Biological Station in 1984.

Scientists worked in groups according to compound classes of acids,
trace elements, organics, and mineral aerosols. Each group was charged
with assessing the present knowledge about emissions, transport, transformation, and deposition; the aim was to focus on long-range transport
(i.e., transport over 1,000 km). Approximately three months before the
workshop, a background paper written by the chairman of each working
group was sent to each participant. At the workshop, background papers
were presented on the atmospheric processes that govern large-scale
transport; case histories were also presented. Four synthesis chapters
containing the summaries of each subgroup's deliberations and consensus,
a concluding chapter, and a chapter on the gaps and recommended research
have been added to these papers to complete this volume.

This book should be of interest to those scientists who wish to
understand some of the complexities of long-range transport and to those
who wish to gain an overall assessment of our present knowledge of the
problems involved. Individuals or agencies concerned with the scientific
and political policy of how one country's emissions can affect another
country should also be interested in reading this publication.

The main support for our workshop was supplied by the Scientific
Affairs Division of the North Atlantic Treaty Organization with
supplemental funding from the U. S. National Oceanic and Atmospheric
Administration, the EXXON Corporation, the Bermuda Biological Station for

Research, Inc., the Bermuda Government, and the W. Alton Jones
Foundation.

As director of this workshop, I sincerely thank the Organizing Com-
mittee--E. L. Atlas, T. M. Church, J. N. Galloway, J. M. Prospero, and
D. M. Whelpdale--for its hard work and diligence. We greatly appreciated
all the help provided by the staff of the Bermuda Biological Station for
Research, Inc., as well as by Brenda Morris and Louise Cruden; their
attention and industry enabled our week to be especially productive. My
thanks to our illustrator, Barbara Wallace, for her major effort. The
Organizing Committee and I also thank Mary-Scott Kaiser for her organ-
izational help and for keeping everything and everyone straight as well
as for her efforts as the technical editor of this book. I am most
indebted to J. E. B. for continual support during this project. One last
point: The weather played as incredible a role this time as it did for
the previous NATO ARW meeting in 1984. It was so stormy that there was
no choice but to meet and write--no one got a Bermuda tan!!

Anthony H. Knap
Ferry Reach, Bermuda

ABSTRACT

Thirty-nine experts in the fields of long-range transport and meteorology agreed to participate actively in a workshop on the long-range atmospheric transport of natural and contaminant substances from continent to ocean and continent to continent. The four major compound classes addressed were trace metals, trace organics, nitrogen and sulfur compounds, and mineral aerosol. After an in-depth briefing covering the meteorological considerations required for an understanding of the long-range atmospheric transport, the state of knowledge concerning the emission, transport, transformation, and deposition of each compound was extensively discussed. Each participant was asked to bring his or her own data to the workshop. These data were discussed in the individual group meetings before being incorporated into the group-discussion chapters--sometimes as case histories, others as a general overview of the field. Gaps in our present knowledge and directions for future research were agreed to in the plenary sessions. A synthesis of the ''state of the art'' at the time was produced at the end of the workshop.

INTRODUCTION

A great deal of evidence supports the idea that long-range, large-scale atmospheric transport is important in moving both natural and man-made substances around the world. That much of the soil in Bermuda has been deposited by large-scale, long-range transport only attests to the scale of this process. The recent findings from the Pacific Asian Dust Network, which illustrate the importance of one major storm to the flux of material to the ocean, clearly demonstrate the magnitude of the research question. The Arctic Haze phenomenon, the transport of sulfur and nitrogen species out of North America to the Atlantic Ocean, and acid rain are all examples of how the emissions of one country or continent affect other areas. And yet, we know only a little about a few compounds!

This workshop was designed to determine the processes important to the large-scale, long-range atmospheric transport of four compound classes and to assess our present knowledge. Our discussions of long-range transport could not proceed without an understanding of the mechanics of emission, transport, transformation, and deposition. Therefore, a background paper distributed before the workshop provided a working document for assessing what was known about each compound class—acids, trace elements, organics, and mineral aerosols. For example, photochemistry is far more important when considering the transport and deposition of trace organics or nitrogen oxides than when considering the transport of mineral aerosols. Meteorology provided the link between different compound classes.

Examples of long-range transport highlighted the relationship of the processes to the four classes under consideration. The nuclear accident at Chernobyl made us all even more aware of how effective meteorological processes are in distributing material throughout the environment. Volcanic emissions provide natural sources of many geochemically important elements and further illustrate how essential it is that the relationships between the natural and anthropogenic sources of any element be resolved. The recent awareness of the ozone hole in the atmosphere of the Antarctic and the increases in atmospheric carbon dioxide further emphasize the importance of the global environment. This book provides the most current information available on the processes involved in the atmospheric transfer of various materials from continent to ocean and from continent to continent.

1. LARGE-SCALE METEOROLOGICAL REGIMES AND TRANSPORT PROCESSES

Douglas M. Whelpdale
Atmospheric Environment Service
4905 Dufferin Street
Downsview, Ontario M3H 5T4, Canada

Jennie L. Moody
Department of Environmental Sciences
Clark Hall, University of Virginia
Charlottesville, VA 22903, USA

1.1. INTRODUCTION

The major environmental issues of the day—acid rain, the Chernobyl accident, global warming—attest to the fact that many natural substances and
waste products from human activities are transported for large distances
through the atmosphere. The atmosphere serves both as a reservoir in
which chemical and physical reactions take place and as a pathway for
material to go from one geochemical reservoir to another. Two main
influences on the behavior of substances in the atmosphere are the properties of the substances themselves and the ambient meteorological
regime. For substances with relatively short atmospheric residence
times, ambient meteorological conditions during release into the atmosphere are important factors in the eventual disposition of the material.
For residence times of several days to a few years, the large-scale
meteorological regime exerts a major influence on the atmospheric distribution and removal of the material. For longer-lived species, the
continuing meteorological role becomes less significant once the material
is hemispherically or globally distributed. In this particular forum,
interest is primarily on the second case since this is the one of importance for the long-range transport from continent to ocean and continent
to continent.
        Several comprehensive reviews of the meteorology of transport and of
the atmosphere-ocean exchange of pollutants and other substances are
available. In 1975 the U. S. National Research Council convened a workshop to examine the ''Tropospheric Transport of Pollutants and Other
Substances to the Oceans,'' a meeting with deliberations highly relevant
to this one. A chapter of the resulting publication (National Academy of
Science 1978) describes meteorological considerations and modeling
approaches appropriate to this problem. In the publication from a NATO
ASI on ''Air-Sea Exchange of Gases and Particles'' (Liss and Slinn 1983),
the first chapter by Hasse provides an extensive review of the meteorology and fluid dynamics of the troposphere, including the near-surface
layer. A chapter by Rodhe and a companion working-group report in the
book from a recent NATO ARW on ''The Biogeochemical Cycling of Sulfur and
Nitrogen in the Remote Atmosphere'' (Galloway et al. 1985) provide a

3

A. H. Knap (ed.), The Long-Range Atmospheric Transport of Natural and Contaminant Substances, 3–36.
© 1990 *Kluwer Academic Publishers.*

review of techniques for studying long-range transport, along with a
number of case studies. The chapter by Merrill in the book from a recent
NATO ASI on ''The Role of Air-Sea Exchange in Geochemical Cycling''
(Buat-Ménard 1986) provides a thorough and highly relevant review of the
meteorological aspects of transport to the oceans.

   The purpose of this chapter is to identify those areas of the world
where long-range, continent-to-ocean and continent-to-continent transport
might occur and to review those portions of the atmospheric pathway of
airborne constituents that control such transport. In the following
section, we review the general circulation patterns over the world oceans
to show which continental areas might serve as potential source regions
for long-range transport. In the subsequent section, for the individual
atmospheric processes that comprise the atmospheric pathway, we assess
the current knowledge of their role in controlling long-range transport.
In the final section we have proposed some of the studies that would be
required to increase our understanding of the atmospheric pathway and
improve our ability to quantify large-scale mass transport of pollutants
and other substances.

1.2.   LONG-RANGE-TRANSPORT OVERVIEW

1.2.1. Transport Eastward From North America

North of 30°N, the landmasses of North America and Greenland create a
steep horizontal temperature gradient that leads to large seasonal variations in the general circulation of the atmosphere over the North Atlantic Ocean. The region of maximum cyclonic development is over this
western perimeter of the North Atlantic Ocean. The most intense atmospheric circulation is during winter when pressure and temperature gradients are the steepest (Fig. 1-1). Depressions forming off the east coast
of North America deepen as they move northward and eastward. The migration of these storms results in the quasi-stationary position of the
Icelandic low. In Figure 1-2, storm tracks indicate the preferred transport paths over the North Atlantic Ocean in January. Most of these storm
centers move at speeds over 12 m/sec and are accompanied by heavy seas
with wave heights frequently higher than 3 m. The frequency of storms in
the North Atlantic and the associated precipitation suggest that material
will be removed rather than transported, unless it is carried above the
boundary layer in a region where entrainment and removal in precipitation
systems would be minimized.

   During the summer months over the North Atlantic Ocean, the pressure
gradient and surface circulation are generally weak (Fig. 1-3). Comparing Figures 1-1 and 1-3, the Icelandic low has dissipated and the
subtropical belt of high pressure, the Bermuda or Azores high, is the
predominant synoptic feature. Surface wind speeds over the North Atlantic Ocean are significantly lower during the summer than the winter
because subsidence occurs over much of the central North Atlantic Ocean.
Cyclogenesis still occurs, but systems develop further poleward and less
frequently than in winter. This is illustrated by the storm tracks,
which generally show a more northerly migration in summer (Fig. 1-4).

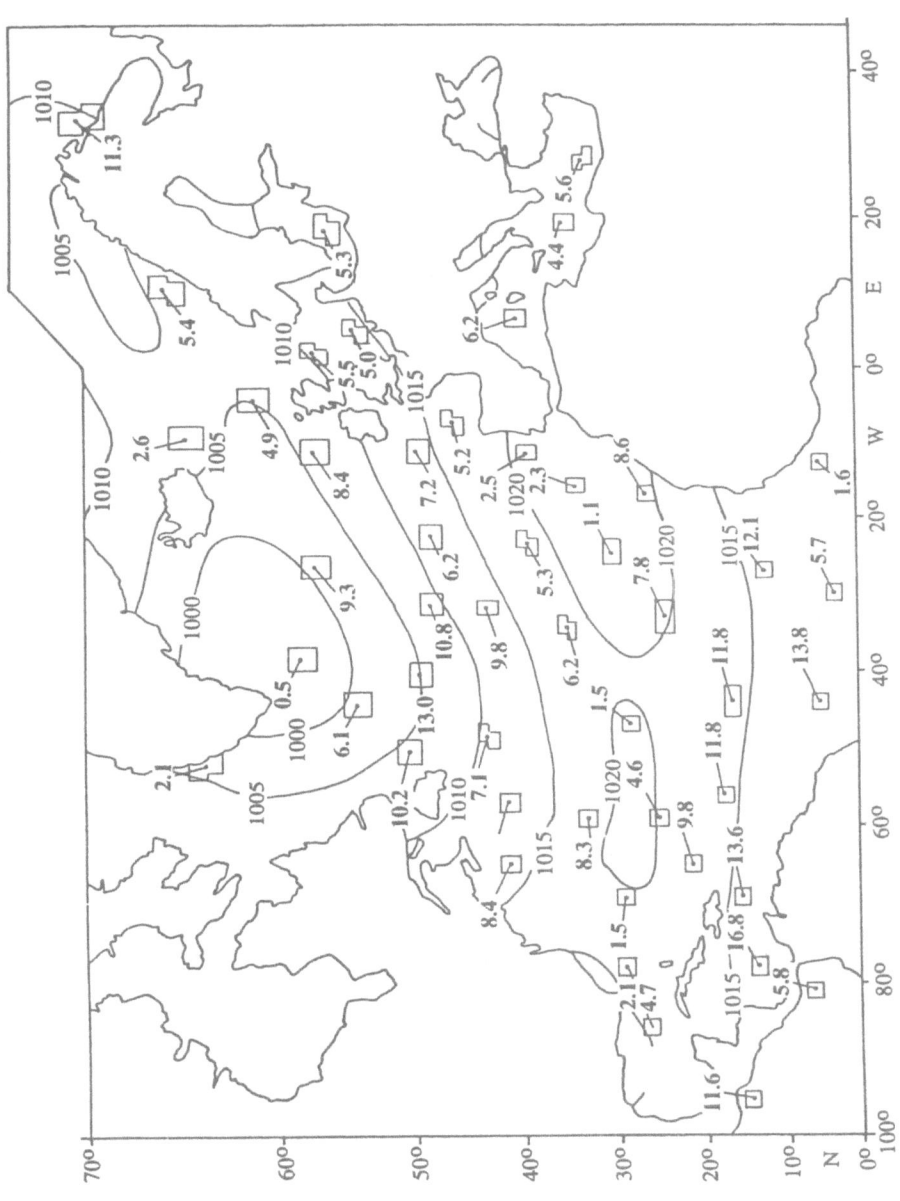

Figure 1-1. Mean sea-level pressures (mb) and mean winds (knots) for the North Atlantic Ocean for January (adapted from Meserve 1974).

6

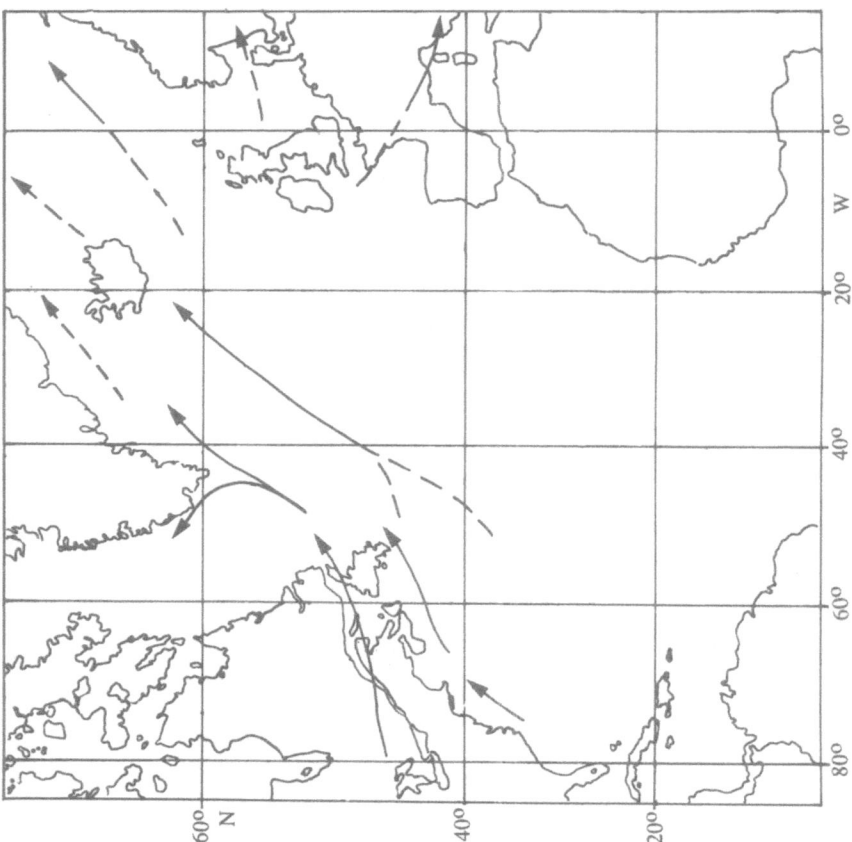

Figure 1-2. Storm tracks for January. Arrows indicate direction; **solid lines, primary tracks; dotted lines, secondary tracks** (adapted from Tucker and Barry 1984).

7

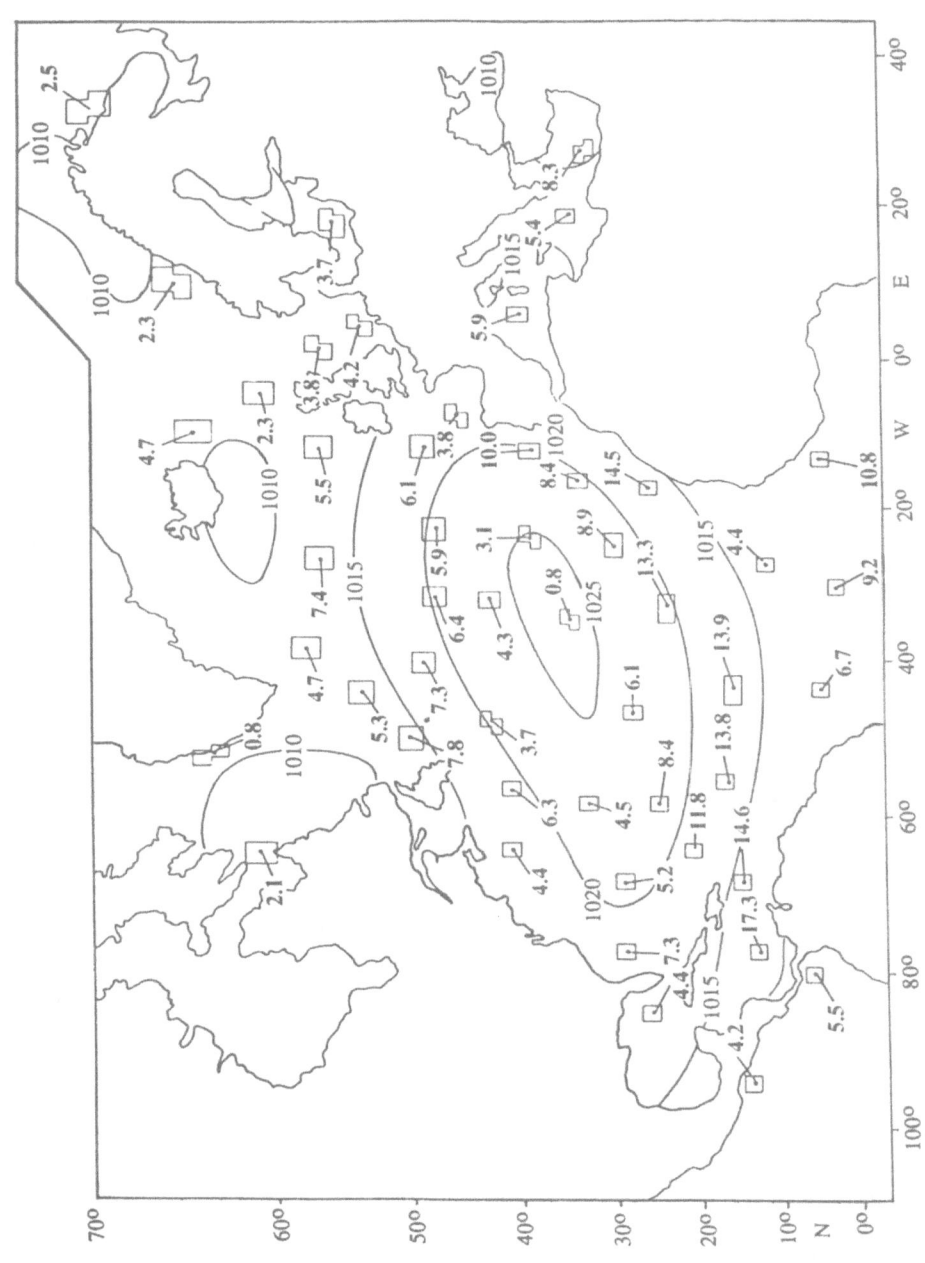

Figure 1-3. Mean sea-level pressures (mb) and mean winds (knots) for the North Atlantic Ocean for July (adapted from Meserve 1974).

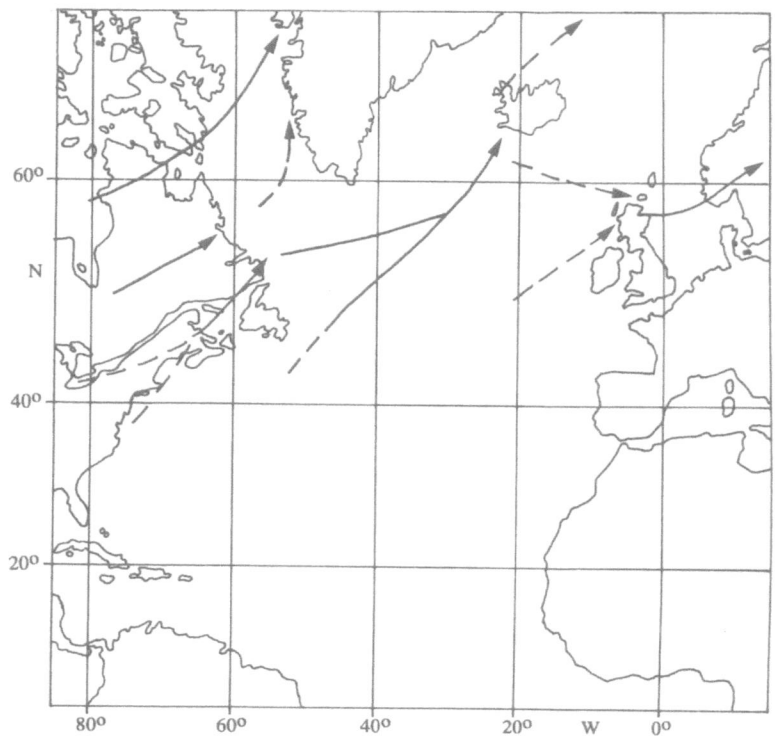

Figure 1-4. Storm tracks for July. **Arrows** indicate direction; **solid
lines**, primary tracks; **dotted lines**, secondary tracks (adapted from
Tucker and Barry 1984).

During both seasons, the upper level winds are predominantly westerly although waves in the upper atmosphere occur in relation to the
migrating systems of cyclones and anticyclones. Figure 1-5 indicates
the resultant 500-mb streamlines for January and July. Wind speeds can
be greater than 20 m/sec between 30-55°N in winter, with a maximum at
40-45°N. In summer, the average speed of westerlies is considerably
lower and the maximum is at 50°N.

Given the mean transport paths described above, material could be
transported eastward from North America over the North Atlantic Ocean.
The conditions in winter would be more likely to result in deposition
over the ocean, unless material were advected above the level of cloud
entrainment. In the summer, transport speeds are somewhat slower, but
the location of the North Atlantic high pressure serves to bring material
advected above the planetary boundary layer down to the surface, toward
the European continent.

1.2.2. Transport Westward From Africa

The westward transport of material from Africa has the potential to occur
in both hemispheres between the band of subtropical high pressure and the

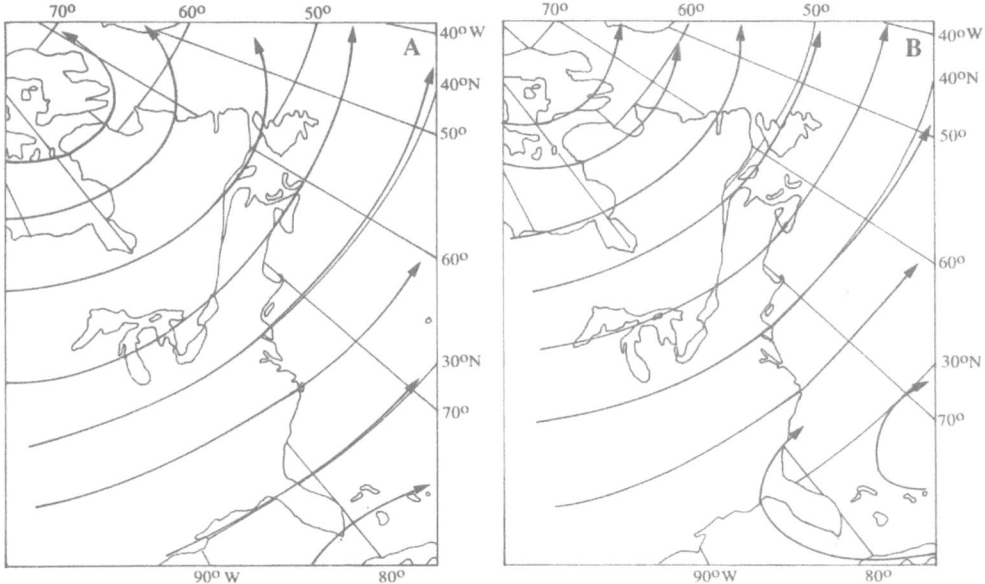

Figure 1-5. Resultant 500-mb streamlines for **A** January and **B** July
(adapted from Whelpdale et al. 1984).

equatorial low. This is the region of the northeast trade winds of the
Northern Hemisphere and the southeast trade winds of the Southern Hemisphere. Compared to the rapidly changing synoptic patterns of the midlatitudes, the trade winds are relatively consistent.

The actual division between the domain of the northeast and southeast trades is the intertropical convergence zone (ITCZ), which can be
thought of as the thermal equator. Because of the lower average temperature of the Southern Hemisphere, which contains the extensive cold
Antarctic landmass, this thermal equator is north of 0°. In the winter, the ITCZ is between 0°N and 5°N. During summer in the Northern
Hemisphere, the ITCZ moves northward, with an east-west axis typically
between 5°N and 10°N.

The trade winds occur in a relatively shallow region of steady surface winds capped by a strong inversion. This temperature increase is
the result of large-scale subsidence around the subtropical high pressure. The strength of the trade winds generally decreases with height.
In winter much of the equatorial Atlantic above the trade inversion is
under the influence of prevailing westerlies. However, in the summer,
easterlies persist throughout the troposphere over tropical latitudes.

This region of surface convergence, along the ITCZ, causes much convective activity and cloudiness. In the winter, however, little organized synoptic activity results. During the summer, synoptic systems form near the region of maximum wind shear (5°-10°N) along the trough of low pressure. These systems can cross the North Atlantic in approximately 4 days traveling at speeds of 5-10 m/sec.

The position and relative steadiness of the trade-wind region suggest that material could be transported westward from Africa in either season. However, given the northward displacement of the ITCZ and the vertical extent of the tropical easterlies, material is more likely to be transported from North Africa westward across the southern North Atlantic Ocean during summer.

1.2.3. Transport Eastward From South America

The general circulation of the South Atlantic Ocean is driven by the large meridional gradient in heating between the cold Antarctic landmass and the equator. The seasonal variations in general circulation are less pronounced than those in the North Atlantic Ocean because of the greater expanse of ocean in the Southern Hemisphere. The low latitudes of the South Atlantic Ocean are dominated by the southeast trades, which are relatively steady. The weak seasonal variation is seen by comparing Figures 1-6 and 1-7. This circulation is driven by the South Atlantic

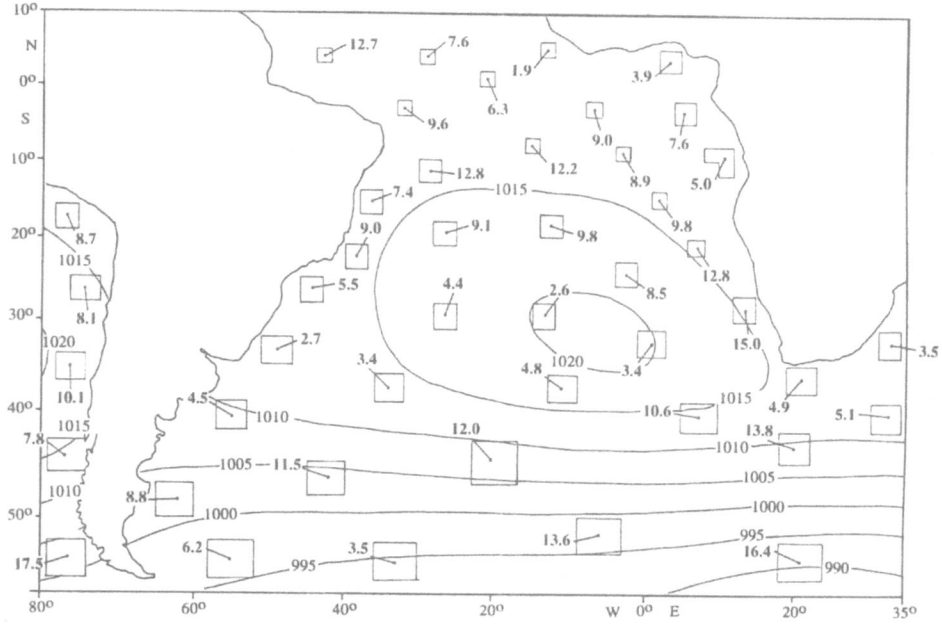

Figure 1-6. Mean sea-level pressures (mb) and mean winds (knots) for the South Atlantic Ocean for January (adapted from Meserve 1974).

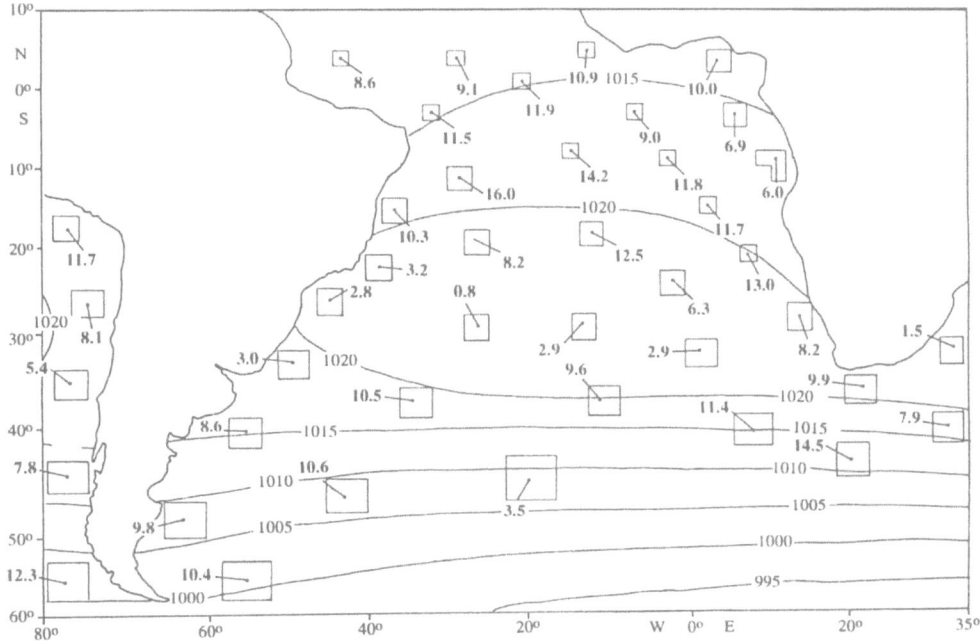

Figure 1-7.   Mean sea-level pressures (mb) and mean winds (knots) for the
    South Atlantic Ocean for July (adapted from Meserve 1974).

subtropical high at its most poleward location (30º-35ºS) in January.   In
winter (July) in the Southern Hemisphere, the climatological mean high
pressure is over 25ºS.   This variation in the location of the subtropical
high affects the location of the prevailing westerlies.

        In both seasons, the maximum westerly winds in the troposphere occur
in the midlatitudes (40º-50ºS).   This is the region of the polar front,
the zone where subtropical and polar air masses meet and midlatitude
synoptic weather systems develop.   Maximum wind speed defines the polar
jet, which steers the movement of cyclonic storms.   The westerly winds
vary in strength from approximately 5 m/sec at the surface to 25 m/sec at
500 mb (see conversion chart in Appendix, p. 311).   During summer, at low
latitudes of the Southern Hemisphere the equatorial easterlies persist up
to a height of 300 mb.   During winter, the easterlies weaken and the
north-south extent of the westerlies expands in the upper troposphere
forming a secondary maximum in wind speed, known as the subtropical jet.
In general, however, the wind speeds of the westerlies do not show a
pronounced seasonal variation.

        This description of the westerlies defines the region in which mate-
rial could be transported eastward from South America.   This climatological mean zonal flow results from the averaging of daily weather systems
that vary greatly.   Figure 1-8 shows the tracks of cyclonic storms
embedded in the westerlies in February and August.   These storm systems

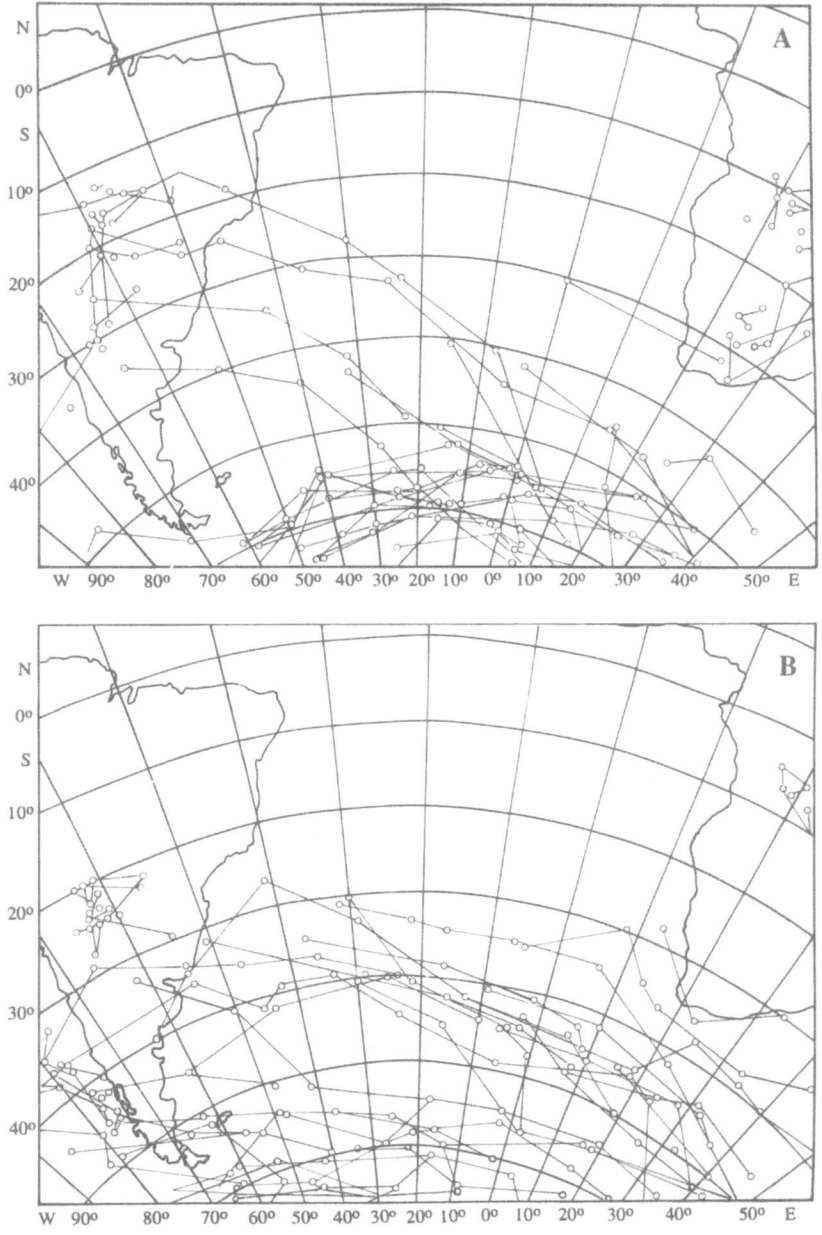

Figure 1-8. Daily positions and tracks of cyclone centers for **A**
February, 1963, and **B** August, 1963 (adapted from Hoflich 1984).

take 5-8 days to cross the South Atlantic Ocean. This region of the
South Atlantic is characteristically very cloudy, with heavy seas (wave
heights of 2-3 m), and gale force winds are relatively common. This
suggests that material in the boundary layer may be readily removed at
the surface or precipitated. However, in situations of great vertical
mixing, material may be detrained through cloud tops into the free troposphere. In these episodic situations, material advected eastward may
have a considerably longer atmospheric lifetime.

1.2.4. Transport Eastward From Asia

The seasonal extremes in the mean surface pressure over the North Pacific
Ocean are driven by large differences in heating over the ocean and the
surrounding landmasses. Figure 1-9 shows the surface pressure distribution and mean winds in January with a low-pressure system centered just
west of the Aleutian Islands (50°N) and the subtropical Pacific high off
the Baja peninsula (30°N). In summer (Fig. 1-10), the low-pressure center dissipates and the Pacific high, moving poleward to 40°N and westward, dominates the general circulation. Spring and fall are transition
months and, accordingly, depict the relative weakening and strengthening
of these surface features.

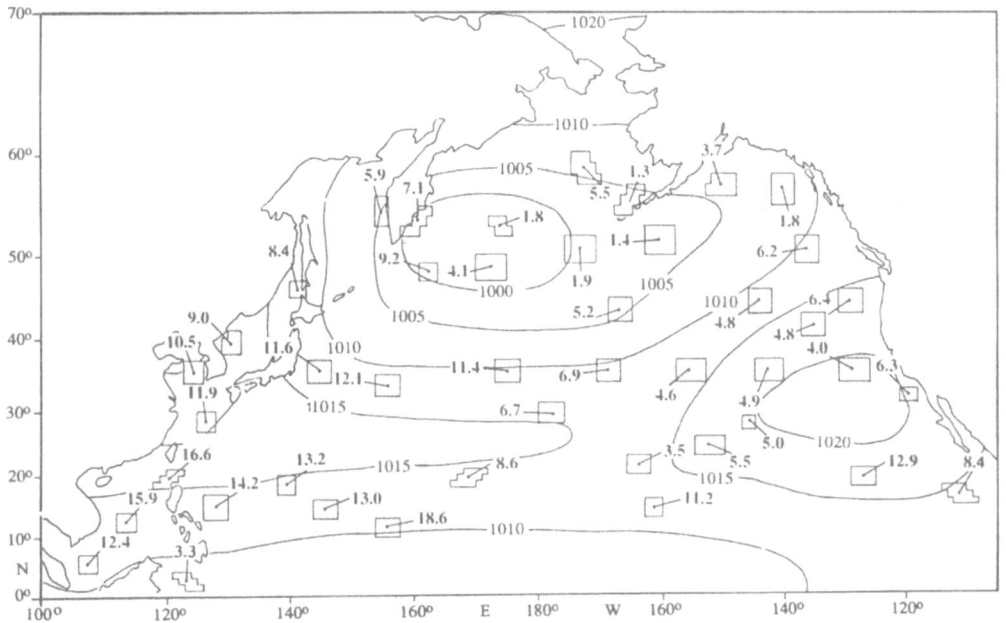

Figure 1-9. Mean sea-level pressures (mb) and mean winds (knots) for the
North Pacific Ocean for January (adapted from Meserve 1974).

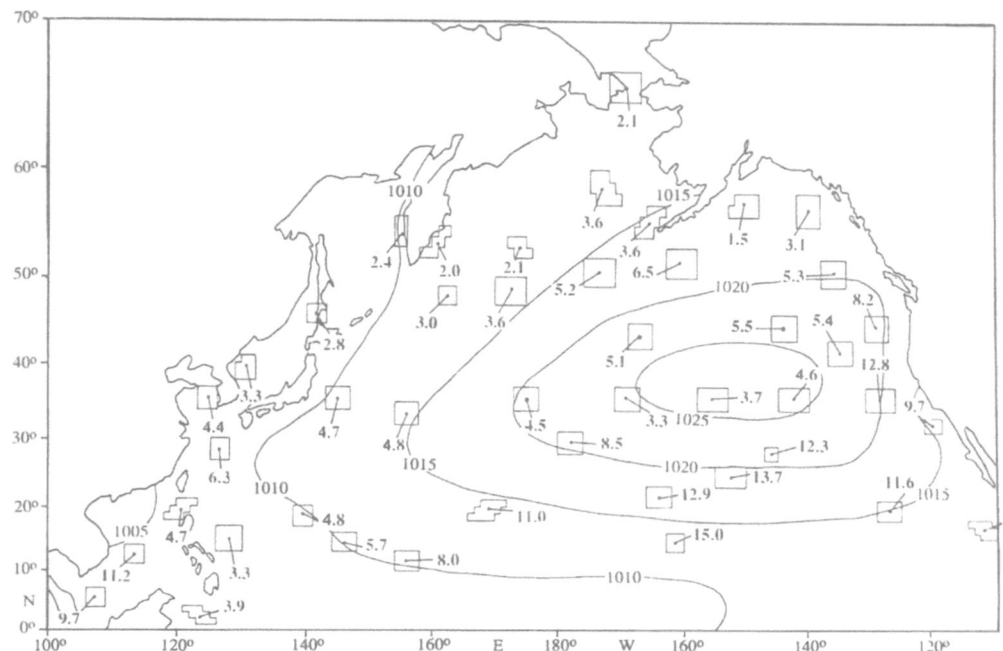

Figure 1-10. Mean sea-level pressures (mb) and mean winds (knots) for the North Pacific Ocean for July (adapted from Meserve 1974).

The large low-pressure region in the cold months results from the frequent passage of cyclones, which then intensify near the Aleutian Islands. These storms form along the polar front, a region where cold, dry Siberian air masses collide with warm, humid maritime air masses. The frequency of westerly winds and thus the potential to transport material off the Asian continent can be seen in Figure 1-11. During winter, the prevailing flow provides a transport path off the east coast of northern China, Korea, and Japan (north of 30°N) roughly 40-60% of the time. In the summer, this transport path is oriented more northeastward toward Alaska and the Arctic.

Wind speeds, which are stronger in the winter, increase with height. During winter, the westerlies at 500 mb have speeds of 12 m/sec or greater between 25°N and 50°N. In the summer, the westerlies weaken to 8 m/sec with a north-south extent between 50°N and 30°N.

Figure 1-12 shows the tracks by month of extratropical depressions or severe storms and tropical cyclones for every year from 1920 to 1940. This figure illustrates that storms crossing the North Pacific Ocean to North America are most frequent in January and the transition months of March and November. Given these transport paths and the relatively high wind speeds, material could be transported from the Asian continent across the North Pacific Ocean to North America. However, these regions of storminess also enhance the possibilities for removal. As with

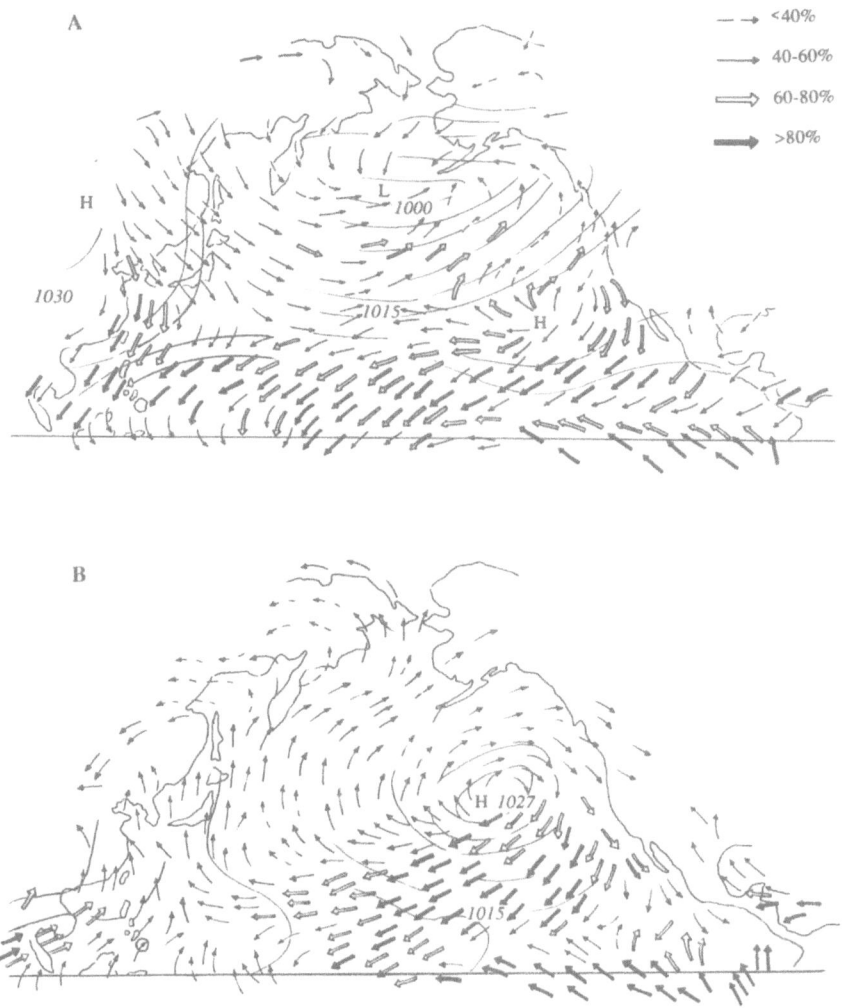

Figure 1-11. Frequency of wind directions over the North Pacific Ocean for **A** winter and **B** summer (adapted from Terada and Hanzawa 1984).

transport across the North Atlantic Ocean, the greatest potential for the eastward transport of material may be during summer months in the mid-latitudes (40-55°N) above the planetary boundary layer, with material advected to the eastern Pacific Ocean being mixed back to the surface in subsiding air.

1.2.5. Transport Westward From South America

Westward advection from South America, analogous to westward transport from Africa, is driven by an easterly flow around subtropical high

16

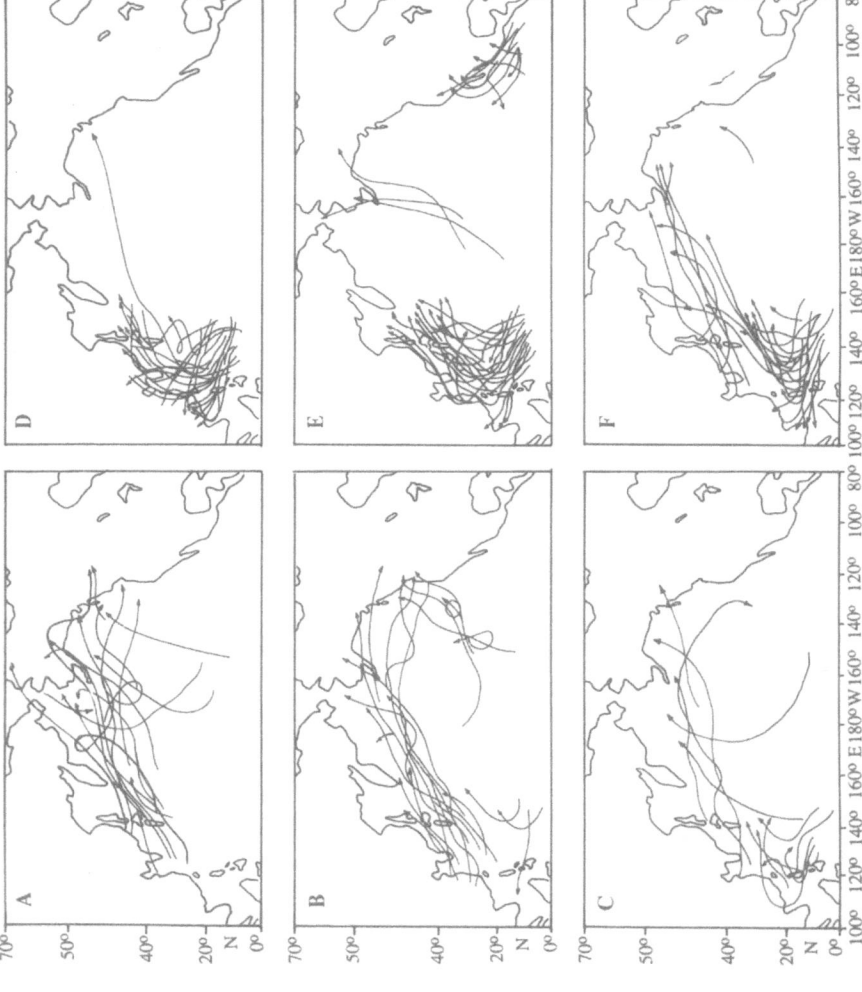

Figure 1-12. Tracks of the main extratropical depressions and tropical cyclones from 1920 to 1940 for A January, B March, C May, D July, E September, and F November (adapted from Terada and Hanzawa 1984).

pressure zones in the North and South Pacific Oceans (30°-35°N and 30°-35°S, respectively). Material at the surface is transported easterly in trade winds and aloft in a narrow band of equatorial easterlies. Figure 1-13 shows the surface pattern of the trade winds that could advect material westward from South America between 10°N and 30°S. Summer transport in the southeasterly trades has the potential to advect material over the equatorial Pacific with wind speeds of 3-6 m/sec (Fig. 1-14). However, in summer, low pressure over the western South Pacific Ocean, relative to the eastern South Pacific Ocean, leads to a breakdown in the southeastern trade pattern. During winter months in the Southern Hemisphere, trade winds are somewhat stronger with speeds averaging 4-7 m/sec (Fig. 1-15).

As indicated in the description of westward transport from Africa, the intertropical convergence zone (ITCZ) is displaced to the north of the equator. This means the southeast trades extend into the Northern Hemisphere. Because of the change in direction of the coriolis force, winds turn northward and eastward. This may act to inhibit the westward flow off the northwest coast of South America. Additionally, most of South America lies in the Southern Hemisphere. Therefore, the greatest potential for material to be transported westward appears to be south of the equator. Circulation around the South Pacific high leads to subsidence over the central Pacific Ocean. Because of its large westward expanse, the South Pacific Ocean is the most likely sink for material advected in the easterlies.

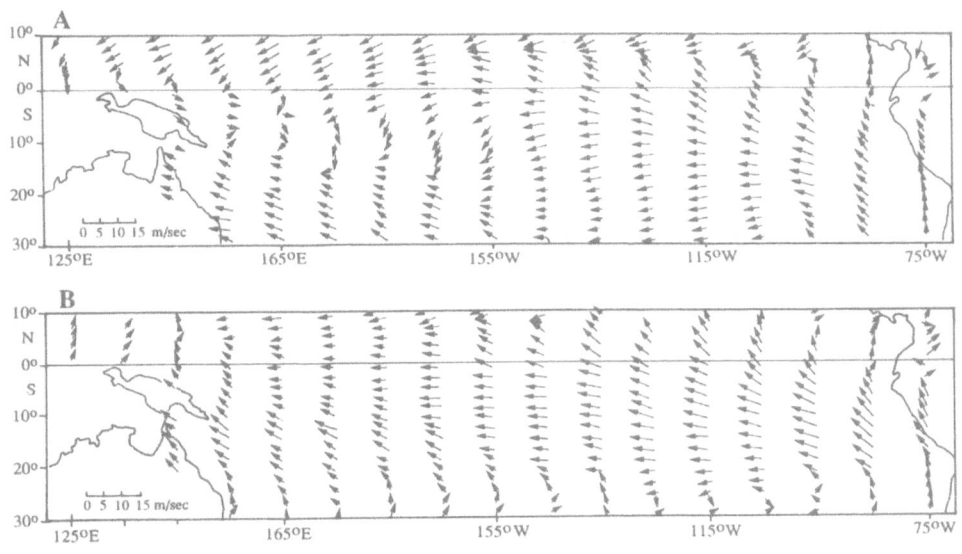

Figure 1-13. Mean surface vector winds at lower latitudes of the South Pacific in A January and B July (adapted from Streten and Zillman 1984).

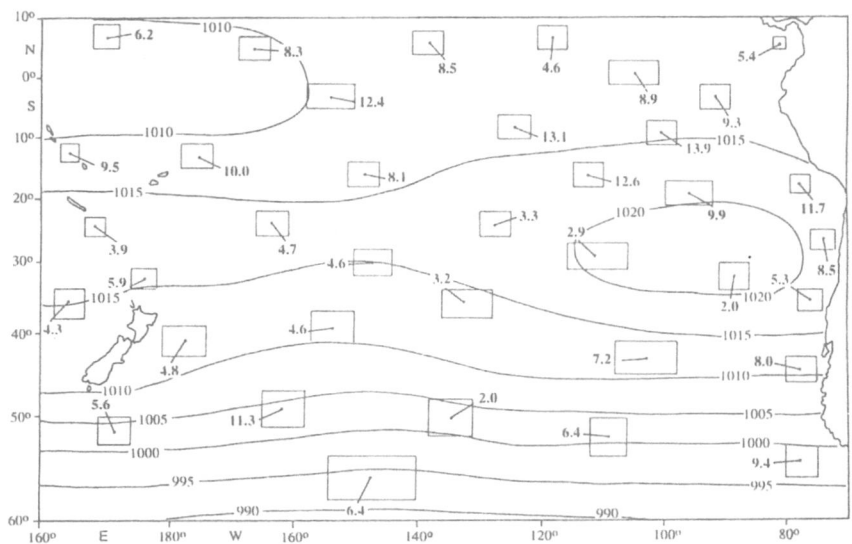

Figure 1-14. Mean sea-level pressures (mb) and mean winds (knots) for the South Pacific Ocean for January (adapted from Meserve 1974).

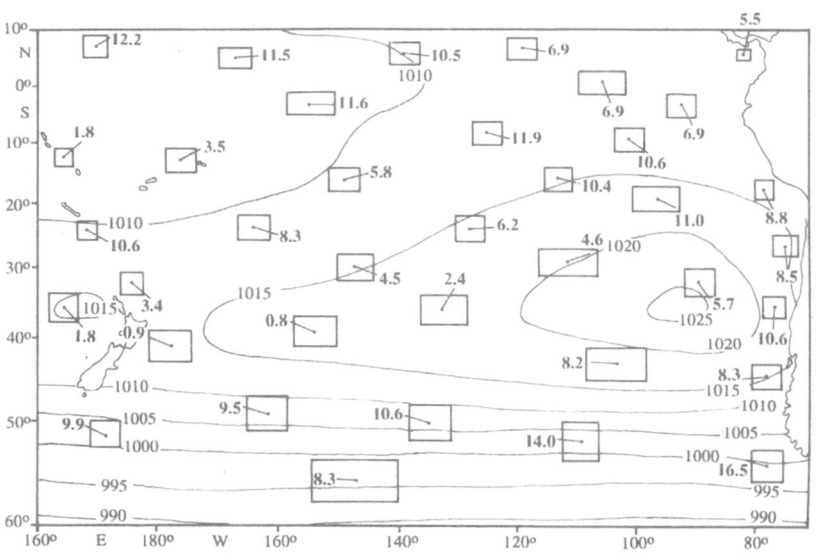

Figure 1-15. Mean sea-level pressures (mb) and mean winds (knots) for the South Pacific Ocean for July (adapted from Meserve 1974).

1.2.6. Transport to the Polar Regions

The general circulation of the polar regions is dominated by polar easterlies, relatively shallow wind layers with westerlies aloft. This westerly flow at higher altitudes is weaker than in midlatitudes because
there is not such a strong meridional temperature gradient. The transport of heat to polar regions, a major feature of earth's general circulation, is accomplished by transient waves in midlatitudes.

In the Northern Hemisphere, poleward transport generally occurs on
the eastern edge of low-pressure systems and the western edge of high-
pressure systems. In a climatological sense, transport toward the north
pole should be favored in the winter between the Icelandic low and the
Siberian high and between the Aleutian low and the less intense and consistent high-pressure cell over northern North America. As was shown in
the sections describing transport over the North Atlantic and North
Pacific Oceans, storminess near the Icelandic and Aleutian lows favors
deposition. Therefore, the greatest potential for long-range transport
to the Arctic is on the backside of strong continental high-pressure
systems during the winter months.

The coincident occurrence of subsidence and enhanced radiational
cooling results in the formation of a strong surface inversion during
winter (polar night) in both hemispheres. These conditions, along with
extremely low temperatures and humidity, do not favor the formation of
precipitation. Material advected into the Arctic under these conditions
has a relatively long residence time. From Barrie's studies of transport
to the Arctic (Chapter 6, p. 137; Barrie 1986, Barrie et al. 1989), it
appears that the most frequent path of material is around the back of the
Siberian high-pressure cell during winter.

Considerably less is known about the circulation of the south polar
region because of its remoteness. In general, flow in the Southern
Hemisphere is much more zonal at high latitudes, as a result of the smaller landmass and the reduced horizontal temperature gradients. In addition, the Antarctic landmass is more symmetrically centered on the pole
and surrounded by ocean; this results in a less variable and less complex
climate than the Arctic region. Polar high pressure is better defined in
the Southern Hemisphere because of the well-defined ice plateau with
down-slope winds draining the continental interior. This air is replaced
by subsiding air from above. This may be a potential transport path for
bringing material into the Antarctic region.

1.2.7. Concluding Remarks

This review of the general circulation of the atmosphere over the oceans
provides a starting point for determining pathways available for long-
range transport. Figure 1-16 illustrates how transport does occur downwind of continental source regions. In this example, the patterns of
deposition of coarse volcanic glass on the ocean floor are controlled by
the prevailing winds and the location of sources. However, this climatological perspective ignores the distribution of atmospheric conditions
that make up the seasonal or annual average. To address adequately the
frequency of long-range transport, it is important to consider the

20

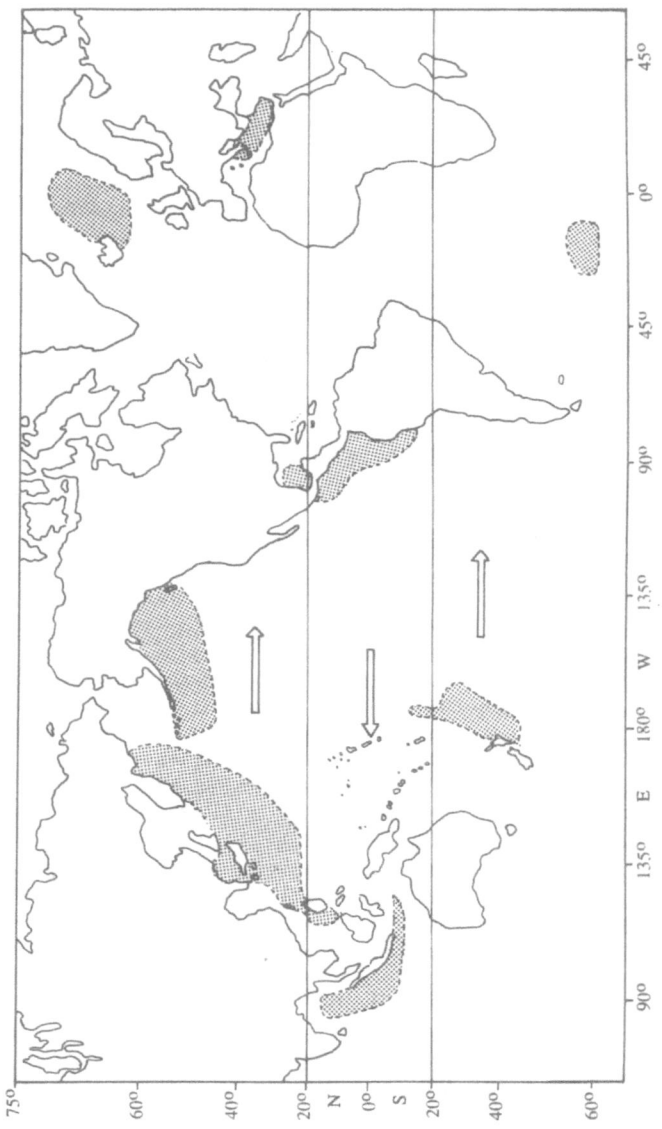

Figure 1-16. **Shaded areas represent abundant, present-day, coarse, volcanic glass distributions found on the ocean floor (adapted from Kennett 1981).**

natural variability in atmospheric conditions and to determine the frequency of those conditions that should actually favor transport. This requires consideration of several important processes, which are discussed in the next section.

1.3. THE ATMOSPHERIC PATHWAY

1.3.1. Introduction

In this section we examine the parts of the atmospheric pathway, i.e., the individual chemical and physical processes, between continental source regions and distant oceans or continents where materials are eventually deposited (Fig. 1-17). Although many individual processes involved can correctly be considered to act on smaller scales, their role in influencing large-scale transport is considered.

In almost every instance it could be said that enough is not known about individual processes to permit a sufficiently precise mathematical description to be used in a model to achieve accurate quantitative predictions of large-scale mass flux. Nevertheless, in most cases, enough is known to permit at least a qualitative estimate of the role of the process in influencing large-scale transport.

The portions of the atmospheric pathway shown in Figure 1-17 are described in varying detail below. We emphasize those that are considered particularly significant in controlling the amount of long-range transport and about which relatively little is known. Some are treated in detail in other sections of this book (for example, emission sources of mineral aerosol in Chapter 3, p. 59) and so we have only noted them here. During the course of discussions at the NATO workshop in Bermuda, it was suggested that we include additional material on some items (for example, a discussion of trajectory models in the transport section). This chapter, along with the references given in the introductory section, should provide a fairly complete introduction to the meteorological aspects of large-scale transport.

1.3.2. Emissions

Traditionally, one thinks of the important characteristics of a source as the identity of the chemical species emitted, the rate of emission (as a function of time), the effective height of injection, and the location of the source. In principle, for most anthropogenic materials and sources, this information can be obtained without an extensive scientific research effort; it is more a data-gathering effort. However, in the context of the importance of emission sources (both anthropogenic and natural) for large-scale transport, other factors must be considered, namely:

1. How to measure emission rates for certain types of sources (e.g. diffuse natural or anthropogenic sources) and highly sporadic sources.

22

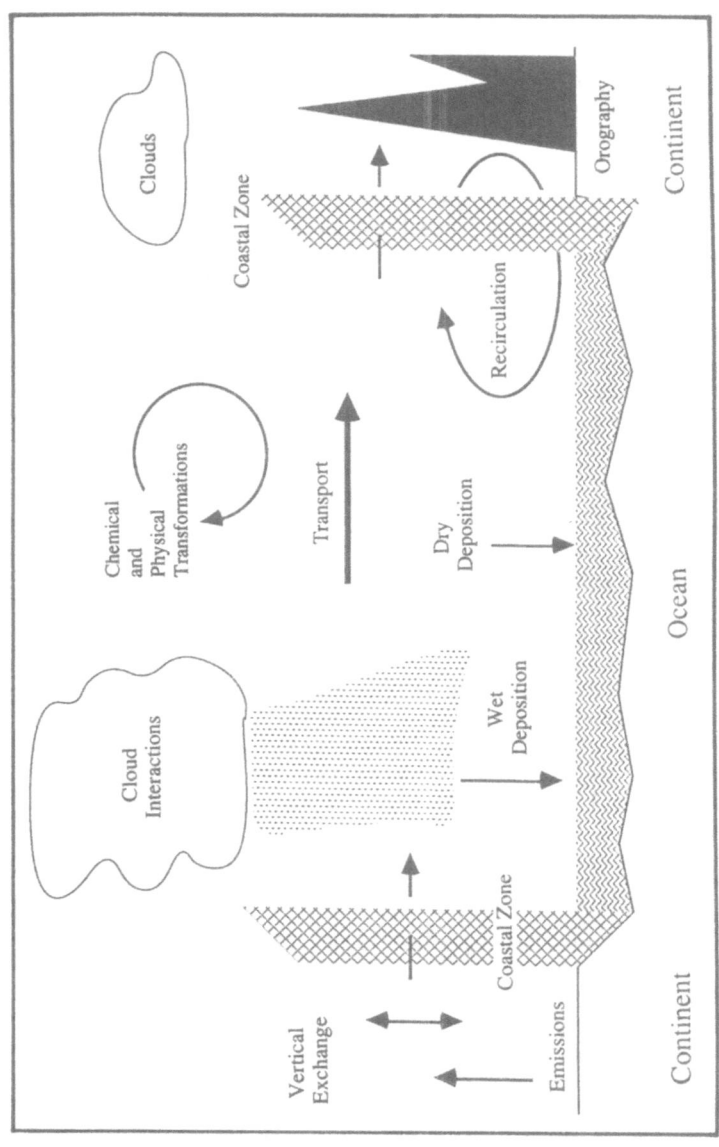

Figure 1-17. Component parts of the atmospheric pathway of materials between a source and a receptor.

2. How to characterize the nature and efficiency of prevailing emission-injection processes and near-source vertical-exchange mechanisms.

3. How to determine and characterize the overall efficiency of a source region; i.e. how much material is transported away from a continental source region.

4. In regions remote from sources, how to disentangle the signal from a distant source from a highly variable local background signal.

A common example may serve to illustrate some of these difficulties. A portion of the emissions injected into the atmosphere in a coastal region may reach the marine environment without being incorporated into the well-mixed surface layer; this material has the potential to travel much further in the marine environment than in the mixed layer where removal processes are more effective. The importance of the source region, from the perspective of long-range transport, is determined by the amount of material that escapes the coastal zone, rather than by total emissions.

Several of the difficulties identified above are addressed in other chapters of this book, where problems specific to families of chemicals are examined. We have given our recommendations for future study of these topics at the end of this chapter.

1.3.3. Vertical Exchange

The height at which material is emitted into the atmosphere (e.g., from a stack) or the height to which it is carried by vertical motions near sources may have a large influence on its eventual range of influence. In addition, at any time during transport, mechanisms that vertically redistribute substances in the atmosphere, in particular between the planetary boundary layer and the free troposphere, have a significant influence on the efficiency of large-scale transport.

Pollutant concentrations near source regions are usually higher in the boundary layer. Vertical-exchange mechanisms are important because they transport pollutants upward out of regions where wet and dry removal processes are effective to regions of greater vertical stability and higher wind speed where the pollutants have a longer atmospheric residence time and greater transport distance.

Over land the strong diurnal cycle of surface heating results in a similar cycle of stability in the near-surface layer of the atmosphere. Daytime instability results in turbulent mixing and rising motions throughout the mixed layer, typically a few hundred to a few thousand meters in depth. When convection is strong enough, buoyant plumes or bubbles of boundary-layer air may penetrate into the free troposphere, introducing boundary-layer contaminants into a region where the potential for transport is much greater. At night, the stability of the lower atmosphere suppresses vertical motion. One result of this is an effective decoupling of the large-scale flow above the surface-based night-time inversion. In this situation, material tends to remain at the level at which it was injected, until the stability changes.

Over the sea the diurnal surface-heating effect is much less pronounced and the atmosphere is generally neutral or near neutral. Exceptions may occur under light wind conditions with an air-sea temperature difference caused by air-mass advection, upwelling, or ocean currents. Downwind of North America, for example, summertime advection of relatively warm air over the colder North Atlantic favors a stable maritime regime and wintertime advection of cold air over the warmer water favors an unstable regime. Although this latter situation is conducive to vertical exchange of material to higher levels, it is also effective in bringing material near the surface where it is more susceptible to removal processes. The net efficiency of such vertical-exchange mechanisms in promoting long-range transport must take account of stability, wind speed, and precipitation.

A second mechanism of vertical exchange is the entrainment of free tropospheric air into the turbulent boundary layer at the interface between the two regions. This may occur when air in the lower free troposphere undergoes radiative cooling or cooling from the evaporation of recently entrained cloudy air. Entrainment serves to reduce boundary-layer concentrations by dilution and to make the boundary layer increase in depth.

A third important vertical-exchange mechanism is frontal uplifting. Advancing fronts displace air both horizontally and vertically. Clouds and precipitation associated with fronts are caused by the uplifting, and boundary layer air is lifted into the free troposphere by this process. This mechanism is most effective with the traveling midlatitude and tropical disturbances.

Downward exchanges typically differ somewhat from upward ones because of different energetics of the transfers. Downward transfer by turbulence is effective on small scales as long as there is a vertical gradient in the property being transferred. For well-established convection, downdrafts usually occur over a much wider area and have lower velocities. An example is the slow sinking motions between clouds. Large-scale subsidence in high-pressure cells is also less energetic than the upward motions.

In summary, the potential for large-scale transport is increased if vertical-exchange processes transport material from levels where wind speeds are low and removal processes are efficient into regions of high winds, vertical stability, and inefficient removal. This often means a vertical transfer to above the surface boundary layer into the free troposphere. Such transfers are most often accomplished by convection and frontal lifting.

1.3.4. Pollutant Interactions with Clouds

Clouds play an important role in both the dynamics and chemistry of pollutants in the atmosphere. Their role depends upon such factors as vertical motion, location, vertical and geographical extent, liquid-water content, and phases present.

Convective clouds with sufficient vertical motion and extent ''pump'' materials upward from the near-surface, below-cloud layers. Relatively little is known quantitatively about how effective this process is in redistributing substances in the vertical and thus in influencing the distance they may be transported. To be effective in vertical transport, clouds must be in regions where there are contaminant materials in the lower atmosphere. To play a role in exchange between the boundary layer and the free troposphere, clouds must also have strong enough vertical motion to penetrate into the free troposphere. Such vigorous convection is sporadic in the midlatitudes but more consistent in the humid tropics.

Clouds are also important because they serve as ''reaction vessels'' for much of the chemistry that goes on in the atmosphere and because they are responsible for precipitation scavenging. Clouds may influence atmospheric transformations in many ways: they change predominant reaction pathways; they mix different species, enhancing reactions or causing dilution; and they may reduce the residence times of soluble species and increase those of others by transporting them to regions of lower temperature where they are more stable (e.g., PAN). In addition, the movement of a cloud through a region may introduce a high degree of temporal and spatial variability into concentrations of soluble and reactive species.

The role of cloud elements in heterogeneous chemical reactions is closely tied in with their role in wet deposition, as discussed below. In the context of large-scale, over-ocean transport, a remaining important gap in our knowledge concerns precipitating clouds and precipitation amounts over the ocean. With the increased use of satellites anticipated for such observations, improved estimates of over-ocean wet deposition may also be expected in the coming decade.

In summary, many details of the dynamics and the chemistry of individual clouds and simple cloud systems are reasonably well known. At the present time, unfortunately, we do not know how efficient these processes are on the large scale in influencing long-range transport; consequently they tend to be simplified in chemical transport models.

1.3.5. The Coastal Zone

Many coastal regions are heavily populated, with the associated sources of air pollution and dumping in the coastal oceans. Contaminants are undoubtedly transported through the atmosphere over relatively short distances from continent to ocean and from ocean to continent. In addition, however, the coastal zone is of particular interest as a portion of the atmospheric pathway for substances susceptible to long-range transport. From an atmospheric point of view, the important changes at the continent/ocean boundary are changes in surface roughness and topography and in thermal properties of the surfaces. The coastal zone is often a region of intense aerosol production. These abrupt changes in surface features are the cause of complex and often intense micro- and mesoscale meteorological phenomena, such as land-sea circulations, vertical stability, and local rainfall enhancement. In turn, one might expect to find a pronounced alteration of the atmospheric chemical regime, at least in the lower levels of the atmosphere, that may be reflected in transport on larger scales.

Land-sea breeze systems are caused by temperature differences over the two surfaces. They occur mainly with high-pressure systems in summer when wind speeds are low and insolation is strong. During the day air warmed by the land surface rises, moves offshore aloft, descends over the cooler sea surface, and returns at a lower level. At night the reverse occurs, with the cool air descending over the land and draining out over the sea, to rise again over the relatively warm sea surface. The lower branch of such a circulation may extend to more than a kilometer into the atmosphere and the upper branch to more than three kilometers. These systems are an important feature of the weather in low latitudes, often penetrating more than a hundred kilometers inland. In the tropics the rising warm air may enhance convection and lead to heavy precipitation. In polluted areas, these mesoscale circulations may result in a significant amount of pollutant ''processing'' or depletion before some fraction escapes to the larger-scale flow.

With an onshore flow, the increased surface roughness of the land intensifies the turbulence, which may result in more dry deposition over land than over the ocean and possibly to convergence some few kilometers inland accompanied by enhanced convection. As well, with an onshore flow, the combination of increasing elevation and surface heating by insolation may result in upslope flows and convective activity. In both situations, local enhancement of precipitation and thus wet deposition may occur. Thus, with an onshore flow, the coastal zone may be a region where pollutants are removed from the atmosphere with increased efficiency.

In the case of offshore flows, intense local circulation systems might cause a ''premature aging'' of pollutants although the enhanced convective activity from the flow of cold air over the warm water would increase deposition over the coastal ocean. Sievering et al. (1989) recently suggested that large aerosol particles from sea spray in the coastal zone may effectively enhance the dry deposition of acidic gases, such as sulfur dioxide. This could be true for both offshore and onshore flows where such gases are present. In all cases, the result would be that less material, particularly primary pollutants, would be available for further transport. Hastie et al. (1988) find that sulfur and nitrogen species in the near-coastal zone are being more rapidly depleted with offshore flow than previously supposed.

Two situations in the coastal zone that are conducive to long-range transport involve flow away from the continent in stable conditions. The first, noted above, occurs when warm air flows over a cold ocean; the increase in stability will reduce the risk of deposition and favor continued transport. The second case occurs when polluted air in the upper boundary layer is entrained into the free troposphere as it passes over the coast into the marine environment where there is a lower, stable boundary layer.

The role of the coastal zone in controlling the fraction of continental emissions passing into the marine environment is not yet well known and is not modeled satisfactorily. As suggested in the final section, it is an area ripe for further detailed field and model investigations.

1.3.6. Chemical and Physical Transformations

Most atmospheric constituents undergo a series of chemical and physical transformations during their lifetimes in the atmosphere. The efficiency of these transformations control the chemical and physical forms in which substances exist in the atmosphere and, therefore, their susceptibilities to subsequent transport or removal. The efficiency of the individual transformation processes depends greatly on the species. Transformation rates depend on the chemical and physical properties of the substance and the ambient radiation and meteorological conditions. The rates of these chemical and physical transformations must be known as a function of ambient conditions encountered during transport if models are to be used to estimate large-scale fluxes. These topics are treated in detail in other chapters of this book and are not pursued further here.

1.3.7. Wet and Dry Deposition

Wet and dry deposition comprise the final portion of the atmospheric pathway of chemical constituents. Deposition rates must be known to determine the eventual impact of continent-to-ocean and continent-to-continent transport.

The individual processes that comprise precipitation scavenging and wet deposition are reasonably well understood, and extensive efforts have been made to incorporate this knowledge into models of varying degrees of complexity. In principle, wet deposition is relatively easy to measure, except over the ocean.

A particular issue raised at the NATO ARW in Bermuda concerned the limitations and strengths of using scavenging ratios to estimate wet deposition. The scavenging ratio is defined as the ratio of the concentration of a species in precipitation from a storm system to that in the air flowing into the storm. The concept is a simple parameterization of a very complex process and is based on the premise that, on average, a species' concentration in precipitation is proportional to that in the inflowing air. The attractiveness of the concept is that one can take advantage of numerous ground-level air-concentration measurements to estimate precipitation concentration and wet deposition.

The concept also has a number of important limitations:

1. Even when based on accurate, representative measurements, the scavenging ratio is highly variable because of differences in storm type and dynamics and in microphysical cloud processes. An ensemble-average scavenging ratio should be based on data from several (>30) storms, and it should be applied to estimate average precipitation concentration from many storms, not from individual storms.

2. Scavenging ratios are often derived from ground-level air concentrations, which do not necessarily represent those in the air entering the storm and on precipitation data from one site. In frontal storms and storms over the ocean where there can be strong vertical gradients in ambient concentration, scavenging ratios may be quite inappropriate.

3. Even under optimum conditions, mean deposition estimates based on scavenging ratios have associated uncertainties of a factor of 2 to 3.

Nevertheless, when the scavenging-ratio is applied within the limitations noted above, it can be very useful. It is often useful when applied for relatively long periods of time and over large geographical regions. Thus, in the case of wet deposition on the scale of continent-to-ocean and continent-to-continent transport, it can be valuable both for deposition calculated from air concentrations and for use in large-scale chemical transport models.

We understand most processes that contribute to dry deposition and to the atmosphere-surface exchange of gases reasonably well. However, these processes are also usually incorporated into large-scale transport and deposition models very simply—by using an exchange coefficient and an ambient concentration. The experimental determination of dry deposition is difficult and, in most cases, is achieved by measurement of ambient concentration and calculation using a deposition velocity.

In summary, in the wet- and dry-deposition portion of the atmospheric pathway, the main measurement limitations are a lack of reliable air-concentration and wet-deposition data over the oceans and an inability to distinguish between local and distant contributions over the continents. From a modeling point of view, a major problem area is the accurate mathematical simulation of the several removal processes as a function of the conditions encountered en route. The determination of these en-route conditions poses a major difficulty.

1.3.8. Transport

Atmospheric transport on the scales considered at the NATO meeting at the Bermuda Biological Station is controlled primarily by the stable global-scale circulation patterns and by large synoptic-scale circulation systems. As described by Hasse in Liss and Slinn (1983), the large-scale circulation in each hemisphere is characterized by a band of fairly stable, light easterly winds between approximately the equator and 30° by a band of westerly winds in the temperate latitudes and by light easterly winds in the polar regions. Imbedded in these flow regimes are the huge, semipermanent high- and low-pressure cells, such as the Azores High and the Aleutian Low. Superimposed on these features are the travelling disturbances, such as the tropical cyclones, the high-pressure cells, and the depressions of the midlatitudes, which bring much of the day-to-day variability of the weather.

The scale and the permanence of these features dictate the average or preferred atmospheric pathways of atmospheric constituents. Average conditions or patterns can be misleading, however, because they do not depict the high degree of temporal variability inherent in the atmospheric system. In addition to the dominant diurnal and seasonal cycles that influence atmospheric behavior in many regions, specific conditions that are conducive to such phenomena as acidic deposition and Saharan dust transport are known to be highly sporadic, or episodic.

As noted in earlier sections, smaller-scale features, such as the intensity of vertical-exchange processes or the occurrence of precipitation, also exert a significant influence on the effectiveness of the large-scale transport of specific substances. In coastal regions, the presence of mesoscale land-sea breeze circulations may control the amount of mixed-layer material transported over the ocean. Although these features are often not explicitly contained in either conceptual or mathematical models of large-scale transport, they must be recognized and incorporated into our understanding of the overall system.

One of the most commonly used techniques for interpreting measurements and evaluating the transport of chemical constituents in the atmosphere is through the application of atmospheric trajectory models. These models can calculate transport pathways forward in time from a source region or backward in time from a receptor. All trajectory calculations rely upon measurements of vertical profiles of pressure, temperature, and wind, which are made discretely in time and space. Miller (1987) has reviewed the various applications of backward trajectory models for the interpretation of chemistry data and has tabulated the advantages and disadvantages of several trajectory-calculation methods for two basic types of models: (1) dynamic models, which employ pressure or temperature fields to calculate isobaric or isentropic trajectories, respectively; and (2) kinematic models, which employ measured wind fields.

Whenever trajectories are used to interpret atmospheric chemical measurements, the assumptions underlying the method used must be understood. For example, isentropic trajectories are based on an assumption of adiabatic vertical motions. However, the adiabatic assumption is violated when clouds or precipitation are encountered (a frequent occurrence in tropical latitudes and midlatitudes) or when significant radiative cooling occurs (a frequent occurrence in the Arctic and Antarctic).

In addition to model assumptions, it is important to keep in mind that the coarse resolution of input data limits the accuracy of all trajectory models. Even in regions of dense meteorological data (e.g., eastern North America), the spatial resolution of data is approximately 400 km, with a temporal resolution of 12 hr. These discrete data must be interpolated in space and time. Several recent studies have estimated the magnitude of interpolation errors in the location of trajectories after a certain number of hours (Draxler 1987, Kahl and Samson 1986, Kuo et al. 1985). The estimates of errors range from 140–290 km after 24 hr to 350–495 km after 72 hr. Results from these studies also suggest that model uncertainty depends on the meteorological characteristics for the period of interest. For example, trajectory errors tend to be smallest under relatively steady wind-flow conditions (Draxler 1987) and to increase for complex situations, such as a developing midlatitude cyclonic system (Kuo et al. 1985).

The Chernobyl case study (Chapter 7, p. 149) provides an excellent illustration of the variability in estimated transport patterns using (1) different forward-trajectory models (e.g., 3-dimensional dynamic models versus isobaric); (2) different interpolation methods; and (3) different meteorological data. The model discrepancies shown in this case study should encourage respect for the inherent uncertainty that exists in attempting to determine precise transport paths through the application of models. Single trajectories, whether forward or back, should not be interpreted as exact representations of transport from point to point, and single trajectories should not be considered as source-receptor models.

Several methods have been developed that attempt to reduce the reliance on single trajectories and their inherent uncertainty. One approach is statistical—to look at an ensemble of trajectories for a long period of measurements. In this way no measurement is critically dependent on the path of a single trajectory. This method has been successfully applied to describe the flow climatology (Miller and Harris 1985) and to identify the chemical climatology (Moody and Galloway 1988) of a site on Bermuda. Another approach is to treat trajectories as mean transport paths with a modeled amount of uncertainty in trajectory location as a function of time. This is similar to modeling a Gaussian plume and results in a 2-dimensional probability representation of the potential transport path (Samson and Moody 1980). These fields can be used in combination with chemical measurements to illustrate a long-term transport bias for a constituent of interest (Keeler and Samson 1989).

Below are our summary observations and recommendations for applying air-mass trajectories to ground-based-aerosol and precipitation data to identify sources and apportion source emissions.

1. Single trajectories should not be used to determine source-receptor relationships.

2. Trajectory accuracy depends on the type of model and the resolution of data employed.

3. Trajectory accuracy depends on the meteorological conditions being modeled.

4. Trajectories provide more reliable estimates where large area sources are involved.

5. Long-term trajectory climatologies are more reliable than single-trajectory calculations.

6. Application of trajectories to bound the geographical extent of probable source regions can help represent inherent trajectory uncertainty (Chapter 7, p. 149).

7. When available, knowledge of conditions along a trajectory (e.g., precipitation) help interpret chemical measurements.

8. The combined use of multiple chemical tracers (such as, radon) and trajectories may help interpret transport (e.g., Levy 1987).

Although the trajectory models briefly described above serve as the basis for many more complex and higher resolution chemical transport models, it was outside the scope of the present chapter, or indeed of the workshop, to discuss more sophisticated modeling techniques.

1.3.9. Recirculation

A special case of atmospheric transport that is particularly important for the introduction of materials into a continental atmosphere involves the large-scale return flows in the stationary or travelling synoptic-scale disturbances. When synoptic-scale circulation patterns are over a coastal region, they may carry continental materials offshore over some stretch along the coast but then return them to the continental atmosphere some several hundred kilometers further along. As an example, one might think of the Azores High off the coast of Africa or the Pacific High in summer. Relatively little attention has been paid to this phenomenon in the past, but the recent modeling work of Brost et al. (1988) and Levy (1989) has shown that such large-scale recirculation, seemingly against the ''annual average'' flow regime, can carry pollutants to unsuspected locations. Such circulation patterns are not easily accommodated in large-scale flux calculations based on mean flow patterns.

As was indicated in the discussion of the coastal zone, even on smaller scales local circulations may bring material back ashore that might have been expected to travel with the larger-scale flow. Intensive measurement programs and the application of chemical transport models of the appropriate scale are required to evaluate and quantify these transport paths.

1.3.10. Remote Sensing of the Long-Distance Transport of Trace Gases

Recent advances in remote-sensing technology and the associated data-analysis techniques have resulted in important new capabilities for measuring the distribution of tropospheric aerosols, carbon monoxide (CO), ozone (O_3), and water vapor (H_2O) over wide time and space scales. The initial applications of these techniques include studies of CO and O_3 produced in association with large-scale agricultural burning in the tropics (Fishman and Browell 1988, Fishman et al. 1986), the formation and transport of regional air pollution (Fishman et al. 1987), and the first measurements of the global distribution of CO in the free troposphere (Reichle et al. 1986).

The remote-sensing methods that have the most potential to contribute to studies of the long-distance transport of natural and contaminant trace gases in the next five to ten years are:

1. Airborne lidar techniques to measure simultaneously the vertical distribution of aerosols, O_3, and water vapor during intensive field campaigns (Browell et al. 1983). These

measurements are essentially continuous (ca. 5 Hz) and sufficiently accurate and precise to be useful over the entire range of concentrations found in the troposphere.

2. Surface-based lidar methods developed for remotely sensing aerosols and O_3 from fixed sites or a mobile laboratory to provide an almost continuous time-series measurement of the vertical distribution of O_3 and aerosols over a surface-monitoring site. These techniques are primarily limited by a relatively high electrical-power requirement and interference from clouds.

3. Space-based measurements applicable to long-distance transport studies include tropospheric O_3 studies using total ozone measurements from the Total Ozone Measurement Sensor (TOMS) on the Nimbus-7 satellite and CO measurements made with the Measurement of Air Pollution Sensor (MAPS), which has been flown on two space shuttle missions. Although the TOMS on Nimbus-7 is not expected to function much longer (it has already operated several years longer than expected), there are plans to put another TOMS on a polar-orbiting NOAA satellite in the early 1990s. MAPS is currently scheduled for another launch once the U.S. Space Shuttle Program resumes regular operations. There are opportunities for both instruments to be placed on future space platforms if a priority can be documented.

These remote-sensing techniques are most useful when they are incorporated into large-scale experiments with coordinated surface, aircraft, and satellite measurements. For transport studies, a sufficiently comprehensive network of meteorological observations for a detailed dynamical analysis is especially important.

1.4. CONCLUDING REMARKS

In this review, we have described the major atmospheric flow regimes of the world, suggesting where the potential for large-scale transport exists, and have identified the main portions of the atmospheric pathway between source and eventual receptor. Taking into consideration both the meteorological regimes and the continental source regions of natural and anthropogenic atmospheric constituents, large-scale atmospheric transport might be expected to occur eastward from North America, westward from northern Africa, eastward from southern South America, eastward from Asia, westward from northern South America, and into the polar regions. Other chapters of this book provide examples where such transport in these areas has been documented.

For large-scale transport to occur, a combination of smaller-scale atmospheric processes must be favorable. We have described the individual portions of the atmospheric pathway, or the physical and chemical processes that comprise them, to show how they may influence large-scale transport. We selected four portions of the atmospheric pathway as being of particular interest for furthering our ability to estimate large-scale mass fluxes: vertical-exchange processes, pollutant interactions with clouds, coastal-zone effects, and recirculation.

As a result of discussions at the NATO workshop, we included a few additional topics, a review of trajectory modeling, and a description of current remote-sensing capabilities. The other participants considered these subjects to be of particular interest, and they also complemented deliberations in other chapters.

Previous workshops on related topics have recommended important future studies. Two, the 1984 NATO ARW and the 1975 Ocean Sciences Board (National Research Council) workshop (Galloway et al. 1985, National Academy of Sciences 1978, respectively), are in large part still relevant. On consideration of these earlier recommendations and the topics included in this chapter, we believe the following three items would be particularly valuable for better quantifying large-scale mass fluxes through the atmosphere.

1. The characterization of major source regions--including their temporal behavior and the atmospheric (and other) processes that control their effectiveness--and the measurement of vertical distributions of emitted materials slightly downwind of the major source regions in the major climatic zones and meteorological regimes. Such information would permit improved estimates of the effectiveness of these source regions and of their potential for contributing to large-scale transport.

2. An intensive experimental and modeling study in the near-coastal zone (i.e., some few hundred kilometers inland and from a few hundred to a few thousand kilometers over the ocean) downwind of a major source region under important offshore meteorological regimes. Such a study would lead to improved estimates of near-coastal deposition and of the amount of material available for more distant transport. A specific component of such an undertaking might be an experiment to study the interaction of fine-particle anthropogenic aerosol with sea-salt particles and the subsequent impact on wet and dry removal. Such an experiment would involve a sophisticated boundary-layer chemical transport model that would include the chemical constituents sulfur dioxide, nitrogen oxides, and aerosols as a function of particle size as well as the dynamics processes of the particles (i.e., coagulation, diffusion, etc.). Measurements would be made in a line away from, for example, the east coast of North America out to sea and would include vertical profiles (up to 2 km) of selected gases and aerosol trace elements and their size distributions. Simultaneous detailed meteorological observations would be made in the boundary layer, and related physical and chemical sea-water parameters would be measured as a function of distance from the coast. Such an experiment would fit well within either the Western Atlantic Ocean Experiment (WATOX) or Atmosphere/Ocean Chemistry Experiment (AEROCE) programs.

3. An intensive modeling and experimental study in the near-
coastal zone of a ''receiving'' continent under appropriate meteorological conditions. Such a study should examine mesoscale
features associated with sea-breeze effects, possible large-
scale recirculation effects, and the local enhancement of deposition in the near-coastal zone.

It is becoming increasingly evident that the understanding necessary
to describe and quantify the large-scale transport of materials through
the atmosphere, and specifically in this case the continent-to-continent
and continent-to-ocean transport, will be achieved more rapidly by using
a combination of techniques. Virtually all studies on this scale will
benefit from the combination of field and modeling components that are
coordinated and interactive. Investigations of the type we have suggested in this chapter and throughout this book are sufficiently complex and
extensive that one institute, agency, or even country could not hope to
conduct one successfully alone. Instead, the cooperation of a number of
groups and agencies, within the context of a coordinated international
program, would be the most effective.

1.5. REFERENCES

Barrie, L. A. 1986. Arctic air pollution: An overview of current knowl-
edge. Atmos. Environ. 20:643-663.
Barrie, L. A., M. P. Olson, and K. K. Oikawa. 1989. The flux of anthropic
sulphur into the Arctic from midlatitudes in 1979/80. Atmos. Environ. 23, in press.
Brost, R. A., R. B. Chatfield, J. P. Greenberg, P. L. Haagenson, B. G.
Heikes, S. Madronich, B. A. Ridley, and P. R. Zimmerman. 1988.
Three-dimensional modeling of transport of chemical species from
continents to the Atlantic Ocean. Tellus 40B:358-379.
Browell, E. V., A. F. Carter, S. T. Shipley, R. J. Allen, C. F. Butler,
M. N. Mayo, J. H. Siviter, Jr., and W. M. Hall. 1983. NASA multi-
purpose airborne DIAL system and measurements of ozone and aerosol
profiles. Appl. Optics 22:552-534.
Buat-Ménard, P. (ed.). 1986. The Role of Air-Sea Exchange in Geochemical
Cycling. NATO ASI Series C, Vol. 185, Dordrecht:Reidel, 549 pp.
Draxler, R. R. 1987. Sensitivity of a trajectory model to the spatial and
temporal resolution of the meteorological data during CAPTEX.
Climate Appl. Meteor. 26:1577-1588.
Fishman, J., and E.V. Browell, 1988. Comparison of satellite total ozone
measurements with the distribution of tropospheric ozone obtained by
an airborne UV-DIAL system over the Amazon Basin. Tellus 40B:393-
407.
Fishman, J., P. Minnis, and H.G. Reichle, Jr. 1986. Use of satellite data
to study tropospheric ozone in the Tropics. J. Geophys. Res. 91:
14,451-14,465.
Fishman, J., F. M. Vukovitch, D. R. Cahoon, and M. C. Shipham. 1987. The
characterization of an air pollution episode using satellite total
ozone measurements. Climate Appl. Meteor. 26:1638-1654.

Galloway, J. N., R. J. Charlson, M. O. Andreae, H. Rodhe, and M. S. Marston (eds.). 1985. The Biogeochemical Cycling of Sulfur and Nitrogen in the Remote Atmosphere. NATO ASI Series C, Vol. 159, Dordrecht:Reidel, 249 pp.

Hasse, L. 1983. Introductory meteorology and fluid dynamics. In Air-Sea Exchange of Gases and Particles (P. W. Liss and W. G. N. Slinn, eds.) NATO ASI Series C, Vol. 108, Dordrecht:Reidel, 1-51.

Hastie, D. R., H. I. Schiff, D. M. Whelpdale, R. E. Peterson, W. H. Zoller, and D. L. Anderson. 1988. Nitrogen and sulfur over the Western Atlantic Ocean. Atmos. Environ. 22:2381-2391.

Hoflich, O. 1984. Climate of the South Atlantic Ocean. In Climates of the Oceans (H. Van Loon, ed.) World Survey of Climatology, Vol. 15, New York:Elsevier.

Kahl, J. D., and P. J. Samson. 1986. Uncertainty in trajectory calculations due to low resolution meteorological data. Climate Appl. Meteor. 25:1816-1831.

Keeler, G., and P. J. Samson. 1989. Testing source-receptor relationships for trace elements. Environ. Sci. Technol., in press.

Kennett, J. P. 1981. Marine tephrachronology. In The Sea (C. Emiliani, ed.) New York:Wiley.

Kuo, Y-H., M. Skumanich, P. L. Haagenson, and J. S. Chang. 1985. The accuracy of trajectory models as revealed by the observing system simulation experiments. Mon. Wea. Rev. 113:1852-1867.

Levy, H. 1987. Tracers of atmospheric transport. Nature 325:761-767.

Levy, H. 1989. U.S. and Canadian combustion nitrogen emissions: A numerical simulation of global and regional transport. Tellus , in press.

Liss, P. W., and W. G. N. Slinn (eds.). 1983. Air-Sea Exchange of Gases and Particles. NATO ASI Series C, Vol. 108, Dordrecht:Reidel, 561 pp.

Meserve, J. M. 1974. North Atlantic Ocean. Vol. 1 (rev. ed.) of U. S. Navy Marine Climatic Atlas of the World. NAVAIR 50-1C-528, Naval Air Systems Command, Washington, D.C., 371 pp.

Miller, J. M. 1987. The use of back air trajectories in interpreting atmospheric chemistry data: A review and bibliography. NOAA Tech. Memo., ERL/ARL-155 (available from NTIS, 5285 Port Royal Road, Springfield, VA 22161), 28 pp.

Miller, J. M., and J. M. Harris. 1985. The flow climatology to Bermuda and its implications for long-range transport. Atmos. Environ. 19: 409-414.

Moody, J. L., and J. N. Galloway. 1988. Quantifying the relationship between atmospheric transport and the chemical composition of precipitation on Bermuda. Tellus 40B:463-479.

National Academy of Sciences (Ocean Sciences Board). 1978. The Tropospheric Transport of Pollutants and Other Substances to the Oceans. (Workshop on Tropospheric Transport of Pollution to the Oceans) Washington:National Academy Press, 243 pp.

Reichle, H. G., Jr., V. S. Conners, J. A. Holland, W. D. Hypes, H. A. Wallio, J. C. Casas, B. B. Gormsen, M. S. Saylor, and W. D. Hasketh. 1986. Middle and upper tropospheric carbon monoxide mixing ratios as measured by a satellite-borne remote sensor during Novermber 1981. J. Geophys. Res. 91:10,865-10,887.

Samson, P. J., and J. L. Moody. 1980. Trajectories as two-dimensional probability fields. In Air Pollution Modeling and its Applications (C.D. Wispelaere, ed.) New York:Plenum Press, 43-54.

Sievering, H., J. Boatman, M. Luria, and C. C. Van Valin. 1989. Sulfur dry deposition over the Western North Atlantic: The role of coarse aerosol particles. Tellus, in press.

Streten, N. A., and J. M. Zillman. 1984. Climate of the South Pacific Ocean. In Climates of the Oceans (H. Van Loon, ed.) New York:Elsevier, 263-429.

Terada, K., and M. Hanzawa. 1984. Climate of the North Pacific Ocean. In Climates of the Oceans (H. Van Loon, ed.) New York:Elsevier, 431-501.

Tucker, G. B., and R. G. Barry. 1984. Climate of the North Atlantic Ocean. In Climates of the Oceans (H. Van Loon, ed.) New York:Elsevier, 193-262.

Whelpdale, D. M., T. B. Low, and R. J. Kolomeychuk. 1984. Advection climatology for the east coast of North America. Atmos. Environ. 18:1311-1327.

2. THE LONG-RANGE ATMOSPHERIC TRANSPORT OF TRACE ELEMENTS A CRITICAL EVALUATION

Thomas M. Church
College of Marine Studies
University of Delaware
Newark, DE 19711

Richard Arimoto
Graduate School of Oceanography
University of Rhode Island
Narragansett, RI 02882-1197

Timothy D. Jickells
School of Environmental Sciences
University of East Anglia
Norwich NR4 7TJ, United Kingdom

Leonard A. Barrie
Atmospheric Environment Service
4905 Dufferin Street
Downsview, Ontario,
Canada M3H 5T4

Leon Mart
KFA, Nuclear Research Center
Chemistry Institute
P. O. Box 1913
D-5170 Juelich
Federal Republic of Germany

Frank Dehairs
ANCH
Vrije Universiteit Brussel
Pleinlaan 2
B-1050 Brussels, Belgium

William T. Sturges
Department of Chemistry
Institute of Aerosol Science
University of Essex
Wivenhoe Park
Colchester, Essex CO4 3SQ
United Kingdom

Francois Dulac
Centre des Faibles
   Radioactivites
Laboratoire mixte CNRS-CEA
Avenue de la Terrasse, B. P. 1
91198 Gif-sur-Yvette Cedex,
France

William H. Zoller
Department of Chemistry, BG-10
University of Washington
Seattle, WA 98195

2.1. OVERVIEW

Throughout the earth's geological history trace elements have been atmospherically transported from continent to continent and from continent to ocean; this long-range atmospheric transport is now recognized as an important, if not the most important, mode of global transport for a variety of trace substances and elements. The anthropogenic trace substances found in remote locations attest to the rapidity and extent of such transport.

Trace elements are of interest in studies of long-range atmospheric transport because of their importance in biogeochemical processes, including global oceanic fluxes. Tracers from particular source regions or industrial processes, including various metals with characteristic origins, have been used to document the dispersal caused by atmospheric transport. Trace elements with magnitudes and sources well documented

37

A. H. Knap (ed.), The Long-Range Atmospheric Transport of Natural and Contaminant Substances, 37–58.
© 1990 Kluwer Academic Publishers.

over geologic time would be the ideal tracers. However, the available
data are restricted either to records of trace-element deposition or to
atmospheric data of limited spatial or temporal extent. Thus, the precise nature of a trace element's origin or its mode of atmospheric
transport is often imperfectly known.

Originally, Dr. Church, as chairman of the working group on the
atmospheric transport of trace elements, wrote this chapter as a background paper. It was subsequently amended by the entire working group
to summarize the current knowledge of the global atmospheric transport
of trace elements. The chapter now includes a literature review of the
sources of trace elements (and the methods of deconvoluting natural and
anthropogenic origins), their modes of atmospheric transport and scavenging, and the effects and significance of their transport. The
questions we sought to address were, ''What evidence is there of the
long-range transport of trace elements?'' and, if convincing evidence
existed, ''What is the significance of such transport to oceanic areas?''
Previous NATO ASI volumes (Buat-Ménard 1986a, 1986b; Liss and Slinn
1983) have focused on the air/ sea exchange process; the focus of this
chapter is restricted to the atmospheric transport of trace elements.
In Chapter 9 (p. 177), our working group discusses four case studies
that we felt illustrate the role of long-range atmospheric transport of
trace elements as a pathway of chemicals to oceans and the gaps in our
current knowledge.

2.2. ATMOSPHERIC TRACE ELEMENTS AND THEIR SOURCES

The trace elements that we considered, their sources, their usefulness
as tracers, and their interference factors are listed in Table 2-1. An
interference factor (IF), as first proposed by Lantzy and Mackenzie
(1979), is calculated as the ratio

$$IF = \frac{\text{total global anthropogenic emissions}}{\text{total global natural emissions}} \times 100.$$

Although IFs are useful in assessing man's impact on the atmosphere, the
geographic variations of anthropogenic and natural emissions and the
limited lifetimes of aerosols in the atmosphere cause the scale of man's
impact to vary regionally. We calculated the global IF using the most
recently available data cited in the references. In some cases, we had
to rely on rather old estimates and, especially for anthropogenic emissions, the data may not be reliable. Nevertheless, the IFs showed that
the anthropogenic emissions of many elements in the periodic chart exceeded the natural atmospheric inputs by 10 to 10,000 times. For some
elements (specifically Na, Mg, Ca, Sc, Al, Si, and natural radionuclides), interference factors are usually near 100 since their concentrations are contributed equally by natural and anthropogenic sources.
These elements are often used as tracers of sea salt or crustal dust.

Trace elements are generally introduced to the atmosphere primarily
by energetic processes at the earth's surface. Cosmic inputs are

Table 2-1. Primary sources of atmospheric tracers and trace elements.

Element	Primary Sources	Group	Interference Factor
Ag	Municiple sewage		
Al	Crustal	1a	
As	Volcanic, anthropogenic	3	300
B	Coal burning, oceanic	1c	26
Ba	Crustal, diesel fuel (isotopes may be useful)	1c	
Be	Crustal, coal burning	1c	
^{7}Be	Stratospheric, cosmic, radio isotope	2	
^{210}Bi	Volcanic	1c	
Br	Sea salt, car exhaust	1c	
^{14}C-	Cosmic, nuclear weapons		
Ca	Sea salt, cement manufacture, crustal	1c	
Cd	Mixed anthropogenic emissions	3, 1c	760
Co	Crustal, anthropogenic	1c	63
^{137}Cs	Anthropogenic fallout, radioisotope	2	
Cu	Crustal, volcanic, anthropogenic	3	296
Dy	Diesel emissions	1b	
Fe	Crustal, anthropogenic	3, 1c	39
Hg		3	100–200
I	Sea salt, oceanic emissions		
^{131}I	Power-plant radioisotope	2	
In	Semlting, incineration	1b	
Ir	Mainly volcanic, meteoritic	1c	
K	Crustal > sea salt, biomass burning	1a	
^{85}Kr	Power-plant radionuclide	2	
Mg	Sea salt > crustal		
Mn	Crustal, anthropogenic	1c	52
Mo	Anthropogenic input	3	4500
Na	Sea salt > crustal	1a	
Ni	Crustal, petroleum burning	1b	180

Table 2-1. (continued)

Element	Primary Sources	Group	Interference Factor
Pb	Gasoline cumbustion, volcanic, smelters	1b	2400
Pd	Possible future tracer of car emissions	1b	
^{210}Po	Volcanic	2	
Pt	Possible future tracer of car emissions	1b	
^{238}Pu	Anthropogenic radionuclide	2	
^{239}Pu	Anthropogenic radionuclide	2	
^{240}Pu	Anthropogenic radionuclide	2	
Rare earths	Crustal, petroleum refining, diesel emissions	1c	
^{222}Rn	Crustal emissions	2	
^{106}Ru	Anthropogenic radionuclide	2	
Sb	Coal combustion, incineration	1c	3900
Sc	Crustal	1a	
Se	Coal burning, oceanic	3	63
Si	Crustal	1a	
Sn	Incineration	1b	550
^{90}Sr	Anthropogenic radioisotope	2	
U (series)	Crustal, atmospheric Rn source	2	
V	Crustal, oil combustion	1c	320
Zn	Anthropogenic, vegetation	3	78,500

Note: Group 1 – Primary sources: 1a. natural, 1b. anthropogenic, 1c. mixed origin.
Group 2 – Radionuclides and isotopes, tracers.
Group 3 – Tracers, biogeochemically important, atmospherically transported

minimal. Natural sources for trace elements include the wind-blown
resuspension of crustal material, volcanic emissions, combustion products from forest fires, and biogenic emissions. Of these, wind-blown
crustal dusts from arid and semiarid areas are particularly important
(see Chapters 3 and 10, pp. 59 and 197, respectively). From areas,
such as the Sahara, dust is a continuous major source that is periodi-
cally augmented by outbreaks of dust storms (d'Almeida 1986, Schutz
1980). The transport of crustal materials to the ocean is well documented for areas downwind of desert regions--especially in the North
Atlantic (Delany et al. 1967, Schutz et al. 1980) and the Mediterranean
(Eriksson 1979, Ganor and Mamane 1982) from the Sahara; in the North
Pacific from Asia (Uematsu et al. 1985); and in the South Pacific from
Australia (Collyer et al. 1984). The transport of Saharan red soils to
the European continent is also well documented (Prodi and Fea 1979,
Reiff et al. 1986).

Volcanic emissions, although difficult to quantify, are another
major natural source of trace elements in the atmosphere. Naturally,
the more volatile species Po (Lambert et al. 1982), Bi (Lee et al.
1986), and perhaps Pb (Patterson and Settle 1987) are the most strongly
affected by volcanic sources. However, Patterson and Settle (1987)
estimate that in preindustrial times volcanic inputs accounted for only
about a quarter of the natural atmospheric lead inputs and now account
for only a negligible fraction of the industrial inputs. Lambert et al.
(1988) have quantified trace-element volcanic fluxes by normalizing the
trace-element concentrations (multiplied by an oxide volatility factor)
to that of sulfur, because the volcanic inputs of sulfur are better
known. Lambert et al. (1988) corroborate their lead volcanic emission
estimates from the volcanic ^{210}Po emission flux, the $^{210}Pb/^{210}Po$ ratio
in volcanic gas and the stable $Pb/^{210}Pb$ ratio in associated lava. A
case study on atmospheric volcanic transport is presented in Chapter 8
(p. 163). An inventory of natural and anthropogenic trace-element
atmospheric emissions is presented in Section 2.3 (p. 46).

Emissions from forest fires and biogenic processes are poorly documented but these could be important sources for certain compounds.
Pacyna (1986) has estimated the magnitude of the trace-element emissions
from forest fires but with a great deal of uncertainty. In the case of
biogenic processes, microbial alkylation is of particular concern. Certain elements, such as the metalloids (As, Se, Te, Sb) and noble metals
(such as, Pd, Au) are alkylated by biologically driven reactions.
Hewitt and Harrison (1987) have reported evidence suggesting naturally
alkylated emissions of lead; Fitzgerald (1986) and Mosher et al. (1987)
have found important biogenic emissions from the sea of mercury and of
selenium, respectively.

Man is primarily responsible for the enrichment of many trace elements now found in the atmosphere (Lantzy and Mackenzie 1979) caused by
the combustion of fossil fuels, including such additives as the lead in
gasoline, roasting of ores for refining metals, processing of crustal
materials for manufacturing cements, and burning of waste materials
(Bertine and Goldberg 1971). Man has also developed greater means to
efficiently inject trace elements into the atmosphere by using increasingly taller industrial stacks.

Because the global enrichment of trace elements in the atmosphere tends to parallel the trend of volatility indices, it is sometimes difficult to ascertain whether a trace element is from a natural or an anthropogenic source (Duce et al. 1975). However, several approaches have been used to evaluate the magnitude of trace-element pollution in the atmosphere by determining the enrichments of trace elements over natural levels.

Ice cores and water-column profiles can be used to establish the increases in fluxes over time by providing historical trends in trace-element fluxes. Although the ice cores have yielded data on Pb (Boutron and Patterson 1987, Wolff and Peel 1985), few reliable values have yet been established for other trace elements because of sample contamination problems and analytical difficulties. Lake sediments (Shirahata et al. 1980), marine sediments (Veron et al. 1987), and ombrotrophic bogs (Schell et al. 1986) have all been used to establish historical trends despite the complications caused by bioturbation, elemental migration, and other disturbances to the profiles. Banded corals have also been used to assess temporal variability (Shen and Boyle 1987). However, it remains difficult to convert measured concentration profiles into accurate time scales that correspond to atmospheric fluxes.

Another approach used to establish the amount by which anthropogenic trace elements exceed natural trace elements is to compare data from an area known to be impacted by long-range transport to data from an apparently ''pristine'' environment. For example, Arimoto et al. (1985) report atmospheric lead concentrations of 120 pg/m^3 from samples at Enewetak in the ''dirty'' Northern Hemisphere compared to 16 pg/m^3 in the samples from American Samoa in the ''clean'' Southern Hemisphere. Sturges and Barrie (1987) recorded mean levels during the Arctic winter (polluted) season of 3.5 ng Pb/m3 but summer values of only 0.39 ng/m3. Such comparisons are, of course, complicated because any two areas could have different natural components. Instead of using spatial comparisons as in the previouse example, concentrations at times of high and low pollution are sometimes compared. Barrie (1986), Lannefors et al. (1983), and Mart (1983) have used this method in their studies of the Arctic atmosphere with its annual cycle of high winter pollution and clean summer conditions. In this approach all comparisons are to background levels, which are not necessarily natural levels.

Enrichment factors (Rahn 1976) can be used to ascertain the expected contributions from crustal and marine sources. The enrichment factor relative to crustal rock is given by

$$EFcrust(X) = \frac{[X/R_c]air}{[X/R_c]crust},$$

where X is the concentration of the trace element of interest and R_c the concentration of crustal reference element (generally Al or Si but occasionally Fe, Ba, or Sc). Although the crustal ratio is taken from mean crustal abundances, crustal materials may differ considerably in their compositions (Bowen 1979) giving an uncertainty up to a factor of 10.

The seawater enrichment factor is similarly defined:

$$EFsea(X) = \frac{[X/R_s]air}{[X/R_s]sea},$$

where R_s, the seawater reference element, is normally Na but may be Mg. Chlorine has also been used as a reference element; however, in most circumstances (except close to the surface of the sea), its volatility makes it unsuitable (Duce and Hoffman 1976). The reference material for calculating EF(sea) is surface seawater, but the concentrations of trace constituents can vary spatially (Church et al. 1984). Tables 2-2 and 2-3 show some values of crustal and seawater enrichment factors in remote areas.

Table 2-2. Crustal enrichment factors at remote continental and marine sites.

Element	Great Smokey Mountains	Olympic Mountains	U. S. Glacier Park	Enewetak Atoll Season Wet	Dry	American Samoa
			Crustal			
Ba	0.7	1.8	20			
Co				0.9	1	13
Fe	4.8	4.1	7	0.9	0.8	1.9
K						560
Mn				0.9	1.1	7
Sc				0.8	0.6	0.9
Ti	0.5	0.6				
			Marine			
Ca	0.6	0.5	6			
I				10,000	73,000	
Mg	2.5	11	2			3,700
Na	1.6	8.2				13,000
			Anthropogenic			
Ag	280	1,300	<2,900	83	1,200	4,900
As	<960	960	780			
Cd	650	2,000	5,400	75	180	
Cu	31	76		1.9	6.7	52
Pb	1,300	130	400	11	110	190
Sb						210
Se				3,700	48,000	210,000
V				1.3	5.5	
Zn	51	94		3.6	33	140

Sources: Davidson et al. (1985) for the Great Smokey Mountains, Olympic Mountains, and U. S. Glacier Park; Duce et al. (1983) for Enewetak Atoll; and Arimoto et al. (1987) for American Samoa.

Table 2-3. Seawater enrichment factors calculated from aerosol data at remote marine locations.

Element	BIMS*	North Atlantic	Hawaii	Enewetak Atoll	American Samoa
			Crustal		
Al	5×10^3	1×10^5	3×10^5	1×10^5	4,600
Co	6×10^1	8×10^3	$< 1 \times 10^3$	1×10^3	540
Fe	1×10^4	4×10^7	6×10^5	3×10^4	1,500
K	1	1.3	1.1	1.3	1.2
Mn	1×10^3	4×10^5	8×10^3	2×10^4	4,000
Sc	1×10^1	1×10^5		4×10^4	490
			Marine		
Ca					1.1
I					140
Mg	1	0.9	1.0	1.1	1.0
Se					20,000
			Anthropogenic		
Ag					1,600
Cu	8×10^2	6×10^4	7×10^3	3×10^3	1,400
Pb	4×10^3	5×10^5	5×10^5	3×10^4	30,000
Sb					4.6
V	1×10^2	9×10^3	4×10^2	2×10^2	
Zn	2×10^4	2×10^6	6×10^5	1×10^5	70,000

Sources: Weisel et al. (1984) for BIMS; Duce et al. (1976) for the North Atlantic; Hoffman and Duce (1972) and Duce and Hoffman (1972) for Hawaii; Duce et al. (1983) for Enewetak; and Arimoto et al. (1987) for American Samoa.

*Bubble Interfacial Microlayer Sampler

An enrichment factor of unity with respect to crustal rock or sea-water suggests that the crust or seawater may be the dominant source. However, some pollutant materials, such as fly ash, contain certain elements the proportions of which are indistinguishable from those in the earth's crust (Wangen 1981). In addition, enrichment factors greater than unity may also be the result of other natural inputs. For example, volcanic emissions may be enriched in many trace elements (Fruchter et al. 1980). Marine aerosols may become enriched in elements during formation by bubble bursting or through gas exchange (Duce and Hoffman 1976, Weisel et al. 1984, Whitehead 1984). Natural combustion may also cause an enrichment in atmospheric trace elements.

Enrichment factors less than unity may arise in some circumstances; for example, where there is a loss of halogens from sea-salt aerosol (Kritz and Rancher 1980). Elements derived from the earth's crust have an EF(crust) close to unity (Table 2-2) with some variability, presum-ably from differences in the nature of the crustal material. Elements derived from seawater (Na, Ca, Mg) may be enriched relative to crustal rock even at remote continental sites. The anthropogenically influenced elements (e.g., Pb, Cd, Zn) all show high enrichment factors. Se and I are highly enriched, partly because of fractionation during oceanic emission in gaseous forms. Thus, although enrichment factors can be useful, the effects of different natural sources can complicate any simple interpretation.

Statistical analyses, such as multiple linear regression, have been used to interpret the variability of the crustal, marine, and anthropogenic components in long-range transport (see Chen et al. [1985] for transport to New Zealand and Schneider [1987] for transport to the Kiel Bight). Factor analyses and cluster analyses group elements together by communalities and these technologies have been used to assess industrial, crustal, and marine contributions to trace-element concentrations. Although some multivariant techniques have been used to derive the relative strength of a source, these techniques are more qualitative than quantitative. Examples of such statistical approachs are given by Slanina et al. (1983), Dutkiewicz et al. (1987), Lowenthal and Rahn (1985), Pacyna et al. (1985), Rahn (1985), and Rahn and Lowenthal (1984).

Another technique that may be used to identify the source of an element is the analysis of stable isotopes. Pb stable isotopes have been routinely used for source identification (Chow 1970, Patterson and Settle 1987, Petit et al. 1984, Settle et al. 1982, Shirahata et al. 1980, Sturges and Barrie 1989); Grousset et al. (1988) have used Nd. In the case of Pb, stable-isotope abundances in aerosol particles vary according to the ore sources used for gasoline additives. By examining air trajectories in conjunction with Pb-isotope data, much insight has been gained on the specific continental origins of lead in aerosols from the Pacific Ocean (Settle et al. 1982) and the Mediterranean Sea (Maring et al. 1987). Evidence of the continent-to-continent atmospheric transport of lead in the Northern and Southern Hemispheres has also been found in the ice-core records from Greenland (Murozumi et al. 1969) and the Antarctic (Boutron and Patterson 1983).

Several case studies of natural and anthropogenic Pb have evaluated trace-element sources. Boutron and Patterson (1983, 1987) report Pb concentrations of 1.6 ng Pb/kg in snow deposited from 1797 to 1801 with an increase to 8.4 ng/kg in snow deposited from 1970 to 1973 in the Antarctic snow cores they analyzed. In Greenland, Murozumi et al. (1969) found lead concentrations of only 1.4 ng/kg for 5500-1500 before present but more than 200 ng/kg in the surface snow of recent origin thereby providing the data necessary to show that, in the Northern Hemisphere, atmospheric pollutant levels of lead far exceed natural ones (here by a factor of 140). As further corroboration of the increased level of anthropogenic Pb, Schell et al. (1986) found EF(crust) values for Pb in ombrotrophic bogs in mountainous areas in the United States of 0.7 ng Pb/kg from pre-1800 samples and 71 ng Pb/kg for post-1981 samples, an increase of about 100 ng Pb/kg in enrichment. These are only a few of the existing depositional records that show man's recent contamination of the global atmosphere through the indiscriminate use of Pb.

Again using stable isotopes from different sources of anthropogenic lead, the atmospheric flux of anthropogenic lead to the waters of the North Atlantic Ocean has been clearly seen in coral records (Shen and Boyle 1987), surface sediments (Veron et al. 1987), and surface-water

profiles (Schaule and Patterson 1981). Fogg and Duce (1985) believe
that isotopes of boron, which is emitted from both the ocean and from
the combustion of coal, may also be used in the future to gain additional insight into transport processes.

2.3. TRACE-ELEMENT EMISSIONS TO THE ATMOSPHERE

Pacyna (1986) recently compiled an inventory of natural and anthropogenic sources of trace elements to the atmosphere for 1975 relying
heavily on the earlier compilations of Lantzy and Mackenzie (1979) and
Nriagu (1979). The significant differences between the anthropogenic-
emission estimates of Pacyna (1984, 1986) and those of Lantzy and
Mackenzie (1979), both nominally from 1975, have raised important uncertainties about the interference factors calculated above. Also, the
emission data of Pacyna (1984, 1986) are limited to a few elements
despite clear evidence from other data of the enrichment of trace elements on aerosol particles (Wiersma and Davidson 1986). More recent
oceanic-emission inventories are available for a few specific elements--
B (Fogg and Duce 1985), Hg (Fitzgerald 1986), Se (Mosher and Duce 1987),
Sn (Byrd and Andreae 1982). Dr. Buat-Ménard's estimates of the global
volcanic-emission rates given in Chapter 8 (p. 163) agree generally with
those of Pacyna (1984, 1986).

Reductions in anthropogenic atmospheric emissions of trace elements
are expected because of improved emission controls and product-use patterns introduced in developed countries (Boyle et al. 1986, Schaule and
Patterson 1981). However, industrial activity and power consumption have
increased in developing countries (see Chapter 4, p. 87). Thus, our
knowledge of global-emission inventories has become adequate to define
the scale of global interference in the atmospheric transport for only a
few elements (Table 2-4). The anthropogenic source of each of the elements listed in Table 2-4 exceeds the natural source, and many of the
elements have significant atmospheric inputs to the ocean whereas in the
past fluvial sources were probably dominant.

Table 2-4. Worldwide trace-metal atmospheric emissions
(10^6 kg/yr) of enriched trace elements.

Element	Sources	
	Natural	Anthropogenic
As	7.8	24
Cd	0.96	7.3
Cu	19	56
Ni	26	47
Pb	19	449
Se	0.4	1.1
Zn	4	314

Sources: Data from Bewers et al. (1988) and Pacyna (1986).

2.4. ATMOSPHERIC TRACE-ELEMENT TRANSPORT

Aeolian dust in the atmosphere is efficiently transported across large distances. Indeed, the iron-rich soils of many remote islands, such as Bermuda, have been attributed to aeolian transport. The soil dust brought to Bermuda by wind systems over pathways of thousands of kilometers arrive circuitously from the south, east, and west (e.g., Chen and Duce 1983). The nature and chemistry of aeolian dust transport is discussed in Chapter 3 (p. 59).

Trace elements from anthropogenic emissions are usually concentrated in much finer aerosol particles than crustal elements. Thus many anomalously enriched trace elements are transported to remote areas on submicrometer aerosols (e.g., Duce et al. 1975). Once such trace-element-enriched aerosols are mixed into the free troposphere and stratosphere, the probability of their removal decreases and they can be transported over considerable distances.

The sequence of volatilization-condensation-coagulation that takes place during industrial processing can transfer trace elements from an original high-density, oxide-mineral phase (density > 3 g/cm^3) to a low-density small-particle phase (density ~ 1.7 g/cm^3), as discussed for Zn and Cu by Sugimae (1984). Such low-density, small-aerosol particles can be rich in elemental carbon (Bradley et al. 1981). The rapid oxidation in ambient air of volatile organometal compounds after industrial or biological emission can also enhance metal concentrations in the particulate phase. By contrast mercury oxidizes slowly in the atmosphere; thus gas-to-particle conversion does not occur (i.e., there is more Hg in the vapor phase than in the particle phase), giving a greater potential for efficient long-range transport (Fitzgerald 1986). Although phase and oxidation-state transformations affect the long-range transport of trace elements, they also tend to mask any original source signals. These complications emphasize the value of discrete particle analysis along trajectories downwind of an emission source (Bruynzeel and Vergrieken 1985) or of statistical analysis of multi-element or elemental source signatures (Dutkiewicz et al. 1987, Lowenthal and Rahn 1985, Pacyna et al. 1985, Rahn and Lowenthal 1984).

There are several secondary effects that can complicate otherwise simple interpretations of atmospheric data. Enrichment factors for Pb in atmospheric aerosols are generally high and sea-salt recycling of lead in remote oceanic areas can have a confounding effect (Settle and Patterson 1982). However, a strong correlation of Br (normally a marine element) with Pb in some areas is indicative of a predominant leaded-gasoline combustion source (Dulac et al. 1987, Lininger et al. 1966, Oblad and Selin 1986, Sturges and Harrison 1986). In their investigations of possible biomethylation contributions to atmospheric lead levels, Hewitt and Harrison (1987) have found that 5% to 50% of the total Pb measured at a remote Scottish island is in the form of organic lead (but not as the gasoline additive tetra alkyl lead). There is still, however, some controversy as to whether this alkylated Pb is from natural or anthropogenic sources.

Probably the best documented tracers of long-range atmospheric transport come from radionuclides associated with nuclear explosions.

The results of studies on radionuclides suggest hemispheric transport
and scavenging on time scales of weeks and interhemispheric exchange on
the order of months. However, the disadvantage of studying bomb debris
is that the emissions are sporadic from restricted areas and from upper
tropospheric and stratospheric testing. Examples of the anthropogenic
tropospheric injection of radionuclides include nuclear power-plant
accidents, such as the one at Chernobyl (see Chapter 7, p. 149; see
also Smith and Clark 1986). The unfortunate accident at Chernobyl did
produce several tracers that Fowler et al. (1987) have recovered from
particle traps in the Mediterranean Sea. Buesseler et al. (1987) and
Kempe and Nies (1987) have conducted similar research using traps in the
Black Sea and the North Sea, respectively. One important finding has
been that tropospheric trajectories can change dramatically over the
Eurasian continent on rather short and unpredictable time scales (see
Chapter 7, p. 149).

2.5. ATMOSPHERIC TRACE-ELEMENT SCAVENGING

Several factors affect the scavenging of atmospheric aerosols and their
associated trace elements during long-range transport. Aerosol trace
elements can be removed from the atmosphere by either wet or dry
deposition.

Dry deposition occurs primarily by gravitational settling and
Brownian diffusion. Choularton et al. (1982) have demonstrated that, in
a stratified atmosphere typical of stable atmospheric conditions
(Richardson number $Ri > 0$), more of the larger particles (> 0.3 μ in
diameter) are removed by gravitational settling. However, during times
of increased turbulence and buoyant motion, such larger particles can
remain suspended and thus be transported over longer distances. In
general, increased turbulence will redistribute aerosols vertically
upward, thereby increasing the possibility of their being transported
over long distances.

Wet deposition, by incorporating particles and their gaseous precursors into precipitation, also affects long-range transport (Schutz
and Kramer 1987). Precipitation, or dew, allows differential scavenging
mainly by virtue of differences in the hygroscopicity and solubility of
the aerosol particle within the cloud and the size of the particle below
the cloud (Barrie and Schemenauer 1986). Relatively large, crustal
aerosol particles (coarse-mode particles) and associated coatings are
removed preferentially by below-cloud impaction processes; more soluble
particles become hygroscopic cloud nuclei and are removed by nucleation
and entrainment within the cloud (Jaffrezo and Colin 1988). For
example, Bergametti (1987) and Tanaka et al. (1980) have seen the preferential removal of large particles by precipitation. However, some
studies suggest that in urban areas small Pb-rich particles have higher
scavenging ratios than large particles because they can function as
nucleation sites during hydrometer formation (Jaffrezo 1987), perhaps
from the increased hygroscopicity when SO_2 oxidizes to sulfate. Over
the open ocean, organosulfur compounds (chiefly DMS which phytoplankton
emit) can be rapidly oxidized to the sulfate aerosol particles that form

the primary nuclei for marine clouds (Charlson et al. 1987). Mineral
aerosols may also be removed over the ocean by in-cloud scavenging of
excess marine sulfate, a significant portion of which is acidic (Charlson and Rhode 1982) and may also solubilize trace elements within the
particle.

In coastal areas, diurnal heating and cooling cycles cause the
convective transport and recirculation of air between sea and land in
the boundary layer (Chapter 1, p. 3). Diurnal cycles could possibly
lead to enhanced deposition in coastal areas but we are not sure.
Although precipitation measurements over the ocean are scarce (Austin
and Geotis 1980, Cambray et al. 1975, Elliot and Reed 1984), some data
from the midlatitudinal North Atlantic (Tucker 1961) and the North Sea
(Dedeurwaerder et al. 1982) do indicate that, in similar climatic
regions, there is less precipitation over the ocean than over land. This
may be because the surface of the ocean is flat compared with the land
where the rougher surface enhances the vertical mixing that ultimately
leads to cloud formation and consequential particle removal via precipitation (Nguyen 1968).

The pluvocity (rate of rainfall per day) of an air mass will affect
such parameters as scavenging ratios (Bergametti 1987, Jaffrezo 1987,
Slinn 1983). Consequently, it will also affect the long-range transport
of trace elements from continents to oceans. In addition, small aerosols may be more efficiently scavenged by internal mixing with sea-salt
aerosols (Andreae et al. 1986). This mixing increases the hygroscopicity of aerosols over the sea and thus the probability that they will
be removed by in-cloud scavenging or by dry deposition. This supports
the theory that, in remote areas where the concentration of aerosols is
low, nucleation is induced independently of particle radii (Buat-Ménard
and Duce 1986).

2.6. TRACE-ELEMENT RECYCLING BETWEEN THE ATMOSPHERE AND OCEANS

During long-range transport, particularly over the ocean, trace elements
can be recycled from the sea surface during the formation of sea-salt
aerosols. One result of this recycling is that aerosol trace elements
over the ocean can be transformed from fine, insoluble particles to
larger, more soluble particles such that (1) trace-element concentrations in rain often decrease while it is raining and (2) insoluble
coarser mineral aerosols are scavenged faster than finer enriched particles (Bergametti 1987). Thus, over the ocean a fraction of the trace
elements in precipitation is derived from recycled sea spray from the
sea surface. The recycled trace-element fraction in oceanic precipitation is difficult to assess. However, because of the distinctive
anthropogenic signatures in the surface microlayer, the use of stable
lead isotopes is again one of the more promising approaches (Settle and
Patterson 1982).

Another approach for assessing trace-element recycling is to take
the measured enrichment factors from the open ocean (Weisel et al. 1984)
and assume that all sea salt in precipitation is similarly enriched
(Church et al. 1984). This would enable rainwater concentrations of

trace elements to be corrected for the recycled component. Both
approaches suggest that in remote marine areas an appreciable fraction
of the trace elements in precipitation can be recycled and that these
recycled fractions may even dominate for some elements, such as Pb, Cd,
Zn, Mn, and V (Church et al. 1984).

2.7. TRACE-ELEMENT DEPOSITION RECORDS

Trace-element deposition from the atmosphere has been inferred from
recent, ground-based aerosol and precipitation data and, over longer
periods of time, from profiles in permanent ice fields and from oceans.
In the former case, ground-based data have been augmented with aircraft
collections; however, sampling has been very limited in space and time.
In the latter case, trace-element data using values from an oceanic
water column must also consider vertical upwelling and horizontal advection (Jickells et al. 1987).

Ice-core records in permanent ice fields probably provide the most
accurate assessment of the extent and magnitude of atmospheric deposition. The samples of Boutron et al. (1988) in the Antarctica ice fields
and of Murozumi et al. (1969) and Peel (1986) in the Greenland ice fields
extend back for centuries beyond the advent of industrialization. New
Soviet efforts in Antarctica (Jouzel et al. 1987) have used deep-ice
drill cores to penetrate beyond one full glacial cycle. These cores are
being used to assess how global climatic cycles affect the atmospheric
dispersal of such elements as Pb (Boutron et al. 1988). Unfortunately,
because ice records are mostly restricted to polar regions, they may not
represent atmospheric concentrations and fluxes in temperate or tropical
regions.

2.8. THE EFFECTS AND SIGNIFICANCE

The effects of the long-range atmospheric transport and deposition of
trace elements are often subtle. For instance, Fe(III) and Mn(II) in
mineral aerosols have recently been found to act as a heterogeneous
catalyst for the oxidation of atmospheric sulfur dioxide (Hoffman and
Boyce 1983, Ibusuki and Takeuchi 1987, Martin and Hill 1987). Thus the
long-range transport of trace elements may have impacts on the long-
range transport of acid rain and may also provide unique tracers of
atmospheric pathways.

Trace-element emissions to the atmosphere have increased dramatically in recent times. It is becoming more and more obvious that the
long-range atmospheric transport of both natural and anthropogenic emissions is having a global impact. The increase in atmospheric fluxes has
changed oceanic profiles of lead; atmospheric deposition has now taken
the place of fluvial transport as the dominant transport pathway (Settle
and Patterson 1982). Although we are still unable to assess accurately
the current atmospheric and fluvial transport rates based on the existing data, we believe that atmospheric deposition is geochemically

significant for trace elements other than Pb, such as Zn and Cd (Bewers
et al. 1988, Jickells et al. 1987, Maring and Duce 1987).

Duce, in his 1986 publication, makes a good case for the long-range
atmospheric transport and deposition of biologically essential trace
elements to oligotrophic oceanic areas. These areas are highly stratified and depend on the upwelling of essential elements to sustain biological production. Martin and Fitzwater (1988) have proposed that
atmospherically derived iron may be biologically important and even a
limiting nutrient under some circumstances. Although atmospheric transport could affect the productivity of oceans in some regions, other
metals, such as copper, could decrease productivity. From a biological
standpoint, total fluxes are not as important as the bioavailability of
the trace elements. The increased acidity of rainwater and cloud water
can be expected to increase the solubilization of most trace elements
(Lindberg and Harriss 1983). The coupling of atmospheric deposition and
biological productivity in surface seawater is perhaps the most important consequence of long-range atmospheric transport of trace elements.
Since much of the trace-element flux occurs by wet deposition, precipitation over the oceans could well be the mechanism that effectively
couples the deposition of atmospherically transported trace elements to
episodes of oceanic productivity in oligotrophic oceans.

2.9. REFERENCES

Andreae, M. O., B. J. Charlson, F. Bruynzeels, H. Storms, R. Vergrieken,
    and W. Maenhaut. 1986. Internal mixture of sea salt, silicates, and
    excess sulfate in marine aerosols. Science 232:1620-1622.
Arimoto, R., R. A. Duce, B. J. Ray, and C. K. Unni. 1985. Atmospheric
    trace elements at Enewetak atoll: 2. Transport to the ocean by wet
    and dry deposition. J. Geophys. Res. 90:2391-2408.
Arimoto, R., R. A. Duce, B. J. Ray, A. D. Hewitt, and J. Williams. 1987.
    Trace elements in the atmosphere of American Samoa: Concentrations
    and deposition to the tropical South Pacific. Geophy. Res. 92:8465-
    8479.
Austin, P. M., and S. G. Geotis. 1980. Precipitation assessment over the
    ocean. In Air Sea Interaction: Instruments and Methods (F. Dolosen,
    L. Hasse, and R. Davis, eds.) New York:Plenum, 523-541.
Barrie, L. A. 1986. Arctic air pollution: An overview of current knowl-
    edge. Atmos. Environ 20:643-663.
Barrie, L. A., and R. S. Schemenauer. 1986. Pollutant wet deposition
    mechanisms in precipitation and fog water. Water Air Soil Pollut.
    30:91-104.
Bergametti, G. 1987. Apports de matiere par voie atmospherique a la
    Mediterranee Occidentale: Aspects geochimiques et meteorologiques.
    Ph.D. dissert., Univ. of Paris, 296 pp.
Bertine, K. K., and E. D. Goldberg. 1971. Fossil fuel combustion and the
    major sedimentary cycle. Science 173:233-235.

52

Buesseler, K. O., H. D. Livingston, S. Honjo, B. J. Hay, S. J. Manganini, E. Degens, V. Ittekkot, E. Izdar, and T. Konuk. 1987. Chernobyl radionuclides in a Black Sea sediment trap. Nature 329:825-828.

Bewers, M., R. Duce, T. Jickells, P. Liss, J. Miller, H. Windom, and R. Wollast. 1988. Land to sea transport of contaminants: Comparison of riverine and atmospheric fluxes. GESAMP Rept. on State of the Marine Environment, United Nations Pub. Series, Geneva:WMO, n.p.

Boutron, C. F., and C. C. Patterson. 1983. The occurrence of lead in Antarctic recent snow, firn, and prehistoric ice deposited over the last two centuries. Geochem. Cosmochim. Acta. 47:1355-1368.

Boutron, C. F., and C. C. Patterson. 1987. Relative levels of natural and anthropogenic lead in recent Antarctic snow. J. Geophys. Res. 92:8454-8464.

Boutron, C. F., C. C. Patterson, C. Lorius, V. N. Petrov, and N. I. Barkov. 1988. Atmospheric lead in Antarctic ice during the last climatic cyle. Ann. Glaciol. 10:5-9.

Bowen, H. J. M. 1979. Environmental Chemistry of the Elements. New York:Academic Press, 333 pp.

Boyle, E. A., S. D. Chapnick, G. T. Shen, and M. P. Bacon. 1986. Temporal variability of lead in the western North Atlantic. J. Geophys. Res. 91:8573-8593.

Bradley, J. P., P. Goodman, I. Y. T. Chan, and P. R. Buseck. 1981. Structure and evolution of fugitive particles from a copper smelter. Environ. Sci. Technol. 15:1208-1212.

Bruynzeel, R., and R. Vergrieken. 1985. Direct detection of sulfate and nitrate layers in sampled marine aerosols by laser microprobe mass analysis. Atmos. Environ. 19:1969-1970.

Buat-Ménard, P. 1986a. Air to sea transfer of atmospheric trace metals. In The Role of Air-Sea Exchange in Geochemical Cycling (P. Buat-Ménard, ed.) NATO ASI Series C, Vol. 185, Dordrecht:Reidel, 477-496.

Buat-Ménard, P. 1986b. The ocean as a sink for atmospheric particles. In The Role of Air-Sea Exchange in Geochemical Cycling (P. Buat-Ménard, ed.) NATO ASI Series C, Vol. 185, Dordrecht:Reidel, 165-183.

Buat-Ménard, P., and R. A. Duce. 1986. Precipitation scavenging of aerosol particles over remote marine regions. Nature 312:508-510.

Byrd, J. T., and M. O. Andreae. 1982. Tin and methyltin species in seawater: Concentration and fluxes. Science 218:565-569.

Cambray, R. S., D. F. Jefferies, and G. Topping. 1975. An estimate of the input of atmospheric trace elements into the North Sea and the Clyde Sea (1972-3). Harwell Rept. AERE-R7733, Harwell (Oxfordshire):United Kingdom Atomic Energy Authority, 26 pp.

Charlson, R. J., and H. Rodhe. 1982. Factors controling the acidity in natural rainwater. Nature 295:683-685.

Charlson, R. J., J. E. Lovelock, M. O. Andreae, and S. G. Warren. 1987. Oceanic phytoplankton, atmospheric sulphur, cloud albedo and climate. Nature 326:655-661.

Chen, L., and R. A. Duce. 1983. The source of sulfate, vanadium and mineral matter in aerosol particles over Bermuda. Atmos. Environ. 17:2055-2064.

Chen, L., R. Arimoto, and R. A. Duce. 1985. The sources and forms of phosphorus in marine aerosol particles and rain from northern New Zealand. Atmos. Environ. 19:779-787.

Choularton, T. W., G. Fullarton, and M. J. Gay. 1982. Some observations on the influence of meteorological variables on the size distribution of natural aerosol particles. Atmos. Environ. 16:315-323.

Chow, T. J. 1970. Isotopic identification of industrial pollutant lead. In Procs., 2nd Int. Clean Air Congress. New York: Academic Press, 348-352.

Church, T.M., J. M. Tramontano, J. R. Scudlark, T. D. Jickells, J. J. Tokos, and A. H. Knap. 1984. The wet deposition of trace metals to the western Atlantic Ocean at the mid-Atlantic Coast and on Bermuda. Atmos. Environ. 18:2657-2664.

Collyer, F. X., B. G. Barnes, G. J. Churchman, T. S. Clarkson, and J. T. Steiner. 1984. A trans-Tasman dust transport event. Weather and Climate 4:42-46.

D'Almeida, G. A. 1986. A model for Saharan dust transport. Climate Appl. Meteorol. 25:903-916.

Dedeurwaerder, H. L., F. A. Dehairs, G. G. Decadt, and W. F. Baeyens. 1982. Estimation of dry and wet deposition and resuspension fluxes of several trace elements in the southern bight of the North Sea. In Precipitaiton Scavenging, Dry Deposition, and Resuspension (H. R. Pruppacher, R. G. Semonin, and W. G. N. Slinn, eds.) New York: Elsevier, 1219-1231.

Delany, A. C., D. W. Parkin, J. J. Griffin, E. D. Goldberg, and B. E. F. Reinmann. 1967. Airborne dust collected at Barbados. Geochim. Cosmochim. Acta 31:885-909.

Duce, R. A. 1986. The impact of atmospheric nitrogen, phosphorus, and iron species on marine biological productivity. In The Role of Air-Sea Exchange in Geochemical Cycling (P. Buat-Ménard, ed.) NATO ASI Series C, Vol. 185, Dordrecht:Reidel, 497-529.

Duce, R. A., and G. L. Hoffman. 1972. Consideration of the chemical fractionation of alkali and alkaline earth metals in the Hawaiian marine atmosphere. J. Geophys. Res. 77:5161-5169.

Duce, R. A., and E. J. Hoffman. 1976. Chemical fractionation at the air/sea interface. Ann. Rev. Earth Planetary Sci. 4:187-228.

Duce, R. A., G. L. Hoffman, and W. H. Zoller. 1975. Atmospheric trace metals at remote Northern and Southern Hemisphere sites: Pollution or natural? Science 187:59-61.

Duce, R. A., G. L. Hoffman, B. J. Ray, I. S. Fletcher, G. T. Wallace, J. L. Fasching, S. E. Piotrowctz, P. R. Walsh, E. J. Hoffman, J. M. Miller, and J. L. Hefter. 1976. Trace metals in the marine atmosphere: Sources and fluxes. In Marine Pollution Transfers (H. L. Windom and R. A. Duce, eds.) Lexington, MA:D. C. Heath, 77-119.

Duce, R. A., R. Arimoto, B. J. Ray, C. K. Unni, and P. J. Harder. 1983. Atmospheric trace elements at Enewetak Atoll: 1. Concentrations, sources, and temporal variability. J. Geophys. Res. 88:5321-5342.

Dulac, F., P. Buat-Ménard, M. Arnold, U. Ezat, and D. Martin. 1987. Atmospheric input of trace metals to the western Mediterranean Sea: 1. Factors controlling the variability of atmospheric concentrations. J. Geophys. Res. 92:8437-8453.

Dutkiewicz, V. A., P. P. Parekh, and L. Hussain. 1987. An evaluation of regional trace element signatures relevant to the northeastern United States. Atmos. Environ. 21:1033-1044.

Elliot, W. P., and R. K. Reed. 1984. A climatological estimate of precipitation for the world oceans. Climate Appl. Meteorol. 23:434-439.

Eriksson, K. G. 1979. Saharan dust sedimentation in the western Mediterranean Sea. In Saharan Dust: Mobilization, Transport, Deposition (C. Morales, ed.) New York:Wiley, 197-210.

Fitzgerald, W. F. 1986. Cycling of mercury between the atmosphere and oceans. In The Role of Air-Sea Exchange in Geochemical Cycling (P. Buat-Ménard, ed.) NATO ASI Series C, Vol. 185, Dordrecht:Reidel, 363-408.

Fogg, T. R., and R. A. Duce. 1985. Boron in the troposphere: Distribution and fluxes. J. Geophys. Res. 90:3781-3796.

Fowler, S. W., P. Buat-Ménard, Y. Yokoyama, S. Ballestra, E. Holm, and H. V. Nguyen. 1987. Rapid removal of Chernobyl fallout from Mediterranean surface waters by biological activity. Nature 329:56-58.

Fruchter, J. S., D. E. Robertson, J. C. Evans, K. B. Olsen, E. A. Lepel, J. C. Laul, K. H. Abel, R. W. Sanders, P. O. Jackson, N. S. Wogman, R. W. Perkins, H. H. Van Tuyl, R. H. Beauchamp, J. W. Slade, J. L. Daniel, R. L. Erikson, G. A. Sehmel, R. N. Lee, A. V. Robinson, O. R. Moss, J. K. Briant, and W. C. Cannon. 1980. Mount St. Helen's ash from the 18 May 1980 eruption: Chemical, physical, mineralogical and biological properties. Science 209:1116-1125.

Ganor, E., and Y. Mamane. 1982. Transport of Saharan dust across the eastern Mediterranean. Atmos. Environ. 16:581-587.

Grousset, F. E., P. E. Biscaye, A. Zindler, J. M. Prospero, R. Chester. 1988. Nd isotopes as tracers in marine sediments and aerosols: North Atlantic. Earth Planetary Sci. Ltrs. 87:367-378.

Hewitt, C.N., and R. M. Harrison. 1987. Atmospheric concentrations and chemistry of alkyllead compounds and environmental alkylation of lead. Environ. Sci. Technol. 21:260-266.

Hoffman, M. R., and S. D. Boyce. 1983. Catalytic autoxidation of aqueous sulfur dioxide in relationship to atmospheric systems. In Trace Atmospheric Constituents: Properties, Transformations and Fates (S. E. Schwartz, ed.) New York:Wiley, 147-189.

Hoffman, G. L., and R. A. Duce. 1972. Trace metals in the Hawaiian marine atmosphere. J. Geophys. Res. 77:5322-5329.

Ibusuki, T., and K. Takeuchi. 1987. Sulfur dioxide oxidation by oxygen catalyzed by mixtures of manganese(II) and iron(II) in aqueous solution at environmental reaction conditions. Atmos. Environ. 21:1555-1560.

Jaffrezo, J. L. 1987. Etude du lessivage des aerosols atmospheriques par les precipitations. Ph.D. dissert., Univ. of Paris.

Jaffrezo, J. L., and J. L. Colin. 1988. Rain-aerosol coupling in urban areas: Scavenging ratio measurement and identification of some transfer processes. Atmos. Environ. 22:929-936.

Jickells, T. D., T. M. Church, and W. G. Deuser. 1987. A comparison of atmospheric inputs and deep-ocean particle fluxes for the Sargasso Sea. Global Biogeochem. Cycles 1:117-130.

Jouzel, J., C. Lourius, J. R. Petit, C. Genthon, N. I. Barkov, V. M. Kotlyakov, and V. M. Petrov. 1987. Vostok ice core: A continuous isotope temperature record over the last climatic cycle (160,000 years). Nature 329:403-408.

Kempe, S., and N. Nies. 1987. Chernobyl nuclide record from a North Sea sediment trap. Nature 329:828-831.

Kritz, M. A., and J. Rancher. 1980. Circulation of Na, Cl, and Br in the tropical marine atmosphere. J. Geophys. Res. 85:1633-1639.

Lambert, G., B. Ardouin, and G. Polian. 1982. Volcanic output of long-lived radon daughters. J. Geophys. Res. 87:11,103-11,108.

Lambert, G., M.-F. LeCloarec, and M. Pennisi. 1988. Volcanic output of SO_2 and trace metals: A new approach. Geochim. Cosmochim. Acta 52:39-42.

Lannefors, H., J. Heintzenberg, and H.-C. Hansson. 1983. A comprehensive study of physical and chemical parameters of the Arctic summer aerosol: Results from the Swedish expedition Ymer 80. Tellus 35B:40-54.

Lantzy, R. J., and F. T. Mackenzie. 1979. Atmospheric trace metals: Global cycles and assessment of man's impact. Geochim. Cosmochim. Acta 43:511-525.

Lee, D. S., J. M. Edmond, and K. W. Bruland. 1986. Bismuth in the Atlantic and Pacific: A natural analogue to plutonium and lead? Earth Planetary Sci. Ltrs. 76:254-262.

Lindberg, S. E., and R. C. Harriss. 1983. Water and acid soluble trace metals in atmospheric particles. J. Geophys. Res. 88:5091-5100.

Lininger, R.L., R. A. Duce, J. W. Winchester, and W. R. Matson. 1966. Chlorine, bromine, iodine and lead in aerosols from Cambridge, Massachusetts. J. Geophys. Res. 71:2457-2463.

Liss, P. S., and W. G. N. Slinn (eds.) 1983. Air-Sea Exchange of Gases and Particles. NATO ASI Series C, Vol. 108, Dordrecht:Reidel, 561 pp.

Lowenthal, D. H., and K. A. Rahn. 1985. Regional sources of pollution aerosol at Barrow, Alaska, during winter 1979-1980 as deduced from elemental tracers. Atmos. Environ. 19:2011-2024.

Maring, H., and R. A. Duce. 1987. The impact of atmospheric aerosols on trace metal chemistry in open ocean surface seawater: 1. Aluminum. Earth Planetary Sci. Ltrs. 84:381-392.

Maring, H., D. M. Settle, P. Buat-Ménard, F. Dulac, and C. C. Patterson. 1987. Stable lead isotope tracers of air-mass trajectories in the Mediterranean region. Nature 330:154-156.

Mart, L. 1983. Seasonal variations of Cd, Pb, Cu, and Ni levels in snow from the eastern Arctic Ocean. Tellus 35B:131-141.

Martin, J. H., and S. E. Fitzwater. 1988. Iron deficiency limits phytoplankton growth in the northeast Pacific sub-Arctic. Nature 331:341-343.

Martin, L. R., and M. W. Hill. 1987. The effect of ionic strength on the manganese catalyzed oxidation of sulphur (IV). Atmos. Environ. 21:2267-2270.

Mosher, B. W., and R. A. Duce. 1987. A global atmospheric selenium budget. J. Geophys. Res. 92:13,289-13,298.

Mosher, B. W., R. A. Duce, J. M. Prospero, and D. L. Savoie. 1987. Atmospheric selenium: Geographical distribution and ocean-to-atmosphere flux in the Pacific. J. Geophys. Res. 92:13,277-13,287.

Murozumi, M., T. J. Chow, and C. Patterson. 1969. Chemical concentrations of pollutant lead aerosols, terrestrial dusts, and sea salts in Greenland and Antarctic snow strata. Geochim. Cosmochim. Acta. 33:1247-1294.

Nguyen, B. C. 1968. Etude par les traceurs radioactifs des echanges entre les diverses zones de l'atmosphere au-dessus des continents et des oceans. Ph.D. dissert., Univ. of Paris, 195 pp.

Nriagu, J. O. 1979. Global inventory of natural and anthropogenic emissions of trace metals to the atmosphere. Nature 279:409-411.

Oblad, M., and E. Selin. 1986. Measurements of elemental composition in background aerosol on the west coast of Sweden. Atmos. Environ. 20:1419-1432.

Pacyna, J. M. 1984. Estimations of the atmospheric emissions of trace elements from anthropogenic sources in Europe. Atmos. Environ. 18:41-50.

Pacyna, J. M. 1986. Atmospheric trace elements from natural and anthropogenic sources. In Toxic Metals in the Atmosphere (J. O. Nriagu and C. I. Davidson, eds.), New York:Wiley, 33-52.

Pacyna, J. M., B. Ottar, U. Tomza, and W. Maenhaut. 1985. Long-range transport of trace elements to Ny Alesund, Spitsbergen. Atmos. Environ. 19:857-865.

Patterson, C. C., and D. M. Settle. 1987. Review of data on eolian fluxes of industrial and natural lead to the lands and seas in remote regions on a global scale. Marine Chemistry 22:137-162.

Peel, D. A. 1986. Is lead pollution of the atmosphere a global problem? Nature 323:200.

Petit, D., J. P. Mennessier, and L. Lamberts. 1984. Stable lead isotopes in pond sediments as tracers of past and present atmospheric lead pollution in Belgium. Atmos. Environ. 18:1189-1193.

Prodi, F., and G. Fea. 1979. A case of transport and deposition of Saharan dust over the Italian peninsula and southern Europe. J. Geophys. Res. 84:6951-6960.

Rahn, K. A. 1976. The chemical composition of the atmospheric aerosl. Tech. Rept., Graduate School of Oceanography, Rhode Island Univ., Kingston, 265 pp.

Rahn, K. A. 1985. Pollution aerosol in the northeast: Northeastern-midwestern contributions. Science 228:275-284.

Rahn, K. A., and D. H. Lowenthal. 1984. Elemental tracers of distant regional pollution aerosols. Science 223:132-139.

Reiff, J., G. S. Forbes, F. T. M. Spieksma, and J. J. Reynders. 1986. African dust reaching northwestern Europe: A case study to verify trajectory calculations. Climate Appl. Meteorol. 25:1543-1567.

Schaule, B. K., and C. C. Patterson. 1981. Lead concentrations in the Northeast Pacific: Evidence for global anthropogenic perturbations. Earth Planetary Sci. Ltrs. 54:97–116.

Schell, W. R., A. L. Sanchez, and C. Granlund. 1986. New data from peat bogs may give an historic perspective on acid deposition. Water Air Soil Pollut. 30:393–409.

Schneider, B. 1987. Source characterization for atmospheric trace metal over Kiel Bight. Atmos. Environ. 21:1275–1283.

Schutz, L. 1980. Long-range transport of desert dust with special emphasis on the Sahara. In Aerosols: Anthropogenic and Natural, Sources and Transport (T. J. Kneip and P. J. Lioy, eds.) Ann. N. Y. Acad. Sci. 338:515–532.

Schutz, L., and M. Kramer. 1987. Rainwater composition over a rural area with specific emphasis on the size distribution of insoluble matter. Atmos. Chemistry 5:173–184.

Schutz, L., R. Jaenicke, and H. Pietrer. 1981. Saharan dust transport over the north Atlantic Ocean. In Desert Dust: Origin, Characteristics, and Effects on Man (T. L. Pewe, ed.), Special Paper 186, Boulder:Geol. Soc. Am., 87–100.

Settle, D. M., and C. C. Patterson. 1982. Magnitudes and sources of precipitation and dry deposition fluxes of industrial and natural leads to the North Pacific at Enewetak. J. Geophys. Res. 87:8857–8869.

Settle, D. M., C. C. Patterson, K. K. Turekian, and J. K. Cochran. 1982. Lead precipitation fluxes at tropical oceanic sites determined from ^{210}Pb measurements. J. Geophys. Res. 87:1239–1245.

Shen, G. T., and E. A. Boyle. 1987. Lead and corals: Reconstruction of historical industrial fluxes to the ocean surface. Earth Planetary Sci. Ltrs. 82:289–304.

Shirahata, H., R. W. Elias, C. C. Patterson, and M. Koide. 1980. Chronological variations in concentrations and isotopic compositions of anthropogenic atmospheric lead in sediments of a remote sub-Alpine pond. Geochim. Cosmochim. Acta 44:149–162.

Slanina, J., J. H. Baard, W. L. Zijp, and W. A. H. Asman. 1983. Tracing the sources of chemical composition of precipitation by cluster analysis. Water Air Soil Pollut. 20:41–45.

Slinn, W. G. N. 1983. Air-to-sea transfer of particles. In Air-Sea Exchange of Gases and Particles (P. S. Liss and W. G. N. Slinn, eds.), NATO ASI Series C, Vol. 108, Dordrecht:Reidel, 299–405.

Smith, F. B., and M. J. Clark. 1986. Radionuclide deposition from the Chernobyl cloud. Nature 322:690–691.

Sturges, W. T., and L. A. Barrie. 1987. Lead 206/207 isotope ratios in the atmosphere of North America: Tracers of American and Canadian emissions. Nature 329:144–146.

Sturges, W. T., and L. A. Barrie. 1989. Stable lead isotope ratios in Arctic aerosols: Evidence for the origin of Arctic air pollution. Atmos. Environ., in press.

Sturges, W.T., and R. M. Harrison. 1986. Bromine:lead ratios in airborne particles from urban and rural sites. Atmos. Environ. 20:577–588.

Sugimae, A. 1984. Elemental constituents of atmospheric particulates and particle density. Nature 307:145–147.

Tanaka, S., M. Darzi, and J. W. Winchester. 1980. Short-term effect of rainfall on elemental composition and size distribution of aerosols in north Florida. Atmos. Environ. 14:1421-1426.

Tucker, G. B. 1961. Precipitation over the North Atlantic Ocean. Quart. Royal Meteorol. Soc. 87:147-158.

Uetmatsu, M., R. A. Duce, and J. M. Prospero. 1985. Deposition of atmospheric mineral particles in the north Pacific Ocean. Atmos. Chemistry 3:123-128.

Veron, A., C. E. Lambert, A. Isley, P. Linet, and F. Grouset. 1987. Evidence of recent lead pollution in deep northeast Atlantic sediments. Nature 326:278-281.

Wangen, L. E. 1981. Elemental composition of size-fractionated aerosols associated with a coal-fired, power plant plume and background. Environ. Sci. Technol. 15:1080-1088.

Weisel, C. P., R. A. Duce, J. L. Fasching, and R. W. Heaton. 1984. Estimates of the transport of trace metals from the ocean to the atmosphere. J. Geophys. Res. 89:11,607-11,618.

Whitehead, D.C. 1984. The distribution and transformations of iodine in the environment. Environ. International. 10:321-339.

Wiersma, G. B., and C. I. Davidson. 1986. Trace metals in the atmosphere of remote areas. In Toxic Metals in the Atmosphere (J. O. Nriagu and C. I. Davidson, eds.) New York:Wiley, 201-266.

Wolff, G.W., and D. A. Peel. 1985. The record of global pollution in polar snow and ice. Nature 313:535-540.

3. MINERAL-AEROSOL TRANSPORT TO THE NORTH ATLANTIC AND NORTH PACIFIC: THE IMPACT OF AFRICAN AND ASIAN SOURCES

Joseph M. Prospero
Division of Marine and Atmospheric Chemistry
Rosenstiel School of Marine and Atmospheric Science
Miami, FL 33149

3.1. INTRODUCTION

In many respects, mineral dust is one of the oldest and longest studied examples of long-range transport. The earliest accounts of this pheno- menon, many of which appeared in the general press, are those describing red rains and snows. Although some observers often ascribed these events to fanciful causes (sometimes describing them as ''blood'' rains), many correctly concluded that the material in the rains was soil dust, which was suspected to have come from distant sources. (For a comprehensive review of the literature on dust falls in Europe, see Bucher 1986, Fett 1958.)

Beginning about 200 years ago, descriptions of dust events have appeared with increasing frequency in the scientific literature. In Europe, the dust in rains and snows was often closely examined and the composition reported in minute detail. Many observers eventually con- cluded that North Africa was a major source of European dust events. Deposition rates were estimated based on dust concentrations found in rain samples and snow layers and scientists assessed the geological significance of the deposits. In 1833 Darwin, aboard the Beagle in the Cape Verde Islands, experienced a Saharan dust outbreak; he was impressed by the extent and severity of the event and correctly specu- lated that this mechanism of transport might be a significant source of material in ocean sediments (Darwin 1846).

Despite this long history of study, of all the substances consid- ered at the NATO workshop, mineral aerosol is still one of the more difficult with which to deal quantitatively, for several reasons.

3.1.1. Sources

The regions containing the major sources of mineral aerosol are remote and inhospitable. These areas are invariably arid, underdeveloped, and sparsely populated as well as difficult to reach. Because there are very few meteorological stations in these regions, there is a dearth of meteorological data. Moreover, dust generation occurs sporadically and is highly dependent on weather and climate. Consequently, dust storms usually occur in places where there are few observers of any kind and about which there are few scientific data.

Although soil dust is a natural material, the rate of deflation is greatly affected by land-use practices. Therefore, dust liberated

A. H. Knap (ed.), The Long-Range Atmospheric Transport of Natural and Contaminant Substances, 59–86.
© 1990 Kluwer Academic Publishers.

because of overgrazing, deforestation, or inappropriate agriculture can
be regarded as anthropogenic dust. Although many pollutant materials
can be identified by some characteristic of their composition or by the
composition of an associated species in the same air mass, the physical,
chemical, or mineralogical properties of ''natural'' dust cannot usually
be distinguished from those of ''anthropogenic'' dust.

In contrast to conventional anthropogenic species (for example,
sulfate and nitrate aerosol and their gaseous precursors), soil-dust
emissions are not being monitored by any large-scale programs. The
conventional estimates of soil erosion by wind are not helpful since
they usually deal with losses from agricultural lands, much of which is
transported close to the ground and falls on adjacent lands. Although
such losses can be large and economically important, this deflated material is not transported far enough to be considered in any study of
long-range transport.

3.1.2. Transport

Data from localized studies of deflation are difficult to use in long-
range-transport studies because the mobilized mass is concentrated in
the large-particle-size range and because removal by sedimentation is so
strongly size-dependent. Near a dust storm, most of the aerosol mass
resides in particles with diameters of 10's to 100's of microns. The
Stokes settling velocity of a 100-µm spherical soil particle is 50 cm/
sec (1.8 km/hr) although that of a 10-µm particle is only 0.6 cm/sec
(0.022 km/hr). Therefore, of the dust-storm particles in this size
range, a large fraction will probably never reach the ocean if they
originate farther inland than a few hundred to a thousand kilometers.
This rapid loss of large particles is one reason why the mass median
diameter of soil aerosols over the open ocean is under 10 µm, typically
only several microns. The importance of this size range to long-range
transport is emphasized by the observation that the mass median diameter
of aeolian material in deep-sea sediments is of the same order.

To estimate transport fluxes, more information is needed than just
surface level measurements of aerosol concentrations and sizes at the
source. To begin with aerosol parameters as a function of altitude need
to be measured. These data should then be coupled with detailed information on the synoptic-scale meteorology of the region. Unfortunately,
as stated previously, such information is rarely available in the arid
regions where much of the soil dust originates.

Transport studies are further complicated because the genesis of
dust storms requires specific meteorological conditions. However, these
conditions can vary from continent to continent and from region to
region. Therefore, because dust-storm genesis can be very sensitive to
changes in weather and climate, quantitative field studies of dust generation and transport can only be planned after these linkages are well
understood.

3.1.3. Deposition

Because of the wide range of particle sizes in mobilized dust, the
removal of mineral aerosol from the atmosphere is much more complicated
than that of other types of aerosols. In contrast to mineral dust, most
natural and anthropogenic aerosols that are derived from gaseous precursors grow only to submicron sizes, and their removal from the atmosphere
by gravitational settling is almost negligible. Once such particles are
generated, they remain in the atmosphere until removed by precipitation.
(For most submicron aerosol species, 80-90% are removed by wet deposition.) Thus, by measuring the concentration of these species in precipitation collectors, the major fraction of the deposition is usually
captured. However, the dry removal of dust particles accounts for a
large fraction of the deposition close to the sources. Thus rain collections alone will not provide a quantitative estimate of dust deposition unless one is collecting precipitation far from the dust source
(i.e., several thousand kilometers or more).
 Because dust concentrations over the ocean vary greatly in time and
space and because rain is sporadic, wet-deposition rates for dust can
change greatly from event to event. As we shall see, the major fraction
of the annual dust deposition occurs in very few rainfall events; thus,
to obtain accurate estimates, deposition rates must be measured over
extended time and space scales.
***

 In this paper I have focused on the transport aspects of the two
best documented dust cases: (1) from North Africa across the North
Atlantic and (2) from Asia across the North Pacific. The oceanic data
from these regions are much more extensive and coherent than those from
any other. By understanding the long-range transport and deposition in
these cases, we obtain a better idea of what needs to be done in other
regions. These studies could also be models for the aerosol-transport
and -deposition studies that are needed in other regions.

3.2. AFRICAN DUST TRANSPORT ACROSS THE TROPICAL NORTH ATLANTIC

3.2.1. Oceanic Processes

3.2.1.1. **Introduction.** The tropical North Atlantic has been the site
of many mineral-aerosol studies over the past 25 years. (For reviews of
various aspects of Saharan dust phenomena, see Prospero 1981, Coude-
Gaussen 1984, Middleton et al. 1986, Pye 1987, Prospero and Carlson
1981.) Data have been obtained from island stations, from ships at sea,
and from aircraft covering the entire region between the coast of Africa
and the Caribbean. This region has also been well covered by various
space systems, including geostationary satellites, during much of the
time so that dust outbreaks have been carefully watched as they moved
across the Atlantic (Prospero 1981, Middleton et al. 1986, Pye 1987; see
also Chapter 10, p. 197).
 Dust outbreaks occur on all scales. In many photos, small-scale
dust plumes streaming from the coast of West Africa are clearly defined

and generally seem to be relatively localized. These events are undoubtedly caused by the deflation of coastal sand and soil deposits. In contrast, large dust outbreaks can cause intense haze conditions along the entire west coast of Africa from about 10°-25°N. The haze areas generally move to the west with a speed comparable to that of the trade winds, taking about five days to a week to reach the western Atlantic and Caribbean. Haze can cover much of the tropical Atlantic during this time. Since dust outbreaks occur with a periodicity of several days to a week during much of the year, a new dust cloud will frequently emerge from the coast of Africa while the previous one is still over the western Atlantic. Thus, high dust concentrations in this region are normal, not rare.

3.2.1.2. **Temporal Record.** The most extensive study of dust transport from North Africa is that conducted on Barbados since 1965 (Prospero and Nees 1986). This data set is the most comprehensive record of mineral-aerosol transport in any ocean region. Dust concentrations on Barbados vary greatly on time scales ranging from days to seasons to years. Day-to-day changes and season-to-season changes of one to two orders of magnitude are common. The short-term variations are caused by the passage of dust clouds, which are often visible on satellite photographs; frequently, dust peaks at Barbados can be traced back to the coast of Africa and, in some cases, even to specific source areas. During the summer, the season of maximum annual concentration, concentrations peak on a fairly regular cycle of about 3-5 days, which corresponds to the frequency of passage of easterly waves in this region.

The seasonal cycle at Barbados appears to be characteristic of a large area of the tropical North Atlantic. From the aerosol measurements taken in Miami since the early 1970s, Glaccum and Prospero (1980) determined that mineral aerosol is the dominant insoluble aerosol component borne by onshore winds during the spring and summer months. The composition of this mineral dust is identical to that of the mineral dust collected on Barbados. The seasonal cycle for the transport of Saharan dust to Miami is identical to the transport to Barbados although dust episodes at Miami are less frequent than at Barbados. This difference in frequency may be because of in-transit removal (since several days added travel time could be required to reach Miami) or because the transport is confined to lower latitudes.

Changes in climate have had a dramatic affect on dust transport. First, the annual cycle is well defined throughout the 1965-1984 record with a maximum in the summer months, usually June or July (Fig. 3-1). Although there are significant year-to-year variations, the period of intensive transport normally begins in April or May and ends in September or October. Second, there are two periods when dust concentrations sharply increased: once in the early 1970s and again in the early 1980s. The annual mean concentrations from 1983 and 1984 (18.7 $\mu g/m^3$ and 16.4 $\mu g/m^3$, respectively) are the highest ever recorded; for comparison, concentrations from the mid-1960s are only 3-4 $\mu g/m^3$. These periods of unusually high dust concentrations in the Barbados record correspond to periods of particularly severe drought in Africa, which began in the late 1960s. Indeed, on a year-to-year basis, dust concentrations at

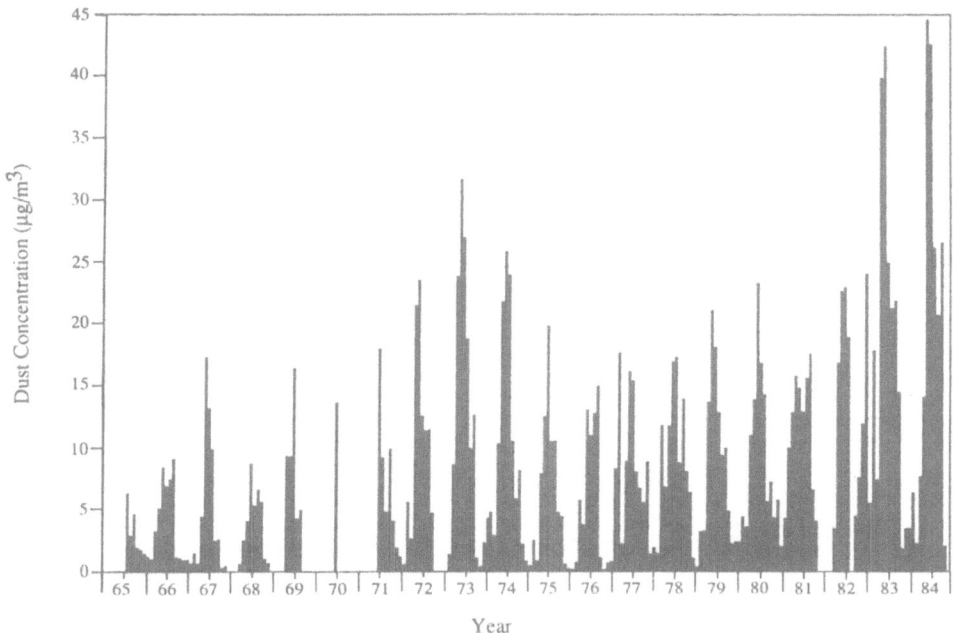

Figure 3-1. Monthly mean trade-wind mineral-aerosol concentrations at
Barbados, West Indies, from 1965 to 1984 (Prospero and Nees 1986).

Barbados are correlated to measured rainfall deficits in the Sahel
(Prospero and Nees 1986). The fact that the dust concentrations in 1983
and 1984 are about four times those of the predrought period shows that
dust mobilization and transport are highly sensitive to meteorological
and climatic factors.

3.2.1.3. <u>Aerosol-Concentration Statistics.</u> Aerosol-concentration frequency distributions can provide information about aerosol sources and
about transport and removal processes. The soil-dust concentrations
found in Barbados are characterized by bimodal, lognormal distributions
(Fig. 3-2). Furthermore, the distributions for severe drought years are
clearly distinguished from those for more moderate years. Figure 3-2A
shows the distribution of daily concentrations for 1979 to 1981 when the
drought had moderated. The lower distribution mode has a geometric mean
of 0.99 $\mu g/m^3$ and contains 42% of the sample set; the upper mode has a
mean of 11.23 $\mu g/m^3$ and contains 58% of the sample set. Figure 3-2B
illustrates the data set for 1984-1985--a time of intense drought. The
mean for the lower mode is 0.96 $\mu g/m^3$; it contains 42% of the sample set.
Thus the lower modes for both sample sets are essentially identical and
apparently constitute the background mineral-aerosol mode for this
region. However, the upper mode mean is 23.1 $\mu g/m^3$, twice that for

64

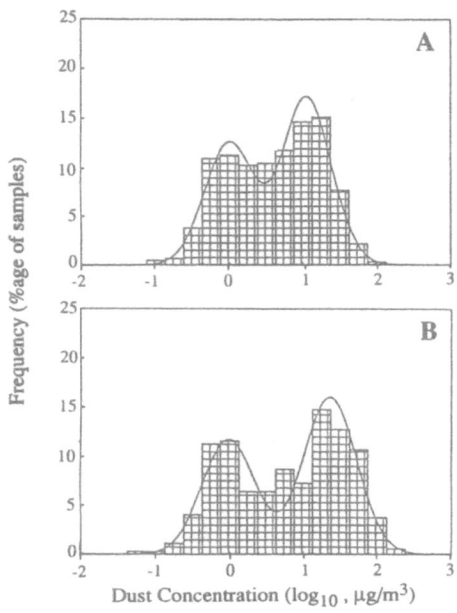

Figure 3-2

Daily dust concentrations of **A** 1979–1981 and **B** 1984–1985 at Barbados. The smooth curves on these frequency histograms represent the best-fit, bimodal distribution as obtained by the moments method. (Savoie et al. 1987)

1979–1981; nonetheless, in both cases they contain identical fractions of measurements. These data suggest that the increased dust transport associated with the drought is attributable either to an increased severity of dust generation or to a decrease in removal rates during transport. The increased dustiness does not appear to be related to an increase in frequency of dust storms. This finding has implications regarding source processes and will be discussed in greater detail later in this chapter.

3.2.1.4. <u>Size Distribution.</u> The most useful data concerning mineral-aerosol size distributions were those from the GARP Atlantic Tropical Experiment in the summer of 1974. This study used identical cascade impactors placed at four locations in the tropical North Atlantic. Using the data from this study, Savoie (1978) reports identical mass median diameters for dust at Miami and Barbados of 2.05 μm and 2.06 μm, respectively. In contrast, the mass median diameters at Sal Island, about 500 km off the west coast of Africa, are more than 6.13 μm and a substantial fraction of the particles in the aerosol mass (about 40%) is above the upper stage cutoff diameter of 11.6 μm. Therefore, the mass median diameters from Sal Island are probably underestimated since impactors cannot efficiently collect particles with diameters larger than about 15 μm. Assuming that the particle sizes are lognormally distributed, the fitted size distributions suggest that a significant fraction of the particles in the samples from Sal Island have diameters larger than 50 μm and some particles are larger than 100 μm in diameter. These distribution figures for Sal Island, reported by Savoie in 1978, agree with those reported by Jaenicke and Schutz (1978) and Schutz (1980) for the same region.

Because of the high settling velocity of particles above about 10 μm in diameter, we can assume that any dust cloud that transits the Sal Island region is still rapidly losing large particles and that the particle spectrum will probably continue to change significantly over the next few days. In contrast, the dust-size spectrum in the western part of the tropical North Atlantic is quite stable, suggesting that sedimentation losses are small and deposition takes place primarily through wet removal. The changes that size distributions undergo between Sal Island and the western part of the tropical North Atlantic illustrate how difficult it is to estimate how much soil dust is removed from the atmosphere without extensive data on aerosol-size distributions.

Although the major portion of the long-range transport flux normally consists of relatively small particles, a significant number of larger particles are also transported. Carder et al. (1986), during their sediment-trap studies in the surface waters around the western Bahamas, found significant numbers of mineral particles with diameters over 20 μm. In fact, large particles constituted a substantial fraction of the downward flux in the water column. This deposition episode occurred in an area of dense haze caused by Saharan dust and they later found that the mineral-aerosol concentrations exceeded 40 μg/m³. Unfortunately, there are few data on the concentration of large-aerosol particles at great distances from their sources. Because it has long been assumed that the long-range transport of such material is not normally possible (Tsoar and Pye 1987), there has been little research on the subject. One other unusual case of the transport of large particles was recorded on Barbados (Prospero et al. 1970) where 25% of the mass of the sample has diameters over 10 μm and 4% over 20 μm. By using satellite imagery, this event is traced to an unusually intense Saharan dust storm.

3.2.1.5. **Vertical Distribution.** The assessment of transport is further complicated because of the nonuniform vertical distribution of dust. Aircraft measurements made at Barbados during the summer of 1969 (Prospero and Carlson 1972) show that the maximum aerosol concentrations are above the marine boundary layer. The dust is confined to an isentropic layer that normally has a top near 550 mb and a base near 750 mb (Carlson and Prospero 1972). This layer is clearly identifiable in meteorological soundings as an unusually warm and dry layer above the cooler and moist marine boundary layer. Dust concentrations in the elevated layer are usually several times those in the marine boundary layer. Recent aerosol measurements by Talbot et al. (1986) yield similar vertical distributions. They measured dust concentrations as high as 350 μg/m³ to 400 μg/m³ on one day when dust concentrations are unusually heavy. Although the dust layer contains many particles with diameters as large as 30 μm, the mass median diameter is 3.2 μm and 85% of the mass is in the 1-μm to 8-μm range.

This elevated layer, called the Saharan air layer, is a persistent feature of the tropical North Atlantic during dust outbreaks. Aircraft studies made off the coast of West Africa in 1974 (Carlson and Caverly 1977, Carlson and Benjamin 1980) reveal the same general structure except

that the top is usually around 500 mb and the base at about 850 mb.
Here, also, dust concentrations in the Saharan air layer are generally at
least several times greater than those in the marine boundary layer.

The structure of the Saharan air layer has its origins in the meteorological events that generate the dust outbreak. In a typical scenario, solar heating of the desert surface produces a strong heating of
the boundary layer air, which induces convective mixing. As the day progresses, the boundary layer deepens and eventually an isentropic mixing
layer develops that can reach an altitude of 6 km and higher. Dust
lifted at the surface can be mixed throughout this isentropic layer.
Dust can be generated by various processes ranging from localized wind
conditions (i.e., gustiness, katabatic winds) to synoptic-scale events.
The general northeasterly flow of air across the Sahara carries this
dusty air mass to the coast where it is undercut by surface-level ocean
winds, which often have a pronounced onshore component. This undercut-
ting produces the layered structure of the Saharan air layer (Prospero
and Carlson 1981). Because of the sharp inversion at the top of the
marine boundary layer in the presence of a Saharan air layer, cloud
development is strongly suppressed; therefore, the dusty air masses are
easily visible in satellite images as hazy, relatively cloud-free areas
(Prospero 1981). Synoptic-scale events associated with easterly waves
can also produce this layered structure (Prospero and Carlson 1981).

Dust from the Saharan air layer is carried into the marine boundary
layer by gravitational settling and by convective mixing across the
inversion of the marine boundary layer. Because of these processes, the
dust concentration in the marine boundary layer will generally be lower
than in the Saharan air layer. Furthermore the size distribution in the
marine boundary layer will not necessarily be representative of that in
the dust outbreak as a whole because surface-level aerosols will contain
a much greater fraction of large particles that have settled out of the
overlying Saharan air layer.

The 3-dimensional character of atmospheric transport is evident for
many species including pollutants. Usually, we can assume that most
aerosol species injected into the atmosphere reside within the original
air parcel until the species are removed, normally by wet processes.
Although this assumption facilitates the atmospheric modeling of many
substances, especially pollutants, it cannot be made for mineral aerosols
because of the high settling velocity of large particles. Consequently,
conventional atmospheric transport models cannot readily incorporate
mineral aerosols. However, modeling techniques are valid for the small
particles in dust-laden air parcels; since it is this size class (less
than about 5-10 μm in diameter) that is of greatest interest in long-
range transport, modeling techniques are important in explaining the
long-range transport of mineral dust as well as other species.

3.2.1.6. Deposition. At the time of our workshop at the Bermuda Biological Station, dust deposition to the tropical North Atlantic had not yet
been studied. All estimates of dust-deposition rates in the current
literature (Chester 1982, Prospero 1981, Schutz 1980), which cover a wide
range of values, are based on models and assumptions and are not necessarily applicable to atmospheric dust.

The only extensive measurements of Saharan dust in rain were made in Miami, Florida, using samples collected over a one-year period (1982–1983) with an automatic collector (Prospero et al. 1987). The deposition rate is highly variable from event to event. In some samples, there are distinct mud deposits in the collector. The major fraction of the annual deposition occurs in a small fraction of the days: 22% in one day; 68% in four days. The total deposition for the year is 10.1 μg Al/cm^2, a value equivalent to a mineral-dust annual deposition rate (based on an Al concentration in soils of 8%) of 126 μg/cm^2 (0.35 μg/ cm^2/day). This rate is comparable to that of the nonbiogenic sediment accumulation rate in the central tropical North Atlantic, 4 mm K/yr or 200 μgm/cm^2/yr. Such high deposition rates suggest that Saharan dust could also be a significant contributor to soil formation, especially on islands built on exposed coral platforms. Recent studies of soils on Barbados show that the trace-metal composition is quite similar to that of Saharan dust and suggests that the dust is the major contributor to the soils (Muhs et al. 1987). Interestingly, another aeolian material, volcanic dust from St. Vincent, comprises the second largest soil component.

The heavy dust deposition to the tropical North Atlantic can increase the concentration of dust-related species in surface waters. The concentrations of Al and Mn are markedly enhanced in the region between 10° and 30°N relative to concentrations in the higher latitudes or south of the equator. The mean concentrations are roughly twice as high as the ''background'' concentrations found in the Atlantic (Kremling 1985).

Deposition fluxes of this magnitude could play an important role in sea-water chemistry (Buat-Ménard 1986). For example, in the Miami study (Prospero et al. 1987), a substantial fraction of the Al in the precipitation samples is in the dissolved state (i.e., passes through a 0.45-μm-diameter pore-size filter). The volume-weighted mean dissolved Al fraction in rain is about 5%. This value is comparable to that obtained by Maring and Duce (1987), who measured the dissolution of Al in aerosol filter samples from the North Pacific. They found that about 5% of the Al is leached when immersed in seawater (pH 8) for periods of up to six hours and about the same amount is leached into distilled water (pH 5.5) after six hours. Therefore, a significant fraction of the Al in soil aerosols can be readily mobilized after brief immersion in different aqueous solutions.

The computed deposition rate of soluble Al from dust (Prospero et al. 1987) ranges from 1–4 × 10^{11} g Al/yr for the tropical North Atlantic. In comparison, the Amazon carries a dissolved Al load of about 2 × 10^{11} g Al/yr. Thus, Saharan dust appears to dominate the Al seawater chemistry in this region. Using similar calculations, the global atmospheric deposition rate of soluble Al should be in the range of 12–40 × 10^{11} g/yr; this flux compares to the river-transport estimates of 4–29 × 10^{11} g/yr of Stoffyn and Mackenzie (1982).

Trace metals and other species found in mineral dust might also be soluble in precipitation or seawater. If so, they could then be important to some marine chemical cycles and consequently to marine biosystems (Duce 1986; see Chapter 10, p. 197).

3.2.1.7. <u>More</u> <u>Evidence</u> <u>of</u> <u>Long-Range</u> <u>Transport</u>. If such quantities of mineral aerosol are transported over such long distances by wind, then we might expect that substantial quantities of other continentally derived materials could also be carried. Savoie et al. (1989a), therefore, analyzed daily aerosol samples taken at Barbados during 1985 for nss-sulfate and nitrate (Fig. 3-3).

High concentrations of mineral aerosol are often associated with unusually high concentrations of aerosol nitrate and nss-sulfate. However, there is no consistent relationship between the dust-concentration peaks and those of the acid species. In some cases, the concentrations of nitrate and nss-sulfate are high when dust concentrations are quite low. From this and other evidence, they conclude that the nitrate and sulfate were not associated with the soil at the time of deflation. Also such high concentrations of acid species are probably not produced by natural or anthropogenic sources in Africa.

Evidence suggests that the acid species must come from anthropogenic sources in the midlatitudes, most probably Europe. The correlation among these species and dust has its roots in the meteorological setting associated with dust outbreaks (Estoque et al. 1974; see also, Middleton et al. 1986, Pye 1987) that normally occur when a high-pressure area is centered near the northwest coast of Africa. Under these conditions, there is a generalized flow from Europe across the Mediterranean and over North Africa. During passage over Africa, dust could be injected into a polluted air mass. However, the amount of dust generated will depend on

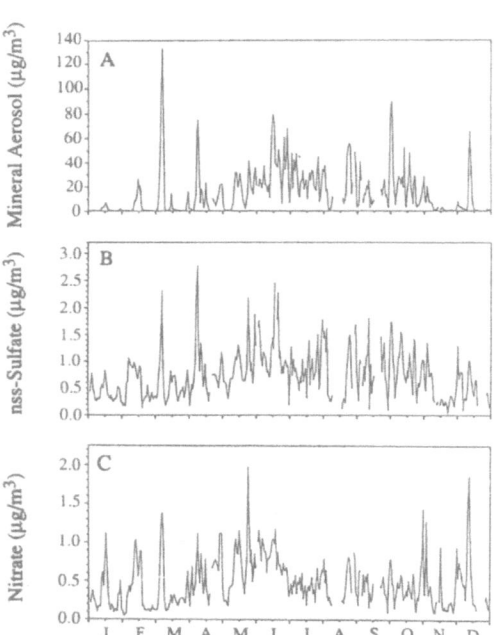

Figure 3-3

Daily averaged concentrations of **A** mineral aerosol, **B** nss-sulfate, and **C** nitrate in trade-wind air at Barbados during 1985 (Savoie et al. 1989a).

the specifics of the synoptic situation; consequently, the correlation between dust and acid species is poor.

As further evidence of the long-range impact of pollutants, Chen and Duce (1983) report finding that vanadium, a pollutant normally associated with fuel-oil combustion, is enriched in Saharan aerosol. Also the concentrations of CO over the tropical North Atlantic are considerably enhanced within the Saharan air layer as compared to concentrations within the marine boundary layer (Gregory et al. 1986).

3.2.2. Continental Processes

3.2.2.1. Sources. Many areas of major dust storms in North Africa are easily identified. Those in the Sahara are the most spectacular and the most frequently reported. For hundreds of years travelers have commented on the immensity of the dust storms in this region. Nonetheless, although these storms are undoubtedly important to the transport process, they are not necessarily the primary sources of the dust that is exported from Africa. Less spectacular deflation events could just as well generate dust that is transported over long distances.

Meteorological reports of dust events and haze are often used to estimate dust-storm activity (Helgren and Prospero 1987, Morales 1979, Morales 1986, Middleton 1986). By using the standard World Meteorological Organization's reporting codes, four separate classes of dust-related conditions from haze to dust storms can be identified in seven degrees of intensity. Other studies (see the review by Goudie 1983) base dust-storm statistics on the occurrence of visibility less than 1 km. Both of these measures of dust activity could be seriously biased. First of all, all observations are made at standard meteorological stations; since the density of stations is roughly related to the population of the area, there are few stations in arid and hyper-arid regions. (In Goudie's review [1983], for instance, he presents no data for North Africa except for Nigeria and the northern Sudan.) Furthermore, the dust reports from the few stations in arid regions are not necessarily representative of the region at large since the stations are generally near population centers. Lands near towns are generally deforested, crossed by roads, and heavily used for agriculture and grazing, all things that greatly increase soil erosion.

Recently, a more quantitative attempt was used to assess dust loads over North Africa by means of a network of sunphotometers (d'Almeida 1986). Because the attenuation of direct solar-radiation intensity (the atmospheric turbidity) is related to the dust load in the atmospheric column, these data can provide an excellent picture of the temporal and spatial variability of dust loadings. However, the data are difficult to interpret because of several factors. First of all, attenuation depends primarily on the concentration of relatively small particles (i.e., roughly ranging from several tenths of a micron to several microns) whereas the dust loading is largely determined by larger particles (roughly tens of microns). Second, because the network is in the Sahel zone, the data must be extrapolated to ascertain values applicable to all of North Africa.

Satellite imagery can be used to identify major dust sources (for examples see Chapter 10, p. 197). Over the desert, dust storms are most visible in the infrared (IR) spectrum. Because of the altitude of a dust cloud, it radiates at a cooler temperature than the surface; hence, in the black-and-white photographs from geostationary satellites, a dust cloud appears as a light-toned area against the black (hot) desert surface. Many spectacular dust storms are identifiable in such imagery, some as large as 1000 km long and several hundred wide. The history of a dust cloud can be followed from the time that it appears until it crosses the west coast of Africa. Over the ocean it can then be followed in the imagery taken at visible-light wavelengths. In the summer months (when most dust is transported across the tropical North Atlantic), most large storms that are clearly identifiable at the time of their inception originate in the central Sahara (Estoque et al. 1974). Unfortunately, dust storms are difficult to identify when water clouds are present and, consequently, large areas to the south of about 20°N could not be routinely monitored. Thus, the relative importance of dust sources cannot be quantitatively assessed on the basis of satellite imagery alone.

Opinions differ on the location of the major dust sources. One school favors sources in the latitude band between 20° N and 30° N, which includes the most arid regions of the Sahara and areas in the Sudan (see, for example, D'Almeida 1986). Another favored source is the Lake Chad basin with its vast expanses of alluvial soils deposited when it was an inland sea. This region is believed to be the source of much of the heavy winter Harmattan haze observed in Nigeria (McTainsh and Walker 1982), on the Ivory Coast, and in adjacent states as well as over the Gulf of Guinea (Bertrand et al. 1974).

The Sahel, the arid area south of the Sahara that has been the hardest hit by drought, is the most difficult to characterize as a dust source. The overall effect of long-term overgrazing, woodcutting, and cash-crop agriculture is greatly exacerbated by the drought. The damage to the soils is tremendous and deflation and erosion have greatly increased.

The increased dust concentrations found at Barbados could come from newly activated sources in the Sahel. First, dust concentrations in Barbados are correlated with rainfall deficits in the Sahel (Prospero and Nees 1986). Second, a drought probably does not affect the stability of the soils within the Sahara since it is already one of the driest regions in the world. If increased dust deflation is not actually related to the changes in rainfall itself but rather to the changes in meteorological conditions associated with a drought—such as, increased wind velocities and gustiness—then the Sahara is indeed the primary source of the ''new'' dust on Barbados (Helgren and Prospero 1987). Indeed, Middleton (1985) reports much more haze at some meteorological stations in both the Sahara and the Sahel. However, to my knowledge, local wind climatology had not yet been studied as a function of drought in the deep Sahara or the Sahel.

3.2.2.2. **Meteorology** of **Dust** **Transport.** The genesis of a major dust event begins with the deflation of a soil surface. For this to occur, certain micrometeorological conditions must be met, the essential one being that the wind speed exceed the threshold velocity for the soil. At the threshold velocity, the aerodynamic drag on the surface dislodges particles from the surface and lifts the dust into the atmospheric boundary layer. The threshold velocity depends on such soil characteristics as particle size and shape, composition, and moisture content as well as the aerodynamics of the surface (Gillette 1981). Because soil properties are quite variable, the threshold wind velocity varies widely from place to place and even from time to time at the same place. Dust generation is also affected by many larger scale characteristics of soils and terrains (see Chapter 10, page 197).

However, for a long-range-transport event to occur, certain meteorological conditions must obtain so that materials deflated at the surface are carried to altitudes where rapid horizontal transport can take place. (For a review of the meteorology of long-range transport, see Merrill 1986 and Chapter 1, page 3.) The long-range transport of dust depends on the coincidence of a specific set of micrometeorological (Tsoar and Pye 1987) and synoptic-scale conditions. These will determine the areal extent of the dust event, the mixing height, and the speed and direction of its transport.

Our group report (Chapter 10, page 197) discusses several papers that have reviewed the meteorology of dust storms in North Africa. These studies agree that, although numerous meteorological conditions can produce dust events, specific synoptic situations are usually associated with transport to a specific region (for example, to Europe, the Middle East, the Gulf of Guinea, the tropical North Atlantic).

Estoque et al. (1974) have published the most detailed study of the synoptic conditions associated with major dust storms in the tropical North Atlantic. Using geostationery satellite (SMS-1) imagery from July and August of 1974, these researchers identify the inception of 17 major dust storms that eventually cover more than several hundred kilometers. These storms are clustered in a region west of the Ahaggar Mountains; the plumes tended to have a northeast-southwest orientation and to move to the southeast. All major dust storms occur when a cyclone is present along the intertropical front to the south of the dust source region. Dust storms often occur when the cyclone appears in conjunction with pressure rises off the northwest coast of Africa. The tightening of the pressure gradient is sufficient to generate geostrophic winds of at least 8.5 m/sec, a velocity that exceeds the deflation wind speeds in some regions (Helgren and Prospero 1987).

Although these findings suggest that dust storms are somewhat predictable and that their sources can be identified, this is not always the case. Major Saharan events are usually associated with large hazy areas that emerge from the cloudy region south of about 20° N. After the first day or two of a dust storm, these two hazy areas often merge so that the soil material cannot be unambiguously attributed to a specific source region. Because of this immense and intense mixing process, the mineralogical composition of African soils collected over a large area of North Africa is relatively uniform (Schutz and Sebert 1987). Element concentrations in Saharan soils (Schutz and Rahn 1982) are also relatively homogeneous. Similarly, the mineralogy of Saharan dust samples collected over the North Atlantic is relatively invariant from one event to another although the concentration of some minerals (e.g., calcite and kaolinite) do show some significant large-scale latitudinal trends (Glaccum and Prospero 1980, Pye 1987).

3.2.2.3. Flux of North African Dust over the North Atlantic. The flux of dust out of Africa is difficult to estimate because of the paucity of information about the sources of the dust and about the temporal and spatial variabilities of aerosol concentrations over North Africa. This is exacerbated by the lack of vertical aerosol measurements. Nonetheless, some values have been estimated based on various assumptions (Table 3-1) and methods.

1. Schutz et al. (1981) input aerosol size and concentration data from the Sahara to a transport model; they then advect the aerosol over the ocean and compute deposition losses during transit. Their size and concentration data fit the experimental data moderately well.

2. Prospero et al. (1979) have computed atmospheric fluxes over the tropical North Atlantic using atmospheric turbidity measurements from a network of ship and island stations.

3. D'Almeida (1986) estimates total column dust loads over North Africa using atmospheric turbidity measurements from a network of stations on the continent. He then incorporates these into a meteorological model and computes the seasonal and directional fluxes.

Table 3-1. Estimates of mineral-aerosol fluxes and deposition to the Tropical North Atlantic (TNA).

Method	Flux (Mt)	Reference
West Coast of North Africa		
Atmospheric turbidity - flux	150	Prospero et al. 1979
Transport model - flux	200	Schutz et al. 1981
Turbidity + model - flux	190	d'Almeida 1986
Central and Western TNA		
Sediment accumulation	30	Prospero et al. 1981
Transport model deposition	33	Schutz et al. 1981.
Western TNA (60°W)		
Aerosol concentration - flux	30	Prospero and Carlson 1972
Atmospheric turbidity - flux	40	Prospero et al. 1979
Transport model - flux	50	Schutz et al. 1981

4. Using the accumulation rate of deep-sea sediments, both Schutz (1980) and Prospero (1981) have computed dust-deposition fluxes on the assumption that the terrigenous material in the sediments is derived from aeolian sources. For a mean accumulation rate of 0.4 g/cm^2/1000 yr for the central and western portions of the tropical North Atlantic, the deposition rate is 30 Mt/yr (Prospero 1981). This value agrees well with the model predictions of Schutz et al. (1981) for ocean areas farther than 1000 km from the coast of Africa (33 Mt).

5. Prospero and Carlson (1972) computed the flux in the trade winds over the western tropical North Atlantic based on aerosol measurements from aboard aircraft and on mean wind flows. This estimate agrees reasonably well with the model calculations of Schutz et al. (1981).

These estimates are not all strictly comparable since the reported fluxes apply to different aspects of the source/transport/deposition cycle. Also, some computations are restricted to a limited size class (e.g., particles with diameters under 5 μm). The problem of comparing these fluxes is further exacerbated by the drought and the large year-to-year fluctuations--a factor of four since the late 1960s (Prospero and Nees 1986). Although the data in Table 3-1 agree surprisingly well, this is probably mostly fortuitous.

We can safely assume that much of the airborne Saharan dust is redeposited onto the continent of North Africa even though there are few measured deposition rates to corroborate this assumption. In his study of the aeolian mantles in Nigeria, McTainsh (1984) obtained deposition rates of about 10 mg/cm^2 (9.1 $μg/cm^2$/day) over three years, which is equivalent to an annual accumulation of a dust layer 74 μm thick. There is a widespread aeolian mantle covering much of Nigeria; near Zaria the depth of this mantle ranges from 10-50 cm at the crests to over 5 m on the lower slopes. These depths are consistent with the measured accumulation rate and the age (40,000 years) of the loess deposits. These aeolian soils have a high chemical fertility.

3.2.3. Conclusions

The Saharan dust cycle is the most thoroughly studied dust event on earth. Nonetheless, there are major gaps in our knowledge of this cycle at all levels from genesis to final deposition. Indeed, we cannot accurately estimate the flux of dust generated in North Africa or the amount of dust leaving African shores, being deposited into the oceans, or being carried over other continents. The few estimates that do exist and that are cited in this paper are as accurate as possible with such limited data. To gain a better understanding of the Saharan dust cycle, much more intensive research is required. Because of the large scale of this phenomenon, active remote-sensing satellite systems are also needed to obtain the necessary data to put flux estimates on a quantitative footing. Such satellite programs are planned for the 1990s (see Chapter 10, page 197).

Finally, although the transport of Saharan aerosol over the tropical North Atlantic has been extensively studied, very little research has been done on mineral-dust transport over the rest of the North Atlantic. Data for this region come primarily from samples taken on occasional ship cruises (Prospero 1979, Chester 1986, Savoie 1984) and from some extended studies on Bermuda (Chen and Duce 1983, Wolff et al. 1986). These data, especially those from Bermuda, show that Saharan dust has a significant impact even at midlatitudes. On Bermuda, winds with trajectories from the southeast sector contain concentrations of crustal elements (Si, Fe, Mn) that are about 10 times greater than the concentrations in trajectories from the northeastern United States. Mean mineral-dust concentrations computed from Si and Fe data (Wolff et al. 1986) are about 4-5 µg/m^3, or about one-third of the value of the concentrations from the Barbados data for the same time. On Bermuda, as on Barbados, summer dust concentrations are about ten times greater than winter concentrations. Nonetheless, the data from the North Atlantic are so limited that the magnitude of this transport or of the quantities of materials deposited cannot be quantitatively assessed.

3.3. TRANSPORT OF ASIAN DUST ACROSS THE NORTH PACIFIC

3.3.1. Oceanic Processes

3.3.1.1. Introduction. We probably have a better understanding of the broader aspects of mineral-aerosol transport over the North Pacific than any other major ocean region considered as a whole, including the North Atlantic. This understanding is primarily the result of an ongoing aerosol-sampling network established in early 1981 as part of the Sea/Air Exchange Program (SEAREX) (Uematsu et al. 1983). The principal stations of this program are set up at Shemya in the Aleutians (52°44'N, 174°06'E), Midway (28°13'N, 177°21'W), Oahu (21°20'N, 157°42'W), Enewetak (11°20'N, 162°20'E), and Fanning (3°55'N, 159°20'W) (Fig. 3-4). The data from this network provide an unparalleled picture of the temporal and spatial characteristics of long-range transport over an immense ocean region.

3.3.1.2. Mineral-Aerosol Concentration Trends. The most striking feature of the SEAREX data is the seasonal periodicity of the concentrations throughout the entire network (Uematsu et al. 1983). The data from Midway seen in Figure 3-5 are typical. As shown in this figure, dust concentrations increase sharply in February and March and remain high until June. They drop very low during the summer, increase briefly in the fall, then decrease again until the next cycle starts in the following year. This pattern is repeated every year over most of the North Pacific. This same seasonal cycle is also evident at Nauru and Fanning, both of which lie predominantly in the Southern Hemisphere circulation; however, the concentration levels at these two sites are much lower.

During the dusty season over the North Pacific, weekly mean concentrations of several micrograms a cubic meter are common. Annual mean concentrations increase with increasing latitude with the lowest at

Figure 3-4

Pacific research sites.

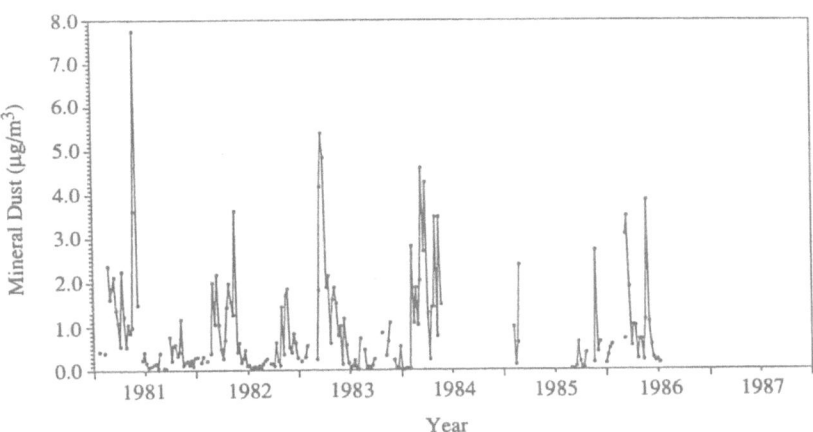

Figure 3-5. Mineral dust found in onshore winds at Midway Island, 1981–
1987 (Prospero et al. 1989).

Fanning (0.05 µg/m^3) and the highest at Shemya (0.89 µg/m^3)(Uematsu et
al. 1983). Even so, the highest mean concentrations, those at Shemya,
are still at least an order of magnitude lower than those at Barbados.
     This annual cycle of dust concentrations parallels that for dust-
storm activity in Asia. It also corresponds to the annual cycle of dust
fall (Kosa) events in Japan. Kosa episodes, which have been well
documented for many years, are derived from soil material transported

from the Asian mainland. During the past 62 years, 85% of the Kosa events observed at Nagasaki occurred between March and June (Tsunogai et al. 1982).

Dust storms in Asia are most prevalent in the spring (Watts 1969) because of the combined effects of low rainfall, more high winds associated with cold fronts, and soil freshly ploughed for spring planting (Ing 1972). During this time, the suspended dust is lifted to very high altitudes, frequently over 5 km, where it is rapidly carried in the wester-lies over vast areas of the North Pacific. Subsequent meteorological events then bring the dust down to lower atmospheric levels and into the easterly wind regime. In this manner, much of the Pacific north of the equator is eventually impacted by Asian dust. The transport of dust from its source to the central Pacific takes one to two weeks (Merrill et al. 1985).

A Japanese network of six stations measured mineral-aerosol concentrations concurrently with the SEAREX study, using identical equipment. A year's data from the principal North Pacific stations in this network are summarized in Table 3-2. The mean concentrations from the Japanese stations at latitudes above 30°N range from 9.7 $\mu g/m^3$ to 32.6 $\mu g/m^3$. These values are over an order of magnitude greater than those from the midlatitude SEAREX stations and are comparable to those in Barbados and the central part of the tropical North Atlantic.

Within the Japanese network, concentrations decrease exponentially with the distance from the Asian coast. The half-decrease distance is about 500 km. The half-decrease distance for the SEAREX network is about

Table 3-2. Mean concentrations of mineral dust in 1981 and the roughly estimated distances from the edge of the Asian continent.

Station	Latitude (°N)	Mean Concentration of Mineral Dust ($\mu g/m^3$)	Distance from the Asian Coast (km)
Okushiri	42.2	32.6	400
Wajima	37.4	17.2*	700
Izumo	35.4	23.5	700
Hachijima	33.2	9.7	1,250
Shemya	52.7	0.89	
Midway	28.2	0.83	4,500
Onna	26.5	11.8	650
Chichijima	27.1	4.4	1,650
Oahu	21.3	0.65	6,500
Enewetak	11.3	0.26	
Fanning	3.9	0.05	

Sources: Tsunogai et al. (1985); for Shemya, Midway, Oahu, Enewetak, and Fanning, Uematsu et al. (1983).

*Calculated from the mean relative concentration given in Table 2 of Tsunogai et al. (1985).

2000 km, which is consistent with other aerosol data and with measurements of radon daughter products.

3.3.1.3. **Aerosol Concentration Statistics.** Frequency histograms of the SEAREX aerosol data for the North Pacific all show bimodal distributions except for Fanning, which is unimodal (Fig. 3-6). The upper mode is comprised of Asian dust episodes. The geometric mean of the high modes range from 4.93 µg/m^3 at Shemya to 1.52 µg/m^3 at Midway to 0.4-0.5 µg/m^3 in the low-latitude easterlies. The lower mode is the background aerosol. The geometric means for the low modes are about an order of magnitude lower, ranging from 0.53 µg/m3 at Shemya and 0.189 µg/m3 at Midway to 0.04 µg/m3 at stations in the easterlies. By way of comparison, the frequency distributions at stations in the central South Pacific are unimodal.

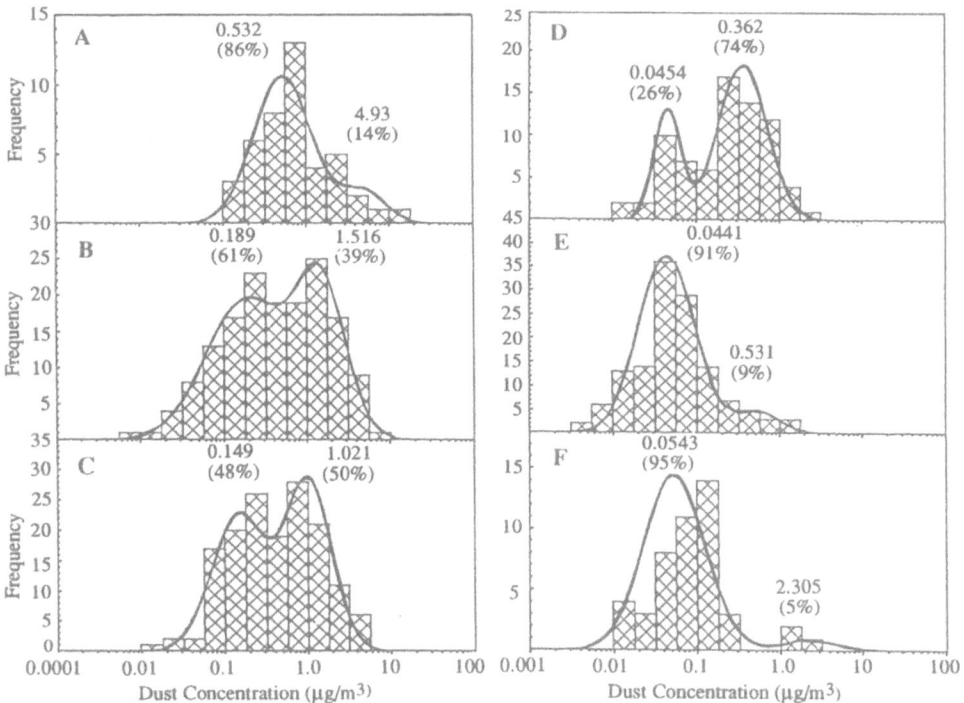

Figure 3-6. Frequency distributions of mineral-aerosol concentrations for A Shemya (n=43), B Midway (n=157), C Oahu (n=153), D Enewetak (n=76), E Fanning (n=138), and F Nauru (n=56) in the North and Equatorial Pacific (unpublished data available from Dr. Dennis Savoie, RSMAS, University of Miami, FL 33149-1098).

The differences in the frequency distributions once again emphasize the complexity of the processes that govern the long-range transport of mineral aerosol. The factor-of-ten differences among the concentrations in the upper mode stress the difficulty of even trying to generalize concentration distributions based on limited temporal and spatial data.

3.3.1.4. Aerosol Size Distributions. The mass median diameters of soil-derived elements measured at Enewetak ranges from 2.0 μm to 2.6 μm (Duce et al. 1983, Arimoto et al. 1985), which is close to the mass median diameters of the Saharan dust aerosol measured at Barbados and Miami. Significantly, the mass median diameter of soil aerosols at American Samoa (1.9 μm) is essentially identical to those at Enewetak and Barbados even though the aerosol concentrations at American Samoa are 10 to 100 times smaller than those at Enewetak and 1000 times smaller than those at Barbados (Arimoto et al. 1987). These data confirm that mineral-aerosol size distributions tend to stabilize after the mineral aerosol is transported over roughly several thousand kilometers.

Unfortunately, there are few data on size distributions of Kosa aerosols in the western Pacific. Kadowaki (1979) reports airborne dust from a Kosa event in Nagoya with a mass median diameter of about 4 μm. Inoue and Naruse (1987) report count median diameters for eight Kosa deposits ranging from 3 μm to 60 μm. Because mass median diameters are larger than count median diameters and, because Inoue and Naruse's values are for deposited material, not for aerosols, their distribution figure is shifted towards larger size particles than would be expected for aerosols. Nonetheless, this wide range of values is attributable to the fact that there are many major dust sources in Asia and that the distances of these sources from Japan range from 1,500 km to 5,000 km.

3.3.1.5. Vertical Distribution. Based on our understanding of the meteorological process of Asian dust storms and the subsequent transport process, we assume that dust is mixed from its source on the surface to as high as 8 km to 10 km into the atmosphere. However, to my knowledge, the vertical distribution of Asian dust over Asia or the Pacific has not been studied. Asian dust was measured at Mauna Loa Observatory in Hawaii from 1979 to 1982 (Parrington et al. 1983, Parrington and Zoller 1984). The seasonal cycle at this station (elevation 3.4 km) is identical to that measured at the SEAREX station on Oahu; the dust concentrations are also comparable. The Mauna Loa data substantiate the concept that Asian dust is primarily transported in the middle troposphere. However, there are no data on which a more detailed picture of the process can be based.

3.3.1.6. Deposition. The most extensive dust-deposition data for the Pacific is from the SEAREX network (Uematsu et al. 1985). The SEAREX researchers collected total deposition (wet and dry) samples for periods of a week at Midway, Oahu, Enewetak, and Fanning during 1981. The average daily deposition during the high-dust season (nominally February through June) at Midway of 0.30 μg/cm^2/day is 3.6 times higher than it is during the clean season (0.084 μg/cm^2/day from July through January). Their deposition rates also show the same latitudinal trend as the aerosol-concentration rates--highest in the midlatitudes and lowest in the low latitudes. The highest annual mean rate is the 0.17 $\mu g/cm^2$/day at Midway and the lowest, the 0.031 $\mu g/cm^2$/day at Fanning.

The short-term temporal variability of dust-deposition from the SEAREX network is quite high. At Midway, about half the annual flux is deposited during two one-week collection periods (i.e., 6% of the annual collection time). If daily precipitation samples, rather than weekly samples, were collected, the actual deposition periods would most likely be much shorter as they are in Miami (Prospero et al. 1987).

The seasonal variability of dust concentrations in the atmosphere is also reflected in the variability of particle concentrations in surface waters. Using data obtained in 1978 during a round-trip cruise from Tokyo to Alaska, Uematsu et al. (1985) report a mean particulate Al concentration of 5.5 $\mu g/kg$ in early July or almost 8 times more than the figure for mid-August of 0.72 $\mu g/kg$. This high water-column concentration in July correlated with the high-dust concentrations measured concurrently over the area. During high-deposition periods, mineral particles made up more than 60% of the suspended particles in the water column. These data emphasize how important short-term deposition is to the total flux.

The SEAREX researchers (Uematsu et al. 1985) report dust-deposition rates of 64 $\mu g/cm^2$/yr and 43 $\mu g/cm^2$/yr for Midway and Oahu, respectively. These figures compare favorably with the values for the mineral-particle flux determined using data from samples taken from sediment traps near the Aleutian Islands (100-200 $\mu g/cm^2$/yr, Tsunogai et al. 1982) and from over the East Hawaii Abyssal Plain (50 $\mu g/cm^2$/yr; Honjo et al. 1982). These deposition rates are also consistent with the deep-sea-sediment accumulation rates of aeolian material over the past 5,000-15,000 years. That the mineralogy of the dust collected aboard ships in the western North Pacific is identical to that of deep-sea sediments (Blank et al. 1985) stresses the importance of Asian dust to sedimentation in the North Pacific.

The literature on Asian dust deposition rates in Japan is reviewed by Inoue and Naruse (1987) who analyzed a variety of types of samples including dust layers in snows. They estimate that recent deposition rates in Japan range from 0.5-1.0 mg/cm^2/yr. These figures are about an order of magnitude greater than the rates measured at Midway, which is at approximately the same latitude as Japan, and are equivalent to a soil accumulation rate of 3.6-7.1 mm/1000 yr. (These rates are 15% to 30% of the Saharan dust-deposition rates measured in Nigeria [McTainsh 1984].) By way of comparison, dust-fall rates in Beijing during dust events are 2.4 mg/cm^2/day, or about three orders of magnitude greater than the deposition rates in Japan (Sheng et al. 1981). Such heavy dust falls are apparently common in Beijing and large areas of China (Middleton et al. 1986). On the basis of previous studies of paleosols in dune deposits, Inoue and Naruse (1987) report that the deposition rate of Asian dust during the last glacial age is 1.9-3.2 mg/cm^2/yr. This is equivalent to 13.5-22.9 mm/1000 yr of deposited soil. This increased dust transport during the glacial age coincides with the period of formation of loess deposits in Asia.

3.3.1.7. Other Evidence of Long-Range Transport to the Pacific.

There is also evidence that species other than mineral aerosols are transported to the North Pacific with the mineral dust. Aerosol nitrate and nss-sulfate have been measured at the same network of stations as that used in the dust studies discussed before (Prospero et al. 1985). The seasonal cycle seen in the nitrate and sulfate concentrations matches that of the mineral-dust cycle although the seasonal changes in the nitrate and sulfate concentrations are not as great as those for dust. It is also significant that the mean concentrations of nitrate and nss-sulfate in the North Pacific are about three times higher than those for the central South Pacific, where continental sources have a minimal impact (Savoie et al. 1989b). These trends suggest that aerosol species transported from continents are important to the chemistry of the atmosphere over the North Pacific (Prospero and Savoie 1989, Savoie and Prospero 1989).

3.3.2. Continental Processes

3.3.2.1. Sources.

Asia has large expanses of land that could be sources of mineral aerosol. First, there are extensive desert regions, especially the Taklamakan, the Gobi, and the Ordos. Second, China has some of the most extensive and deepest loess deposits in the world (Pye 1987), most of which are used extensively for agriculture. Because of the fine texture of these soils, they are easily deflated.

Dust events are quite frequent in Asia. Meteorological stations in northwest China, especially in the Taklamakan and the Kansu Corridor, report that over 30 days a year the visibility is less than 1000 m because of the dust (Middleton et al. 1986). In contrast, other regions of China report less than three days a year of dust events (Goudie 1983).

3.3.2.2. Meteorology of Dust Transport.

There are few studies of dust-storm meteorology in Asia (Ing 1972, Watts 1969). The most detailed investigation is that of Merrill et al. (1985) who studied the conditions associated with a major Chinese dust outbreak that took place in April 1979. Their storm scenario begins with the development of a low-pressure system over Mongolia and the passage of a cold front over the arid regions of western China. After the passage of the cold front, a strong (1050-mb) high-pressure system is established, producing northerly surface winds of 15-18 m/s. These winds are accompanied by many reports of severe dust storms. Two days later, a second cold front moves across the area and almost all stations in the interior of China report dust. The advance of the second front behind the first produces strong uplift; the isentropes indicate that mixing extends from the surface to as high as 400 mb, or about 7 km. Imagery from the Japanese Geostationary Meteorological Satellite shows the dust-laden air moving across the coast of China and over Japan. The dust moves toward the east to the central North Pacific and gradually sinks and turns south, eventually entering the easterly wind regime. Once in the easterlies, the dust backtracks to such Pacific stations as Enewetak.

Although Merrill et al. (1985) find that isentropic trajectory analyses provide insights to the transport process, such analyses are

difficult because of the length of the trajectory paths and transit
times. Trajectory paths to the central Pacific are typically 10,000-
12,000-km long and transport times are generally about 8-14 days. Trajectory analyses using the SEAREX data show that the general scenario in
the preceding paragraph is common during the high-dust season. In contrast, during the low-dust season in July and August, a mixture of
sources could have an impact on the central Pacific; open ocean trajectories suggest sources in North and Central America with transit times of
17-21 days over distances of 9,000-12,000 km. The summer transport takes
place primarily within the trade-wind layer, which does not favor the
efficient transport of aerosols. This transport scenario (and the apparent absence of major dust sources in the Americas) is consistent with the
low mineral-aerosol concentrations reported during the summer months.

3.3.2.3. The Flux of Asian Dust over the North Pacific. Uematsu et al.
(1985) have made the only quantitative estimates of fluxes to the North
Pacific. Based on actual deposition measurements, they compute an annual
flux of 20 Mt/yr for the region between 0°N and 50°N and between 150°E
and 130°W. In their calculations, they assume that the concentrations
measured at the SEAREX stations apply to entire latitudinal bands across
the region. Their value is, therefore, an underestimate since it does
not account for the much higher dust concentrations in the midlatitudes
closer to the Asian and Japanese coasts. Because, as shown earlier,
Asian dust concentrations and deposition rates over Japan are over ten
times greater than those at Midway or Oahu, the total flux out of Asia
must be at least several times the 20-Mt deposition flux Uematsu et al.
calculated.

3.3.3. Conclusion

Despite its large size, the North Pacific is significantly impacted by
continental sources. This impact is most evident in the mineral-aerosol
data and in the nitrate- and sulfate-aerosol data collected concurrently.
Asia is the dominant source for this material. Meteorological conditions
preclude the possibility that North American or Central American sources
could play a major role except near coasts.

3.4. DISCUSSION AND CONCLUSIONS

In this chapter, I have assessed the impact of mineral-aerosol transport
on the two major oceans in the Northern Hemisphere—the Atlantic and the
Pacific. In each region, the mineral-aerosol concentrations are determined by the transport of materials from arid regions located in only one
of the bordering continental land masses: from North Africa for the
Atlantic and from Asia for the Pacific. North America and Europe appear
to play minor roles as sources of dust for oceans in the Northern Hemisphere. It is noteworthy that in one case, the transport takes place
primarily in the tropics although in the other case it occurs in the
midlatitudes. The differences provide an opportunity to study dust

generation and transport in two distinctly different climatic and meteorological environments.

Although this discussion has emphasized material that is transported to the oceans, much material is also exported in other directions. In the case of Saharan dust, there is considerable documentation of substantial fluxes to the south, over the Gulf of Guinea; to the north, over the Mediterranean; and to the east, over the Middle East (Middleton et al. 1986). With Asian dust events, the data are too limited to assess the magnitude of transport to regions other than the Pacific. Nonetheless, ancillary transports are probably large and significant.

The study of dust is important because of its impact on primary geophysical and biogeochemical systems. Some of these impacts are discussed in Chapter 10 (page 197). However, an additional important consideration is that dust, because of its ubiquitousness and its inert chemical properties, can serve as an ideal tool for the study of long-range transport and for the development of transport models.

3.5. REFERENCES

Arimoto, R., R. A. Duce, B. J. Ray, and C. K. Unni. 1985. Atmospheric trace elements at Enewetak Atoll: 2. Transport to the ocean by wet and dry deposition. J. Geophys. Res. 90:2391-2408.

Arimoto, R., R. A. Duce, B. J. Ray, A. D. Hewitt, and J. Williams. 1987. Trace elements in the atmosphere of American Samoa: Concentrations and deposition to the tropical South Pacific. J. Geophys. Res. 92: 8465-8479.

Bertrand, J., J. Baudet, and A. Drochon. 1974. Importance des aerosols naturels en Afrique de l'Ouest. Rech. Atmos. 8:846-860.

Blank, M., M. Leinen, and J. M. Prospero. 1985. Major Asian aeolian inputs indicated by the mineralogy of aerosols and sediments in the western North Pacific. Nature 314:84-86.

Buat-Ménard, P. 1986. The ocean as a sink for atmospheric particles. In The Role of Air-Sea Exchange in Geochemical Cycling (P. Buat-Ménard, ed.) NATO ASI Series C, Vol. 185, Dordrecht:Reidel, 165-183.

Bucher, A. 1986. Recherches sur les Poussieres Minerales d'Origine Saharienne. M.S. thesis, Univ. of Reims-Champagne-Ardenne, 226 pp.

Carder, K. L., R. G. Steward, P. R. Betzer, D. L. Johnson, and J. M. Prospero. 1986. Dynamics and composition of particles from an aeolian input event to the Sargasso Sea. J. Geophys. Res. 91:1055-1066.

Carlson, T. N., and S. G. Benjamin. 1980. Radiative heating rates for Saharan dust. Atmos. Sci. 37:193-213.

Carlson, T. N., and R. S. Caverly. 1977. Radiative characteristics of Saharan dust at solar wavelengths. J. Geophys. Res. 82:3141-3152.

Carlson, T. N., and J. M. Prospero. 1972. The large-scale movement of Saharan air outbreaks over the northern equatorial Atlantic. Appl. Meteorol. 11:283-297.

Chen, L., and R. A. Duce. 1983. The sources of sulfate, vanadium and mineral matter in aerosol particles over Bermuda. Atmos. Environ. 17:2055-2064.

Chester, R. 1982. Particulate aluminum fluxes in the eastern Atlantic. Marine Chemistry 11:1-16.

Chester, R. 1986. The marine mineral aerosol. In The Role of Air-Sea Exchange in Geochemical Cycling (P. Buat-Ménard, ed.) NATO ASI Series C, Vol. 185, Dordrecht:Reidel, 443-476.

Coude-Gaussen, G. 1984. Le cycle des poussieres eoliennes desertiques actualles et la sedimentation des loess peridesertiques quarter-naires. Bull. Centre Rech. Explor. Product. Elf-Aquitaine 8: 167-182.

D'Almeida, G. A. 1986. A model for Saharan dust transport. Climate Appl. Meteorol. 25:903-916.

Darwin, C. 1846. An account of the fine dust which often falls on ves-sels in the Atlantic Ocean. Geol. Soc. London, Q. J. 2:26-30.

Duce, R. A. 1986. The impact of atmospheric nitrogen, phosphorous and iron species on marine biological productivity. In The Role of Air-Sea Exchange in Geochemical Cycling (P. Buat-Ménard, ed.) NATO ASI Series C, Vol. 185, Dordrecht:Reidel, 497-529.

Duce, R. A., R. Arimoto, B. J. Ray, C. K. Unni, and P. J. Harder. 1983. Atmospheric trace elements at Enewetak: I. Concentrations, sources and temporal variability. J. Geophys. Res. 88:5321-5342.

Estoque, M., J. Fernandez-Partagas, D. M. Helgren, and J. M. Prospero. 1974. Genesis of major dust storms in West Africa during the summer of 1974. U. S. Army Res. Office Tech. Rept., Contract No. DAAG29-83-k-0082, University of Miami, Fl., 17 pp.

Fett, W. 1958. Der Atmospharische Staub. Berlin:Deutches Verlag Wissen, 309 pp.

Gillette, D. A. 1981. Production of dust that may be carried great dis-tances. In Desert Dust: Origin, Characteristics, and Effects on Man (T. L. Pewe, ed.) Spec. Paper 186, Boulder:Geol. Soc. Am., 11-26.

Glaccum, R. A., and J. M. Prospero. 1980. Saharan aerosols over the tro-pical North Atlantic--mineralogy. Marine Geol. 37:295-321.

Goudie, A. S. 1983. Dust storms in space and time. Progress Physical Geography 7:502-530.

Gregory, G. L., R. C. Harriss, R. W. Talbot, R. A. Rasmussen, M. Gar-stang, M. O. Andreae, R. R. Hinton, E. V. Browell, S. M. Beck, D. I. Sebacher, M. A. K. Khalil, R. J. Ferek, and S. V. Harriss. 1986. Air chemistry over the tropical forest of Guyana. J. Geophys. Res. 91:8603-8612.

Helgren, D. M., and J. M. Prospero. 1987. Wind velocities associated with dust deflation events in the western Sahara. Climate Appl. Meteorol. 26:1147-1151.

Honjo, S., S. J. Manganini, and J. J. Cole. 1982. Sedimentation of bio-genic matter in the deep ocean. Deep-Sea Res. 29:609-625.

Ing, G. K. T. 1972. A dust storm over central China, April 1969. Weather 27:136-145.

Inoue, K., and T. Naruse. 1987. Physical, chemical, and mineralogical characteristics of modern aeolian dust in Japan and rate of dust deposition. Soil Sci. Plant Nutrition 33:327-345.

Jaenicke, R., and L. Schutz. 1978. Comprehensive study of physical and chemical properties of the surface aerosols in the Cape Verde Islands region. J. Geophys. Res. 83:3585-3599.

Kadowaki, S. 1979. Silicon and aluminum in urban aerosols for characterization of atmospheric soil particles in the Nagoya area. Environ. Sci. Technol. 13:1130-1133.

Kremling, K. 1985. The distribution of cadmium, copper, nickel, manganese, and aluminum in surface waters of the open Atlantic and European shelf area. Deep-Sea Res. 32:531-555.

Maring, H. B., and R. A. Duce. 1987. The impact of atmospheric aerosols on trace metal chemistry in open ocean surface seawater: 1. Aluminum. Earth Planetary Sci. Ltrs. 84:381-392.

McTainsh, G. 1984. The nature and origin of aeolian mantles in central northern Nigeria. Geoderma 33:13-37.

McTainsh, G., and P. H. Walker. 1982. Nature and distribution of Harmattan dust. Z. Geomorphol. N. F. 26:417-435 (Z).

Merrill, J. T., 1986. Atmospheric pathways to the oceans. In The Role of Air-Sea Exchange in Geochemical Cycling (P. Buat-Ménard, ed.) NATO ASI Series C, Vol. 185, Dordrecht:Reidel, 35-63.

Merrill, J. T., R. Bleck, and L. Avila. 1985. Modeling atmospheric transport to the Marshall Islands. J. Geophys. Res. 90:12,927-12,936.

Middleton, N. J. 1985. Effect of drought on dust production in the Sahel. Nature 316:431-434.

Middleton, N. J. 1986. A geography of dust storms in southwest Asia. Climatology 6:183-196.

Middleton, N. J., A. S. Goudie, and G. L. Wells. 1986. The frequency and source areas of dust storms. In Aeolian Geomorphology (W. G. Nickling, ed.) New York:Allen and Unwin, 237-259.

Morales, C. 1979. The use of meteorological observations for studies of the mobilization, transport and deposition of Saharan soil dust. In Saharan Dust:Mobilization, Transport, Deposition (C. Morales, ed.) New York:Wiley, 119-131.

Morales, C. 1986. The airborne transport of Saharan dust: A review. Climatic Change 9:219-241.

Muhs, D. R., R. C. Crittenden, J. H. Rosholt, C. A. Bush, and K. C. Stewart. 1987. Genesis of marine terrace soils, Barbados, West Indies: Evidence from mineralogy and geochemistry. Earth Surface Processes, Landform 12:605-618.

Parrington, J. R., and W. H. Zoller. 1984. Diurnal and longer-term temporal changes in the composition of atmospheric particles at Mauna Loa, Hawaii. J. Geophys. Res. 89:2522-2534.

Parrington, J. R., W. H. Zoller, and N. K. Aras. 1983. Asian dust: Seasonal transport to the Hawaiian Islands. Science 220:195-198.

Prospero, J. M. 1979. Mineral and sea salt aerosol concentrations in various ocean regions. J. Geophys. Res. 84:725-731.

Prospero, J. M. 1981. Aeolian transport to the world ocean. In The Oceanic Lithosphere (C. Emiliani, ed.) New York:Wiley, 801-874.

Prospero, J. M., and T. N. Carlson, 1972. Vertical and areal distribution of Saharan dust over the western equatorial North Atlantic Ocean. J. Geophys. Res. 77:5255-5265.

Prospero, J. M., and T. N. Carlson. 1981. Saharan air outbreaks over the tropical North Atlantic. Pure Appl. Geophysics 119:677-691.

Prospero, J. M., and R. T. Nees. 1986. Impact of the North African drought and El Nino on mineral dust in the Barbados trade winds. Nature 320:735–738.

Prospero, J. M., and D. L. Savoie. 1989. Effect of continental sources on nitrate concentrations over the Pacific Ocean. Nature, in press.

Prospero, J. M., E. Bonatti, C. Schubert, and T. N. Carlson. 1970. Dust in the Caribbean atmosphere traced to an African dust storm. Earth Planetary Sci. Ltrs. 9:287–293.

Prospero, J. M., D. L. Savoie, T. N. Carlson, and R. T. Nees. 1979. Monitoring Saharan aerosol transport by means of atmospheric turbidity measurements, In Saharan Dust: Mobilization, Transport, Deposition (C. Morales, ed.) New York:Wiley, 171–186.

Prospero, J. M., R. A. Glaccum, and R. T. Nees. 1981. Atmospheric transport of soil dust from Africa to South America. Nature 289:570–572.

Prospero, J. M., D. L. Savoie, R. T. Nees, R. A. Duce, and J. Merrill. 1985. Particulate sulfate and nitrate in the boundary layer over the North Pacific Ocean. J. Geophys. Res. 90:10,586–10,596.

Prospero, J. M., R. T. Nees, and M. Uematsu. 1987. Deposition rate of particulate and dissolved aluminum derived from Saharan dust in precipitation at Miami, Florida. J. Geophys. Res. 92:14,723–14,731.

Prospero, J. M., M. Uematsu, and D. L. Savoie. 1989. Mineral aerosol transport to the Pacific Ocean. In Chemical Oceanography, Vol. 10 (J. P. Riley, ed.) New York:Academic Press, in press.

Pye, K. 1987. Aeolian Dust and Dust Deposits. New York:Academic Press, 334 pp.

Savoie, D. L. 1978. Physical and chemical characteristics of the Saharan aerosols over the tropical northern Atlantic. M.S. thesis, Univ. of Miami, FL, 137 pp.

Savoie, D. L. 1984. Nitrate and non–sea–salt sulfate aerosols over major regions of the world ocean: Concentrations, sources, and fluxes. Ph.D. dissert., Univ. of Miami, FL, 432 pp.

Savoie, D. L., and J. M. Prospero. 1989. Comparison of oceanic and continental sources of non–seasalt sulphate over the Pacific Ocean. Nature, in press.

Savoie, D. L., J. M. Prospero, and R. T. Nees. 1987. Frequency distribution of dust concentration in Barbados as a function of averaging time. Atmos. Environ. 21:1659–1664.

Savoie, D. L., J. M. Prospero, and E. S. Saltzman. 1989a. Non–seasalt sulfate and nitrate in trade wind aerosols at Barbados: Evidence for long–range transport. J. Geophys. Res. 94:5069–5080.

Savoie, D. L., J. M. Prospero, and E. S. Saltzman. 1989b. Nitrate, non–seasalt sulfate and methanesulfonate over the Pacific Ocean. In Chemical Oceanography, Vol. 10 (J. P. Riley, ed.) New York:Academic Press, in press.

Schutz, L. 1980. Long–range transport of desert dust with special emphasis on the Sahara. In Aerosols: Anthropogenic and Natural, Sources and Transport (T. J. Kneip and P. J. Lioy, eds.) New York:Ann. N.Y. Academy Sci., 338:515–532.

Schutz, L., and K. A. Rahn. 1982. Trace–element concentrations in erodible soils. Atmos. Environ. 16:171–176.

Schutz, L., and M. Sebert. 1987. Mineral aerosols and source identification. Aerosol Sci. 18:1-10.

Schutz, L., R. Jaenicke, and H. Pietrer. 1981. Saharan dust transport over the North Atlantic Ocean. In Desert Dust: Origin, Characteristics, and Effects on Man (T. L. Pewe, ed.) Special Paper 186, Boulder:Geol. Soc. Am., 87-100.

Sheng, L. T., G. X. Fei, A. Z. Sheng, and F. Y. Xiang. 1981. The dust fall in Beijing, China, on April 18, 1980. In Desert Dust: Origin, Characteristics, and Effect on Man (T. L. Pewe, ed.) Special Paper 186, Boulder:Geol. Soc. Am., 149-157.

Stoffyn, M., and F. T. Mackenzie. 1982. Fate of dissolved aluminium in the oceans. Marine Chemistry 11:105-127.

Talbot, R. W., R. C. Harriss, E. V. Browell, G. L. Gregory, D. I. Sebacher, and S. M. Beck. 1986. Distribution and geochemistry of aerosols in the tropical North Atlantic troposhere: Relationship to Saharan dust. J. Geophys. Res. 91:5173-5182.

Tsoar, H., and K. Pye. 1987. Dust transport and the question of desert loess formation. Sedimentol. 34:139-153.

Tsunogai, S., M. Uematsu, S. Noriki, N. Tanaka, and M. Yamada. 1982. Sediment trap experiment in the northern North Pacific: Undulation of settling particles. Geochem. 16:129-147.

Tsunogai, S., T. Suzuki, T. Kurata, and M. Uematsu. 1985. Seasonal and areal variation of continental aerosol in the surface air over the western North Pacific region. Oceanog. Soc. Japan 41:427-434.

Uematsu, M., R. A. Duce, J. M. Prospero, L. Chen, J. T. Merrill, and R. L. McDonald. 1983. Transport of mineral aerosol from Asia over the North Pacific Ocean. J. Geophys. Res. 88:5343-5352.

Uematsu, M., R. A. Duce, and J. M. Prospero. 1985. Deposition of atmospheric mineral particles in the North Pacific Ocean. Atmos. Chemistry 3:123-138.

Watts, I. E. M. 1969. Climates of China and Korea. In World Survey of Climatology (H. Arakawa, ed.) Amsterdam:Elsevier, 8:1-117.

Wolff, G. T., M. S. Ruthkosky, D. P. Stroup, P. E. Korsog, M. A. Ferman, G. J. Wendel, and D. H. Stedman. 1986. Measurement of SO_x, NO_x, and aerosol species on Bermuda. Atmos. Environ. 20:1229-1239.

4. THE INTERCONTINENTAL TRANSPORT OF SULFUR AND NITROGEN

James N. Galloway
Department of Environmental Sciences
University of Virginia
Charlottesville, VA 22903

4.1. INTRODUCTION

With the increasing industrialization and population of the world, the
impact of anthropogenic activities on the atmosphere is evident on
several spatial scales. Population centers in developed and developing
countries are being affected locally (e.g., visibility reduction) and
regionally (e.g., acid deposition) by the release of combustion products
into the atmosphere. Through studying these local and regional effects,
it is now realized that the contributing materials are not entirely
deposited in the region in which they are emitted. The materials can be
transported over thousands of kilometers (e.g., Levy and Moxim 1987,
Whelpdale et al. 1984). Thus, hemispheric and global scales of atmospheric contamination are now realities.

    The goal of the NATO Advanced Research Workshop held in Bermuda in
early 1988 was to assess the state of knowledge of the transport of various materials from continent to ocean and continent to continent. As a
contribution to the workshop, this paper was specifically concerned with
the large-scale transport of sulfur and nitrogen compounds. The approach
I chose for this preparatory paper was to examine the species of concern
and to discuss the physical and chemical characteristics that could
affect atmospheric residence times of these species and hence their
potential for long-range transport. I have presented evidence supporting
long-range transport in five regions of the world (see Chapter 1, p. 3)
and assessed the extent of our knowledge at the time of the workshop.

4.2. SULFUR AND NITROGEN CHARACTERISTICS

4.2.1. Sulfur

The sulfur compounds that I considered in this assessment are the gases
H_2S, $(CH_3)_2S$ [DMS], and SO_2 and the sulfate aerosols. The interrelation-
ships of these compounds in the atmosphere are depicted in the atmospheric sulfur cycle in Figure 4-1; their concentration ranges and
residence times are in Tables 4-1 and 4-2. The two estimates of the
concentrations of sulfur compounds in remote marine and continental
atmospheres agreed relatively well. Because DMS and SO_2 can be oxidized
to $SO_4^=$ aerosol, the aerosol is the most likely candidate for long-range
transport.

A. H. Knap (ed.), The Long-Range Atmospheric Transport of Natural and Contaminant Substances, 87–104.
© 1990 Kluwer Academic Publishers.

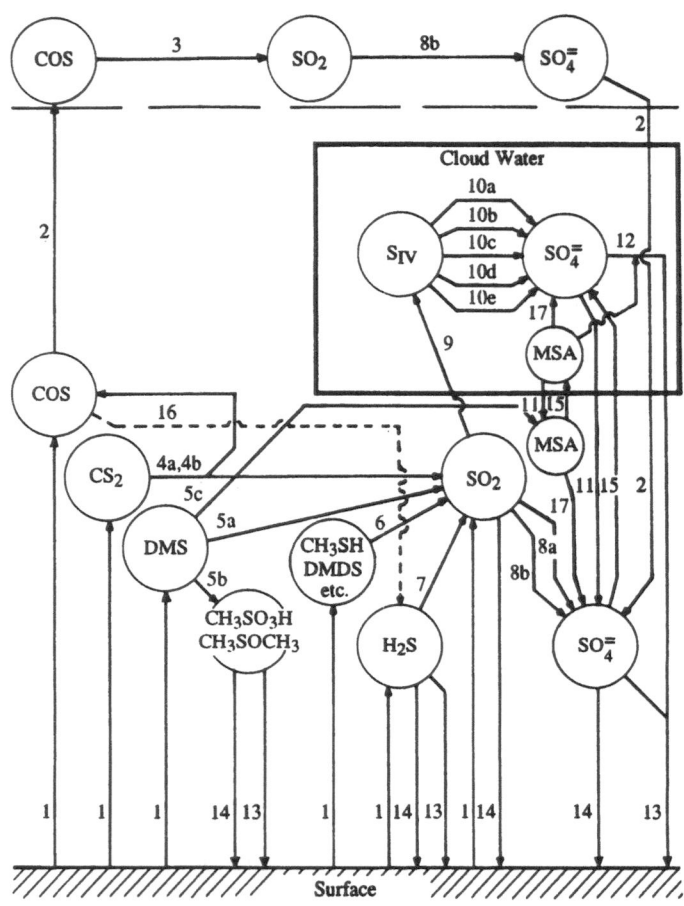

Figure 4-1. The chemical transformations of sulfur in the atmospheric
cycle (Charlson et al. 1985). Circles are chemical species, the box
represents cloud-liquid phase; DMS = CH_3SCH_3, DMDS = CH_3SSCH_3, S_{IV} =
$(SO_2)_{aq}$ + HSO_3^- + $SO_3^=$ + $CH_2OHSO_3^-$, and MSA (methane sulfonic acid)
= CH_3SO_3H. The chemical transformations are:

1: Surface emissions

2: Tropospheric/stratospheric
   exchange

3: $COS + h\gamma \longrightarrow S + S + CO$
   $S + O_2 \longrightarrow SO + O$
   $SO + O_2 \longrightarrow SO_2 + O$

4a: $CS_2 + OH \longrightarrow CS_2OH$
    $CS_2OH \longrightarrow$ multistep $\longrightarrow COS + SO_2$

4b: $CS_2 + h\gamma \longrightarrow CS_2*$
     $CS_2* + O_2 \longrightarrow CS + SO_2$
     $CS + O_2 \longrightarrow COS + O$
     $CS + O_3 \longrightarrow COS + O_2$

5a: $CH_3SCH_3 + OH \longrightarrow$ multistep $\longrightarrow SO_2$

5b: $CH_3SCH_3 + OH \longrightarrow$ multistep $\longrightarrow CH_3SOCH_3$

5c: $CH_3SCH_3 + OH \longrightarrow$ multistep $\longrightarrow CH_3SO_3H$

6:   $CH_3SH + OH \longrightarrow$ multistep $\longrightarrow SO_2$

7:   $H_2S + OH \longrightarrow$ multistep $\longrightarrow SO_2$

8a: $SO_2 + OH \longrightarrow HSO_3$
     $HSO_3 + O_3 \longrightarrow HO_2 + SO_3$
     $SO_3 + H_2O \longrightarrow H_2SO_4$

8b: $SO_2 \xrightarrow{\text{Heterogeneous catalysis}} SO_4^=$

9:   $(SO_2)_g \longleftrightarrow (SO_2)_{aq}$
     $(SO_2)_{aq} + H_2O \longrightarrow HSO_3^- + H^+$
     $HSO_3^- \longleftrightarrow H^+ + SO_3^=$
     $CH_2(OH)_2 + HSO_3^- \longleftrightarrow H_2O + CH_2OHSO_3^-$

10a: $(H_2O_2)_g \longleftrightarrow (H_2O_2)_{aq}$
     $HSO_3^- + (H_2O_2)_{aq} \longrightarrow$ multistep $\longrightarrow H^+ + SO_4^=$

10b: $(O_3)_g \longleftrightarrow (O_3)_{aq}$
     $HSO_3^- + (O_3)_{aq} \longrightarrow$ multistep $\longrightarrow H^+ + SO_4^=$

10c: $(HO_2)_g \longleftrightarrow (HO_2)_{aq}$
     $(HO_2)_{aq} \longrightarrow H^+ + O_2^-$
     $(HO_2)_{aq} + O_2 \xrightarrow{H_2O} (H_2O_2)_{aq} + OH^-$
     $HSO_3^- + (H_2O_2)_{aq} \longrightarrow$ multistep $\longrightarrow 2H^+ + SO_4^=$

10d: $(OH)_g \longleftrightarrow (OH)_{aq}$
     $HSO_3^- + (OH)_{aq} \longrightarrow$ multistep $\longrightarrow 2H^+ + SO_4^=$

10e: $HSO_3^- + O_2 \longrightarrow$ multistep $\longrightarrow H^+ + SO_4^=$

11: Evaporation

12: $SO_4^=$ in cloud water $\longrightarrow SO_4^=$ in rainwater

13: Washout, rainout

14: Dry deposition

15: Cloud nucleation

16: $COS + OH \longrightarrow$ multistep $\longrightarrow H_2S$

17: $MSA \longrightarrow SO_4^=$ by some mechanism

Table 4-1. Concentrations of sulfur compounds in the atmosphere (μg S/m-3)

Species	Remote Marine	Continental	Populated Continental	Reference
SO_2	0.05 - 0.2	0.2 - 1.0		Galloway 1985
	0.2 ± 0.1	0.2 ± 0.1	5.0 ± 2.0	Ryaboshapko et al. 1987
$SO_4^=$	0.02 - 0.35	0.5 - 0.8		Galloway 1985
	0.35 ± 0.15	0.6 ± 0.2	3.0 ± 0.5	Ryaboshapko et al. 1987
DMS	0.3 ± 0.2	0.2 ± 0.1	0.2 ± 0.1	Ryaboshapko et al. 1987

    To estimate the significance of the transport between continents
and between continents and oceans, background concentrations and the
amount of material transported must be known as well as the residence
time of the material.

    How much material is transported from continent to continent depends
on the amount advected from the initiating continent. Since anthropogenic activities are major sources of the sulfur and nitrogen emitted to
the atmosphere over populated continents, continental export from a natural environment must be distinguished from export from an environment
influenced by anthropogenic activities. On a global average, Ryaboshapko
et al. (1987) estimate that 27 Tg S/yr are transported from a natural
continental environment to the marine atmosphere; however, this figure
increases to about 100 Tg S/yr when the human contribution is added.
Thus, even globally, the continent-to-ocean transport of material is
substantial. But, does this material stay in the atmosphere long enough
to be deposited onto a continent downwind of an ocean?

    To answer this question, synoptic meteorological conditions and
residence times need to be known. The residence times of sulfur compounds in the atmosphere range from 10-100 hr (Table 4-2). At an average
wind speed of 8 m/sec and with 5000-8000 km to cover, an air parcel would
take approximately 7-12 days to cross the Atlantic or Pacific Ocean,
respectively. Since the residence lifetimes of sulfur compounds in the
atmosphere average from 0.5-4 days, only a small fraction of the sulfur

Table 4-2. Residence times of sulfur compounds in the marine boundary
    layer and marine free troposphere (hrs).

Species	Boundary Layer Midlatitude Winter	Summer	Subtropic	Equator	Free Troposphere Midlatitude	Tropic
SO_2	≤50	≤50	≤10	≤10	20-60	10-30
$SO_4^=$	50	100	200	50	?	?
DMS	≥150	~100	~50	~50	25-100	~30

Source: Galloway et al. (1985).

advected to a marine atmosphere from an initiating continent would make it ¨to a downwind continent. However, episodic transport above the boundary layer could bypass the typical removal mechanisms of wet and dry deposition (see Chapter 1, p. 3).

4.2.2. Nitrogen

The continents, especially those with large populations, can be major sources of NO in the atmosphere. Once in the atmosphere, NO undergoes a complex series of reactions (Fig. 4-2). Given the correct atmospheric conditions (see Chapter 1, p. 3), the products of these reactions can be transported over large distances. The species of nitrogen that are stable enough to survive long-distance transport are the gases HNO_3, NH_4^+, and NO_3^- aerosol. The concentrations of these species in the remote atmosphere are listed in Table 4-3. I have used nitrate aerosol to document the existence of long-range transport because there was more data available for it than for the other species. However, other nitrogen species could also contribute to the long-range transport of nitrogen. Chapter 11 (p. 231) discusses the possible contributions of other species.

On the average, some nitrogen oxides reside in the atmosphere less time than do sulfur oxides (Summers and Fricke 1989). However, if subjected to episodic transport above the boundary layer, the oxides of nitrogen and of sulfur could undergo similar degrees of dispersion when transported over great distances.

Table 4-3. Concentrations of nitrogen compounds in the remote atmosphere (μg N/m3).

Species	Marine	Continental
HNO_3	0.02	0.10
NH_4^+	0.3	0.1
NO_3^-	0.06	0.06

Source: Galloway (1985)

4.3. THE LONG-RANGE TRANSPORT OF SULFUR AND NITROGEN

To determine the degree that sulfur and nitrogen compounds are transported from one continent to another, the following information is required:

1. The prevailing transport path of air advected from the continent, the seasonality of the transport, and the possibility for transport in the free troposphere.

2. The composition of the marine atmosphere downwind of the continent (to determine if there is evidence of transport).

92

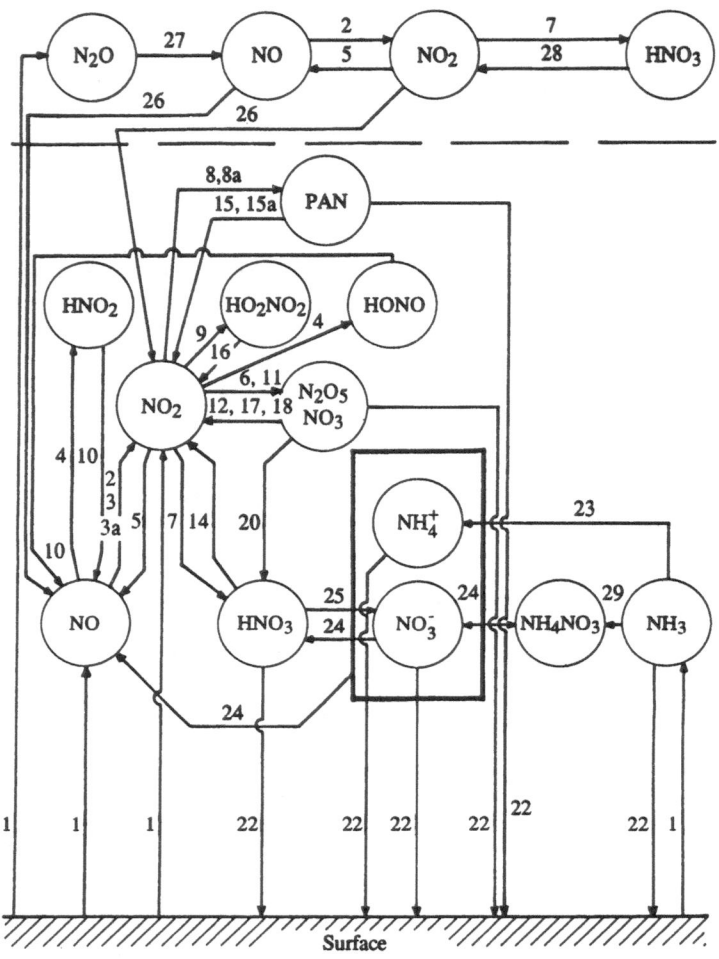

Figure 4-2. The chemical transformations of nitrogen in the atmospheric
cycle (Charlson 1985). Circles are chemical species, the box represents cloud-liquid phase.

1: Sources

2: $NO+O_3 \longrightarrow NO_2+O_2$

3: $NO+HO_2 \longrightarrow NO_2+OH$

3a: $NO+RO_2 \longrightarrow NO_2+RO$

4: $NO+OH+M \longrightarrow HONO+M$

5: $NO_2+h\gamma \longrightarrow NO+O$

6: $NO_2+O_3 \longrightarrow NO_3+O_2$

7: $NO_2+OH+M \longrightarrow HNO_3+M$

8: $NO_2+CH_3C(O)O_2+M \longrightarrow$
    $CH_3C(O)O_2NO_2+M$

8a: $NO_2+R(O)O_2+M \longrightarrow$
    $R(O)O_2NO_2+M$

9: $NO_2+HO_2+M \longrightarrow HO_2NO_2+M$

10: $HONO+h\gamma \longrightarrow OH+NO$

11: $NO_3+NO_2+M \longrightarrow N_2O_5+M$

12: $NO_3+h\gamma \longrightarrow NO_2+O$

13: $NO_3 + x \longrightarrow$ products

14: $HNO_3 + h\gamma \longrightarrow NO_2 + OH$

15: $CH_3C(O)O_2NO_2 + h\gamma \longrightarrow$
     $CH_3C(O)O_2 + NO_2$

15a: $R(O)O_2NO_2 \longrightarrow R(O)O_2 + NO_2$

16: $HO_2NO_2 \longrightarrow HO_2 + NO_2$

17: $N_2O_5 + h \longrightarrow NO_3 + NO_2$

18: $N_2O_5 \longrightarrow NO_3 + NO_2$

19: $CH_3C(O)O_2 + NO \longrightarrow CH_3 + NO_2 + CO_2$

20: $N_2O_5 \longrightarrow HNO_3$

21: $NO_3 \longrightarrow$ products

22: Sinks to surface

23: NH_3 (g) $\longleftrightarrow NH_3$ (aq)
    $H_2O + NH_3$ (aq) $\longrightarrow NH_4^+ + OH^-$

24: Evaporation of cloud water

25: HNO_3 (g) $\longleftrightarrow HNO_3$ (aq)
    HNO_3 (aq) $\longrightarrow NO_3^- + H^+$

26: Tropospheric/stratospheric
    exchange

27: $N_2O + O^1D \longrightarrow NO + NO$

28: $HNO_3 + h\gamma \longrightarrow NO_2 + OH$

29: $NH_3 + HNO_3 \longrightarrow NH_4NO_3$

30: $N_2 + O_2 \xrightarrow{\text{lightning}} 2NO$

3. The estimated advection rates from the source continent and to the receiving continent.

In the following section I use the transport regimes identified by Whelpdale and Moody in Chapter 1 (p. 3) and my review of the published evidence to identify cases of long-range transport. In the working-group report on sulfur and nitrogen (Chapter 11, page 231), Levy and his group further discuss the amount of information that was available on the meteorological and chemical control of transport when our workshop met and present additional case studies.

4.3.1. Eastward Advection of S and N from North America to Europe

The significant amounts of S and N emitted in North America (Fig. 4-3), especially in eastern North America, enhance the possibility of large-scale transport eastward (Husar and Holloway 1982). Because of the magnitude of the emissions, there are numerous examples of this continental signal being found in gas, aerosol, and precipitation samples collected over the Atlantic Ocean. Historical records from Greenland ice cores show increasingly higher concentrations of $SO_4^=$ and NO_3^- (Busenberg and Langway 1979, Finkel et al. 1986, Mayewski et al. 1986, Neftel et al. 1985). Precipitation collected on Bermuda has higher concentrations of $SO_4^=$ and NO_3^- when the airflow is westerly (Galloway et al. 1987). Sulfur and nitrogen in aerosol samples from Bermuda show the same pattern (Chen and Duce 1983), as do the concentrations of SO_2, NO_x and HNO_3 (Wolff et al. 1986). Ryaboshapko et al. (1987) report higher concentrations of SO_2 and $SO_4^=$ aerosol in shipboard samples collected at many different locations on the northern Atlantic Ocean. These findings corroborate the earlier work of Meszaros (1978), who is perhaps the first to report that the atmosphere over the northern Atlantic Ocean is heavily impacted by continental emissions.

94

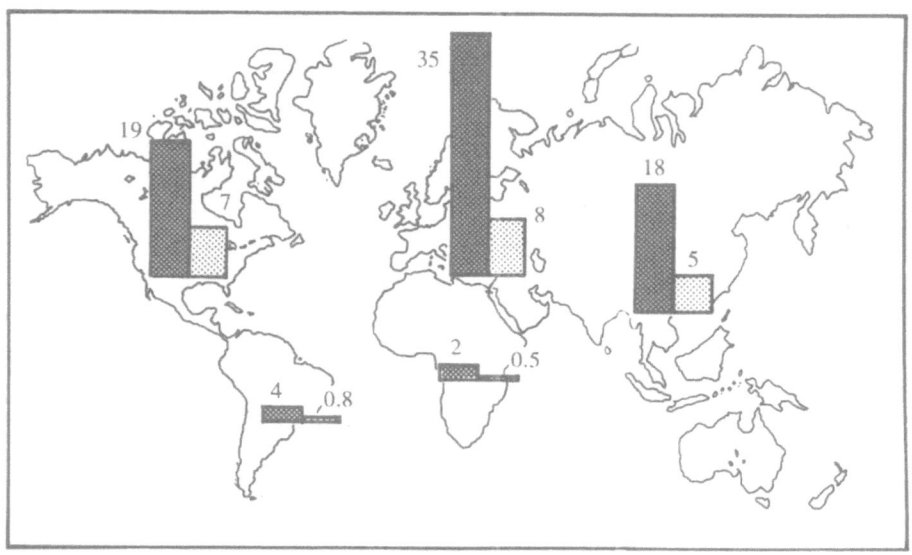

Figure 4-3. Estimated 1980 emissions (Tg/yr) of sulfur (**darker columns**)
and nitrogen (**lighter columns**) to the atmosphere of North America,
South America, Europe, Africa, and Asia (Rodhe and Herrera 1988).

Approximately 12 Tg S/yr are injected into the eastern North American atmosphere, primarily from anthropogenic sources (Galloway et al.
1984). Of the 12 Tg S/yr, estimates of the amount advected eastward to
the North Atlantic Ocean include 2.1 Tg S/yr (Fay et al. 1986), 3.0 Tg
S/yr (Bischoff et al. 1984), 3.5 Tg S/yr (Husar and Holloway 1982),
3.9 Tg S/yr (Galloway and Whelpdale 1980), 4.3 Tg S/yr (Galloway et al.
1984), and 5.8 Tg S/yr (Shannon and Lesht 1986) (Fig. 4-4A). Based on an
analysis of these estimates and a review of the surface and above-ground
data on sulfur concentrations found in the atmosphere of the western northern Atlantic Ocean, Galloway and Whelpdale (1987) estimate that the
most likely amount of sulfur advected eastward is 3-4 Tg S/yr--or about
25% to 30% of the total sulfur emitted to the atmosphere of eastern North
America. We also estimate that 1.9 ± 1.2 Tg S/yr is removed from the
atmosphere over the western Atlantic Ocean by wet and dry deposition and,
by difference, that 1-2 Tg S/yr is transported east of 60° longitude.
    There are two other investigations that address the question of sul-
fur transport from North America to Europe. Nyberg (1977) makes a qualitative observation of transport using air-mass trajectories to identify
source regions of sulfur found in precipitation. In a more rigorous
analysis using empirical data and an independent model, Whelpdale et al.
(1988) estimate that only a few tenths of a teragram of sulfur reach
Europe from North American emissions.

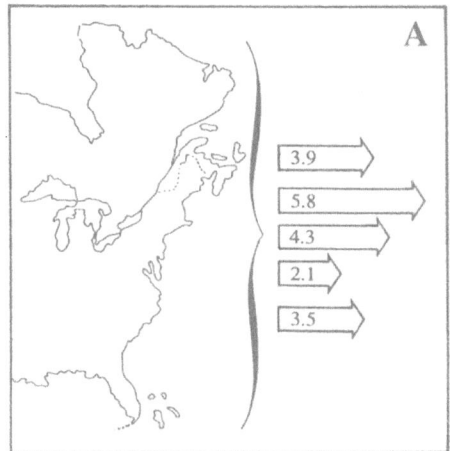 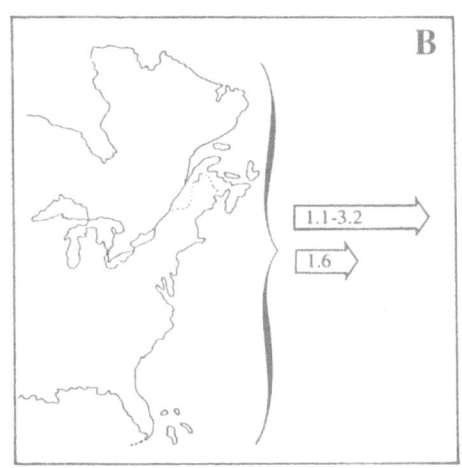

Figure 4-4. Estimates of the advection rate of **A** sulfur and **B** nitrogen
oxides eastward from North America. Sulfur values are from Fay et
al. (1986, 2.1 Tg S/yr), Husar and Holloway (1982, 3.5 Tg S/yr),
Galloway and Whelpdale (1980, 3.9 Tg S/yr), Galloway et al. (1984,
4.3 Tg S/yr), and Shannon and Lesht (1986, 5.8 Tg S/yr); nitrogen-
oxide values are from Galloway et al. (1984, 1.1-3.2 Tg N/yr),
Bischoff et al. (1984, 1.6 Tg N/yr), and Logan (1983, 1.6 Tg N/yr).
(Galloway and Whelpdale 1987)

In the case of nitrogen oxides, Logan (1983) estimates that 4.5 Tg
N/yr are emitted into the atmosphere over eastern North America as NO_x.
As with sulfur, most of this material is of anthropogenic origin. The
estimates of the amount of nitrogen advected eastward from North America
are 1.1-3.2 Tg N/yr (Galloway et al. 1984), and 1.6 Tg N/yr (Logan 1983,
Bischoff et al. 1984) (Fig. 4-4B). Based on an analysis of these estimates and a review of the surface and above-ground data on NO_x concentrations in the atmosphere of the western northern Atlantic Ocean, Galloway
and Whelpdale (1987) estimate that the most likely amount of NO_x
advected eastward is 0.8-1.2 Tg N/yr. This range represents about 18-27%
of the total NO_x emitted to the atmosphere over eastern North America.
We also estimate that less then 10% of the nitrogen oxides advected to
the atmosphere over the northern Atlantic Ocean is transported east of
60°W.

However, precipitation deposited on Bermuda associated with airflow
from the southeast Atlantic Ocean (i.e., with a minimal North American
influence) have concentrations of NO_3^- (and $nss-SO_4^=$) that are about a
factor of two higher than in precipitation in remote marine areas (Galloway et al. 1989). This implies that the remote marine areas used to
determine the natural background of the northern Atlantic Ocean are not
suitable, that there is another long-distance transport source of S and
N, or that the entire atmosphere of the northern Atlantic Ocean is

impacted by continental emissions from North America, Africa, and Europe (Galloway et al. 1989).

4.3.2. The Transport of S and N from South America to the Pacific and Atlantic Oceans

The westward and eastward transport of S and NO_x from South America to the Pacific and Atlantic Oceans is influenced by the northern and extreme southern areas of South America, respectively (see Chapter 1, p. 3). Because these land areas are small and because significantly less S and NO_x are emitted into the atmosphere over South America than over North America (Fig. 4-3), probably too little S and NO_x is transported from South America to detect at any distance, except on an episodic basis.

Unfortunately, few data are available that test this speculation. However, some limited evidence from analyses of ^{210}Pb and lipid-class organic compounds in the atmosphere west of Peru indicates that continental material is transported westward of South America to the Pacific (Schneider and Gagosian 1985, Schneider et al. 1983). Savoie (1984) also presents results from aerosol sampling over the South Pacific Ocean including off the west coast of South America. The highest NO_3^- and $SO_4^=$ concentrations he has found are off the northwest coast of South America. He concludes that the most likely sources of these concentrations are the anthropogenic activities in Chile and Peru although he does not rule out a natural source. Crutzen et al. (1985) conclude that biomass burning during the dry season in northern South America could be a significant source of atmospheric NO_x over the eastern South Pacific Ocean.

4.3.3. Transport East of Asia to the North Pacific Ocean

The population of Asia is about 2.7 billion people of which over one billion are in China, mostly in the eastern portion (Fig. 4-5). Thus, much of Asia's emissions of S and NO_x (Fig. 4-3) could be advected eastward out of China across the eastern and western Pacific. In doing their research in the western Pacific, Fukuda and Tsunogai (1975) and Tsunogai et al. (1985) have used atmospheric aerosol samples from Japan to illustrate the transport of ^{210}Pb and sulfate from Asia. Ito et al. (1986) used atmospheric concentrations taken over two Japanese islands in their evaluation of the residence times of gaseous and aerosol sulfur species over the western Pacific. They concluded that the residence time of SO_2 is 15 hours and that two-thirds of the SO_2 is deposited to the sea while the remainder is oxidized to $SO_4^=$ aerosol either in the boundary layer or in the free troposphere. The aerosol is available for long-range transport.

Prospero et al. (1985) have reported on sulfur and nitrogen transport east of Japan. Their research from several island sites in the middle of the northern Pacific Ocean found concentrations of nss-$SO_4^=$ in covariance with mineral aerosol, indicating that the material was transported from Asia. Prospero's research team also found similar evidence of continental NO_3^- at Fanning Island.

Investigations of methanesulfonic acid (MSA) have also found evidence of a continental influence on atmospheric sulfur over the northern

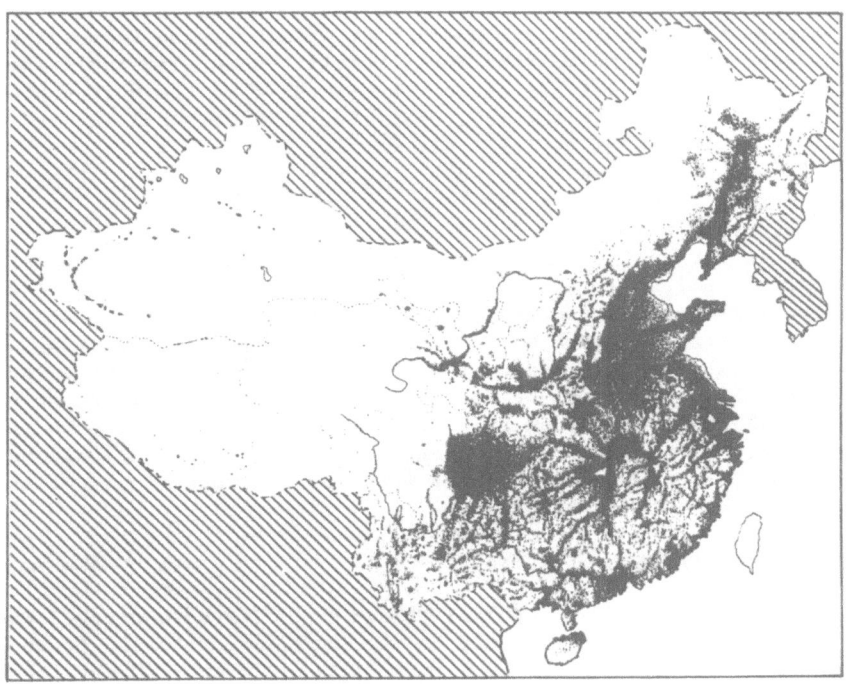

Figure 4-5.   Population distribution in China; each dot represents 5,000
     people (Clayre 1984).

Pacific Ocean.   Saltzman et al. (1986) report that the ratios of the MSA
concentrations to those of nss-$SO_4^=$ at Midway Island could not be
explained by DMS oxidation.   The lower MSA:nss-$SO_4^=$ ratios at Midway may
be related to variations in the input of continentally derived $SO_4^=$, the
composition of oceanic organosulfur emissions, and atmospheric reaction
pathways.   This observation is consistent with the possibility of transport from Asia to the North Pacific Ocean.
      Researchers working on data from Hawaii report similar results.
Winchester and Wang (1989) collected samples at a mountain site in Xin-
jiang in China, at a coastal site in western Japan, and at Mauna Loa
Observatory (MLO) in Hawaii.   In their study Winchester and Wang clearly
show that, during the spring, aerosol sulfur that could be from either an
anthropogenic or a natural source is associated with soil-derived dust
from the Asian continent and is carried across the Pacific to Hawaii.   In
a previous study, Darzi and Winchester (1982) speculated that much of the
sulfuric acid believed to cause the relatively high acidity in rain
samples from high altitudes in Hawaii may originate in Asia.   Based on
3.5 years of aerosol sampling, Parrington et al. (1983) conclude that
nss-$SO_4^=$ is not only transported from Asia during dust outbreaks but also
during nondust periods.

Andreae et al. (1988) have studied the possibility of material being transported from Asia to North America by examining the composition of the atmosphere over the western North Pacific Ocean. In analyzing air samples from an aircraft off the northwest coast of the United States, they have determined the vertical distribution of several gas and aerosol species in post-frontal maritime air masses. Isentropic trajectory calculations and measurements of radon-daughter concentrations showed that the air masses they sampled had been over the Pacific Ocean for 4–8 days. The DMS concentrations they found are consistent with the low levels of DMS found in the surface waters of the northeastern Pacific during their study period, which implies that DMS concentrations are controlled by marine rather than continental processes. The vertical profiles of SO_2 and nss-$SO_4^=$ from the work of Andreae et al. (1988) suggest that, under the conditions prevalent during the flights, (1) significant amounts of SO_2 and nss-$SO_4^=$ are produced from DMS oxidation only within the boundary layer and (2) transport from Asia dominates the sulfur cycle in the free troposphere. These authors also concluded that the concentrations they found of NO_3^- are influenced by the long-range transport of pollutant nitrogen, presumably from Asia.

4.3.4. Transport West of Africa

Because of its unique geographical position relative to synoptic meteorological conditions, Africa could be the source of material found in the atmosphere over both the northern and southern Atlantic Ocean (see Chapter 1, p. 3). The transport of mineral matter from Africa to the Atlantic Ocean has been reported at least since the nineteenth century (Darwin 1846). S and NO_x could be transported not only from Africa but also from Europe since not much S and NO_x is being emitted to the African atmosphere relative to the European atmosphere (Fig. 4–3). With appropriate meteorological conditions, S and NO_x from the European atmosphere could pass through the African atmosphere and be transported westward. Savoie (1984) has thoroughly analyzed the transport of Saharan dust westward from Africa. His primary conclusions are:

1. Large amounts of NO_3^- and nss-$SO_4^=$ are transported into the southern portion of the trade-wind region of the northern Atlantic Ocean during Saharan dust storms.

2. Most NO_3^- and nss-$SO_4^=$ associated with the episodic transport of Saharan dust seems to come from African and European anthropogenic activities.

3. NO_3^- and $SO_4^=$ from Africa and Europe are transported to the east coast of North America.

4. Most NO_3^- and nss-$SO_4^=$ found over the northern Atlantic Ocean come from continents.

Prospero et al. (1989) analyzed twelve months of aerosol sampling from Barbados. They concluded that 60% of the nss-$SO_4^=$ and about 70% to

80% of the NO_3^- transported across the tropical northern Atlantic Ocean are from North Africa or Europe with the higher figures coinciding with Saharan dust storms.

These conclusions are partly supported by the results of an aircraft sampling program held during June 1984 near Barbados (Talbot et al. 1986). At approximately 1.5-3.5 km, high concentrations of Saharan dust were found. Talbot et al. (1986) calculated that Saharan dust is a significant source of airborne NO_3^-, $nss-SO_4^=$, and PO_4^{-3} in the boundary layer air over the tropical northern Atlantic Ocean. Their estimates of the annual input of Saharan dust-related chemical species to the equatorial Atlantic suggests that the deposition of mineral aerosols may be an important source of NO_3^- and PO_4^{-3} in surface waters. Reichholf (1986), based on the work of Prospero and Nees (1986) and Prospero et al. (1981), speculated that Saharan dust may be a major source of nutrients for the Amazonian rain forest.

4.4. OUTLOOK FOR THE FUTURE

The long-range transport of S and NO_x over long distances is most likely from areas with strong sources of S and N. Because of the diffuse characteristics of natural emissions, these influential sources are probably in North America, Europe, or Asia in highly populated areas. It is not, therefore, surprising that S and NO_x have been observed in areas far downwind of Asia, North America, Africa, and Europe.

Since the study of long-range transport is so closely linked to the study of anthropogenic emissions, projections of the population and industrial growth of the world should be closely examined. In this section of the paper, I have used population estimates for five regions of the world for the year 2020 together with reasonable estimates of the per capita S and NO_x emissions to speculate how regional S and NO_x emissions could increase over the next three decades. During this period of time, the populations of the developing areas of Asia, Africa, and South America are projected to increase substantially; in the developed regions (North America and Europe), there are projected to be only slight population increases (Fig. 4-6).

To estimate what the S and N emissions will be in 2020, the rate of increase in per capita S and N emissions must first be ascertained and must be ascertained separately for developing and for developed areas. For developing countries boundary conditions must first be set. A minimum can be set by assuming that there will be no increase in the per capita S and N emissions in 2020, which implies that the industrial base remains the same. A more reasonable estimate would assume a factor of four increase over the 1980 figure (Galloway 1989). Figure 4-7 compares these two estimates to the 1980s rates of S and N emissions.

In developed areas, population growth is expected to be low and there is an increasing awareness of the environmental problems associated with S and N emissions. Therefore, it is probable that by 2020 the total and per capita emissions of S and N will only increase slightly (Fig. 4-7).

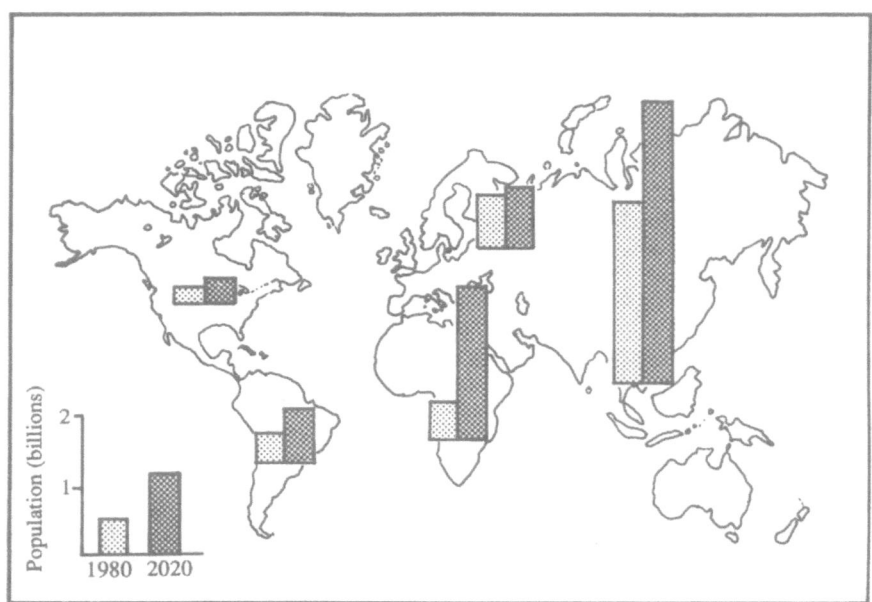

Figure 4-6. World population in 1980 (lighter columns) versus the
estimated world-population figure for 2020 (darker columns).

    These projections show small increases for S and N emissions in
North America and Europe. Thus the North American and European contributions to long-range transport can be expected to change little. However, in developing areas, emissions of S and N to the atmosphere should
increase substantially. The lower limit of the increase in emissions
will probably be about 50%, which will increase the contribution of emissions from developing regions to the long-range transport of S and NO_x.
This increase will primarily affect the atmospheres over the northern
Pacific Ocean and the Southern Hemisphere.
    Because of the strong possibility of these emission increases of S
and N and the potential effect on the marine atmosphere, more measurement
programs are needed to determine baseline and current conditions and to
track future changes in the composition of the marine atmosphere. The
climatic and ecological implications of these projected increases must
also be thoroughly investigated.

4.5. CONCLUSIONS

We already know a great deal about the long-range movements of the air in
the atmosphere. However, we know little about the composition of the air
that is being transported. Although we know that sulfur and nitrogen
compounds can be transported from North America to Europe, Asia to North

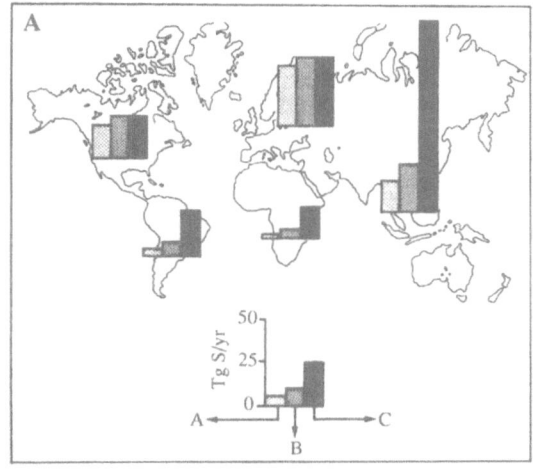

Figure 4-7A

Worldwide sulfur emissions (Tg S/yr) in 1980 (**A, lightest columns**) compared to estimates for 2020 that reflect a population growth (see Fig. 4-6) but no increase in per-capita sulfur emissions (**B, medium columns**) and for 2020 that reflect both a population growth and an increase in per-capita sulfur emissions (**C, darkest columns**). (Galloway 1989)

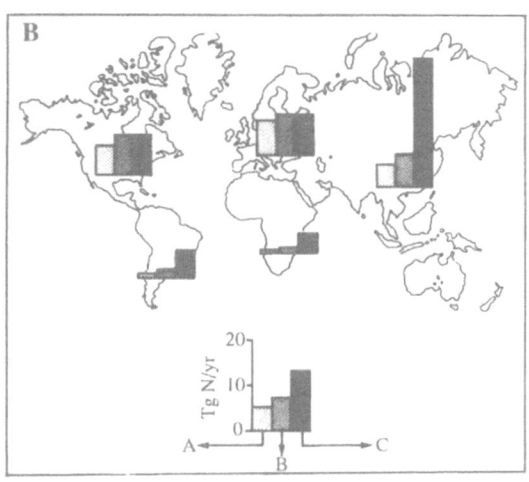

Figure 4-7B

Worldwide nitrogen-oxide emissions (Tg N/yr) in 1980 (**A, lightest columns**) compared to estimates for 2020 that reflect a population growth (see Fig. 4-6) but no increase in per-capita nitrogen-oxide emissions (**B, medium columns**) and for 2020 that reflect both a population growth and an increase in per-capita nitrogen-oxide emissions (**C, darkest columns**). (Galloway 1989)

America, and Africa to the Americas, we do not know with any great precision the magnitude of the transport. In Chapter 4 (p. 87) the working group on the long-range transport of sulfur and nitrogen discuss specific deficiencies in our knowledge. Because of the increased amounts of sulfur and nitrogen projected to be emitted into the atmosphere in the future and the long-range effect of these increases, it is paramount that the deficiencies in our knowledge of long-range transport be eliminated.

102

4.6. REFERENCES

Andreae, M. O., H. Berresheim, T. W. Andreae, M. A. Kritz, T. S. Bates, and J. T. Merrill. 1988. Vertical distribution of dimethylsulfide, sulfur dioxide, formic acid, aerosol ions, and radon over the northeast Pacific Ocean. Atmos. Chemistry 6:149-173.

Bischoff, W. D., V. L. Patterson, and F. T. McKenzie. 1984. Geochemical mass balance for sulfur-, and nitrogen-bearing components: Eastern United States. In Geological Aspects of Acid Deposition (O. P. Bricker, ed.) Boston:Butterworth 7:1-21.

Busenberg, E., and C. C. Langway, Jr. 1979. Levels of ammonium, sulfate, chloride, calcium,, and sodium in snow and ice from Southern Greenland. Geophys. Res. 84:1705-1709.

Charlson, R. J., W. L. Chameides, and D. Kley. 1985. The transformations of sulfur and nitrogen in the remote atmosphere: Background paper. In The Biogeochemical Cycling of Sulfur and Nitrogen in the Remote Atmosphere (J. N. Galloway. R. J. Charlson, M. O. Andreae, H. Rodhe, M. S. Marston, eds.) NATO ASI Series C, Vol. 159, Dordrecht:Reidel, 67-80.

Chen, L., and R. A. Duce. 1983. The sources of sulfate, vanadium and mineral matter in aerosol particles over Bermuda. Atmos. Environ. 17:2055-2064.

Clayre, A. 1984. The Heart of the Dragon. Boston:Houghton Mifflin Co., 280 pp.

Crutzen, P. J., A. C. Delany, J. Greenberg, P. Haagenson, L. Heidt, R. Lueb, W. Pollock, W. Seiler, A. Wartburg, and P. Zimmerman. 1985. Tropospheric chemical composition measurements in Brazil during the dry season. Atmos. Chem. 2:233-256.

Darwin, C. 1846. An account of the fine dust which often falls on vessels in the Atlantic Ocean. Geol. Soc. London Quart. 2:26-30

Darzi, M., and J. W. Winchester. 1982. Aerosol characteristics at Mauna Loa Observatory, Hawaii, after east Asian dust storm episodes. J. Geophys. Res. 87:1251-1258.

Fay, J. A., S. Kumar, and D. Golomb. 1986. Annual and semi-annual anthropogenic sulfur budget for eastern North America. Atmos. Environ. 20:1497-1500.

Finkel, R. C., C. C. Langway, Jr., and H. B. Clausen. 1986. Changes in precipitation chemistry at Dye 3, Greenland. J. Geophys. Res. 91:9849-9855.

Fukuda, K., and S. Tsunogai. 1975. Pb-210 in precipitation in Japan and its implication for the transport of continental aerosols across the ocean. Tellus 27:514-521.

Galloway, J. N. 1985. The deposition of sulfur and nitrogen from the remote atmosphere: Background paper. In The Biogeochemical Cycling of Sulfur and Nitrogen in the Remote Atmosphere (J. Galloway, N., R. J. Charlson, M. O. Andreae, H. Rodhe, and M. S. Marston, eds.) NATO ASI Series C, Vol. 159, Dordrecht:Reidel, 143-175.

Galloway, J. N. 1989. Atmospheric acidification: Projections for the future. Ambio 18:161-166.

Galloway, J. N., and D. M. Whelpdale. 1980. An atmospheric sulfur budget for eastern North America. Atmos. Environ. 14:409-417.

Galloway, J. N., and D. M. Whelpdale. 1987. WATOX-86 overview and western North Atlantic Ocean S and N atmospheric budgets. Global Biogeochem. Cycles 1:261-281.

Galloway, J. N., D. M. Whelpdale, and G. T. Wolff. 1984. The flux of S and N eastward from North America. Atmos. Environ. 18:2595-2607.

Galloway, J. N., R. J. Charlson, M. O. Andreae, H. Rodhe, and M. S. Marston (eds.). 1985. The Biogeochemical Cycling of Sulfur and Nitrogen in the Remote Atmosphere. NATO ASI Series C, Vol. 159, Dordrecht:Reidel, 249 pp.

Galloway, J. N., T. M. Church, A. H. Knap, D. M. Whelpdale, and J. M. Miller. 1987. The Western Atlantic Ocean Experiment (WATOX). In The Chemistry of Acid Rain: Sources and Atmospheric Processes (R. W. Johnson, G. E. Gordon, W. Calkins, and A. Z. Elzerman, eds.) ACS Symp. Series 349, American Chemical Society:Washington, 39-55.

Galloway, J. N., R. S. Artz, W. C. Keene, T. M. Church, and A. H. Knap. 1989. Processes controlling the concentration of SO_4, NO_3^-, NH_4^+, H^+, $HCOO^-$, and CH_3COO^- in Bermuda precipitation. Tellus, in press.

Husar, R. B., and J. M. Holloway. 1982. Sulfur and nitrogen over North America. In Ecological Effects of Acid Deposition, Rpt. PM 1636, Stockholm:National Swedish Environment Protection Board, 95-115.

Ito, T., T. Okita, M. Ikegami, and I. Kanazawa. 1986. The characterization and distribution of aerosol and gaseous species in the winter monsoon over the western Pacific Ocean. II. The residence time of aerosols and SO_2 in the long-range transport over the ocean. Atmos. Chemistry 4:401-411.

Levy, H., II, and W. J. Moxim. 1987. Fate of US and Canadian combustion nitrogen emissions. Nature 328:414-416.

Logan, J. A. 1983. Nitrogen oxides in the troposphere: Global and regional budgets. J. Geophys. Res. 88:10,785-10,807.

Mayewski, P. A., W. B. Lyons, M. J. Spencer, M. Twickler, W. Dansgaard, B. Koci, C. I. Davidson, and R. E. Honrath. 1986. Sulfate and nitrate concentrations from a south Greenland ice core. Nature 232:975-977.

Meszaros, E. 1978. Concentration of sulfur compounds in remote continental and oceanic areas. Atmos. Environ. 12:699-705.

Neftel, A., J. Beer, H. Oeschger, F. Zurcher, and R. C. Finkel. 1985. Sulphate and nitrate concentrations in snow from South Greenland 1895-1978. Nature 314:611-613.

Nyberg, A. 1977. On air-borne transport of sulphur over the North Atlantic. Quart. Royal Meteorol. Soc. 103:607-615.

Parrington, J. R., W. H. Zoller, and N. K. Aras. 1983. Asian dust: Seasonal transport to the Hawaiian Islands. Science 220:195-197.

Prospero, J. M., and R. T. Nees. 1986. Mineral aerosols in the trade winds at Barbados: Impact of the North African drought and El Nino. Nature 320:735-737.

Prospero, J. M., R. A. Glaccum, and R. T. Nees. 1981. Atmospheric transport of soil dust from Africa to South America. Nature 289:570-572.

Prospero, J. M., D. L. Savoie, R. T. Nees, R. A. Duce, and J. Merrill. 1985. Particulate sulfate and nitrate in the boundary layer over the North Pacific Ocean. J. Geophys. Res. 90:10,586-10,596.

Prospero, J. M., M. Uematsu, and D. L. Savoie. 1989. Mineral aerosol transport to the Pacific Ocean. In Chemical Oceanography, Vol. 10 (J. D. Riley, ed.) New York:Academic Press, in press.

Reichholf, J. H. 1986. Is Saharan dust a major source of nutrients for the Amazonian rain forest? Studies on Neotropical Fauna and Environment 21:251-255.

Rodhe, H., and R. Herrera. 1988. Acidification in Tropical Countries. New York:Wiley, 405 pp.

Ryaboshapko, A. G., V. I. Lepeshkin, E. D. Podgurskaya, and V. I. Medinets. 1987. Air pollution monitoring over the North Atlantic. In Procs., Intern. Symp., Intergrated Global Monitoring of the State of the Biosphere (U. A. Izrael, F. Y. Rovinsky, A. V. Tsyban, S. M. Semenov, V. A. Abakumov, eds.), Vol. 2 (WMO Tech. Document 151), Geneva:World Meteorol. Org., 261-282.

Saltzman, E. S., D. L. Savoie, J. M. Prospero, and R. G. Zika. 1986. Methanesulfonic acid and non-sea-salt sulfate in Pacific air: Regional and seasonal variations. Atmos. Chemistry 4:227-240.

Savoie, D. L. 1984. Nitrate and non-sea-salt sulfate aerosols over major regions of the world ocean: Concentrations, sources and fluxes. Ph.D. dissert., Univ. of Miami, FL, 432 pp.

Schneider, J. K., and R. B. Gagosian. 1985. Particle size distribution of lipids in aerosols off the coast of Peru. J. Geophys. Res. 90:7889-7898.

Schneider, J. K., R. B. Gagosian, J. K. Cochran, and T. W. Trull. 1983. Particle size distributions of n-alkanes and ^{210}Pb in aerosols off the coast of Peru. Nature 304:429-432.

Shannon, J. D., and M. M. Lesht. 1986. Modelled trends and climatological variability of the net transboundary mass flux of airborne sulfur between the United States and Canada. In Trans., 2nd Intern. Air Pollut. Specialty Conf., Meteorology of Acidic Deposition, Pittsburg:Air Pollut. Control Assoc., 28-35.

Summers, P. W., and W. Fricke. 1989. Atmospheric decay distances and times for sulphur and nitrogen oxides estimated from air and precipitation monitoring in eastern Canada. Tellus, in press.

Talbot, R. W., R. C. Harriss, E. V. Browell, G. L. Gregory, E. I. Sebacher, and S. M. Beck. 1986. Distribution and geochemistry of aerosols in the tropical North Atlantic troposphere: Relationship to Saharan dust. J. Geophys. Res. 91:5173-5182.

Tsunogai, S., T. Shinagawa, and T. Kurata. 1985. Deposition of anthropogenic sulfate and Pb-210 in the western North Pacific area. Geochem. 19:77-90.

Whelpdale, D. M., T. B. Low, and R. J. Kolomeychuk. 1984. Advection climatology for the east coast of North America. Atmos. Environ. 18:1131-1327.

Whelpdale, D., A. Eliassen, J. Galloway, H. Dovland, and J. Miller. 1988. The transatlantic transport of sulfur. Tellus 40B:1-15.

Winchester, J. W., and M.-X. Wang. 1989. Acid-base balance in aerosol components of the Asia-Pacific region. Tellus, in press.

Wolff, G. T., M. S. Ruthkosky, D. P. Stroup, P. E. Korsog, M. A. Ferman, G. J. Wendel, and D. H. Stedman. 1986. Measurements of SO_x, NO_x, and aerosol species on Bermuda. Atmos. Environ. 20:1229-1239.

5. THE LONG-RANGE TRANSPORT OF ORGANIC COMPOUNDS

Elliot Atlas
Department of Oceanography
Texas A & M University
College Station, Texas   77843

5.1. INTRODUCTION

Studies over the last several decades have revealed that the chemical
composition of organic matter in the atmosphere is astoundingly complex.
The complex composition of atmospheric organic compounds reflects the
variety of their sources--plant emissions, industrial and combustion
effluents, agricultural sources, etc.--as well as secondary products that
result from chemical and photochemical reactions in the troposphere.
Some organic compounds are sufficiently stable to be transported far from
their sources in continental regions to remote areas of the ocean or to
other continents. Such long-range transport is clearly evident, for
example, in the global distribution of high-molecular-weight chlorinated
hydrocarbons. As this chapter shows, pesticide and PCB residues have
been detected in air samples from both the Northern and Southern Hemispheres. Furthermore, the pesticides measured in various organisms around
the world--from fish and seals in the Antarctic to benthic amphipods in
the Arctic Ocean--are ample proof that these compounds are being deposited far from their source and subsequently accumulated by marine organisms (Hargrave et al. 1988; Hidaka et al. 1983; Subramanian et al. 1986a,
1986b; Tanabe et al. 1984, Zell and Ballschmiter 1980).
  Atmospheric transport and surface deposition are significant processes not only for synthetic compounds but also for many naturally
occurring organic compounds (Farmer and Wade 1986, Gagosian 1986, Kawamura and Kaplan 1986, Likens et al. 1983, Zafiriou et al. 1985). Studies
in the North Pacific, for example, have suggested that continentally
derived organic matter may be a significant component of sediments in
deep-ocean basins (Gagosian and Peltzer 1986). Also, careful analysis
has shown that continentally derived organics in remote ocean regions can
be useful as air-mass tracers in conjunction with meteorological models
of long-range transport (e.g., Gagosian et al. 1987).
  The long-range transport of organic compounds is also recognized as
potentially important in major photochemical cycles. The transport of
relatively reactive chemical species from natural or anthropogenic emissions influences the global ozone budget, and other long-lived trace
organics are possible ''greenhouse'' gases.
  Thus there are at least three areas of interest directly related to
the long-range atmospheric transport of organic compounds. These are:

1.   The global budgets and geochemical cycles of both pollutant and
     natural compounds.

A. H. Knap (ed.), The Long-Range Atmospheric Transport of Natural and Contaminant Substances, 105–135.

2. The transformations and reactions of organic compounds in the remote atmosphere and their effect on major photochemical cycles.

3. The use of unique continental markers as tracers of long-range transport to remote areas.

My initial concept as chairman of the working group on organics was to focus on high-molecular-weight synthetic compounds. However, considering the problems and processes involved, I decided to expand the discussion to other chemical species. Thus in this chapter I briefly review several classes of organic compounds in the atmosphere and discuss their concentrations, transport, and deposition and aspects of their chemical transformation. Unfortunately this discussion is limited by the scarcity of data, particularly from remote areas, for many of the compound classes I address. Since not all organic compounds can be reviewed here, I deal only with a relative few organic compounds that are representative of gaseous and particulate organic matter of primarily continental origin.

5.2. ANALYTICAL ASPECTS

As noted, there are relatively few data on organics in remote areas, but the situation is improving. The increasing information available on organics in remote areas is related to increasingly sensitive and specific methods of analysis. In addition, the problems of background contamination and sampling artifacts, which have always been a problem in trace analysis, are being recognized and evaluated. For most of the compounds discussed below, analysis involves some preconcentration of the organic fraction (either particles or gases); extraction and separation of the organics from the interfering compounds (for high-molecular-weight compounds); high-resolution, chromatographic separation of the individual compounds; and detection with the appropriate sensors, e.g., mass spectrometers, flame ionization, photo-ionization, or electron capture detectors. For most species, these analyses are neither routine nor trivial. One reason for the paucity of data on organics is the difficulty (and cost) of analytical protocols. The ability to perform such tasks is linked to the improvement and simplification of the present analytical technologies. Improved methodologies will allow more intensive, and extensive, data collections in remote areas that are needed to advance our understanding of the processes affecting atmospheric organics.

5.3. CONCENTRATION DISTRIBUTION

5.3.1. Volatile Organic Gases (excluding methane)

One technique that is used to identify major organics species in marine air is cryogenic preconcentration followed by GC/MS identification. This technique allows most abundant organics to be identified, depending on the mass range of the instrument, the boiling point, and thermal lability of the samples. Penkett (1982) reports the composition of Atlantic air, excluding C_2-C_5 hydrocarbons and certain aldehydes, using this technique (Table 5-1). The compounds he identifies are mostly halogenated hydrocarbons, but he also identifies two sulfur species, two aromatic hydrocarbons, and acetone. Interestingly, acetone is the most abundant. The total concentration of all species he measures is only ~ 2.3 ppbv.

Of the major compounds in Table 5-1, most have a clear anthropogenic origin. The most stable of these, the fluorocarbons, have been studied in detail at remote sites and their long-term temporal increases are being closely monitored (Prinn et al. 1983, 1987; Simmonds et al. 1983). The main removal processes for these compounds occur in the stratosphere. Two anthropogenic compounds that are degraded in the troposphere are C_2HCl_3 and C_2Cl_4. Because these compounds have a primary source in the Northern Hemisphere and are degraded relatively fast in the troposphere, C_2HCl_3 and C_2Cl_4 have a strong interhemispheric gradient (e.g., see Fig. 5-1). This N/S gradient is used to estimate average OH radical concentrations that are within reason (Class and Ballschmiter 1987, Penkett 1982) although there are some differences in the concentrations of C_2Cl_4 used to characterize the remote troposphere in each hemisphere.

Table 5-1. Concentrations of abundant organic species (excluding NMHC) in Atlantic air (derived from a GC/MS analysis of a whole-air sample).

Compound	Mass Number	Estimated Concentration (pptv)
CCl_3F	101	184
CCl_2F_2	85	288
$CFCl_2CF_2Cl$	101	25
CF_2ClCF_2Cl	135	19
CHF_2Cl	51	45
CH_3CCl_3	97	84
CCl_4	117	87
CH_3Cl	50	341
CH_2Cl_2	84	60
$CHCl_3$	83	27
CH_3Br	96	11
CH_3I	142	2.4
C_2HCl_3	130	6.3
C_2Cl_4	129	35
C_6H_6	78	95
C_7H_8	91	7
$(CH_3)_2CO$	58	480
COS	60	400
CS_2	76	150

Source: Penkett (1982)

108

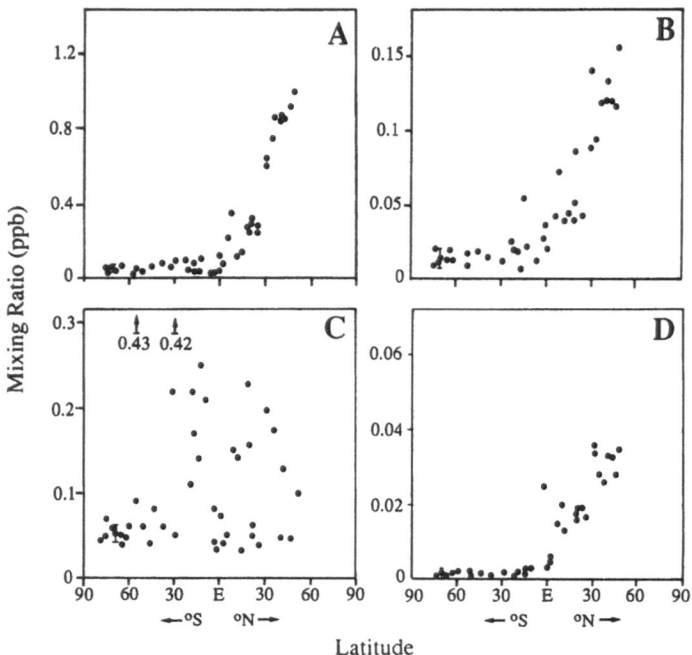

Figure 5-1. Distribution of selected hydrocarbons and perchloretylene (**A** = C_3H_8, **B** = C_6H_6, **C** = C_3H_6, **D** = C_2Cl_4) in the Atlantic Ocean atmosphere (Rudolph et al. 1984).

For example, the average background concentrations of C_2Cl_4 in the Southern Hemisphere reported by different researchers can be an order of magnitude different (1.5 ppt versus 14 ppt) although in the Northern Hemisphere the averages differ by a factor of 3 (15 ppt versus 40 ppt) (Class and Ballschmiter 1986, 1987; Hov et al. 1984; Penkett 1982; Singh et al. 1983; and unpublished data available from the author). In any case, differences in the concentration and composition of light halogenated hydrocarbons are used not only to estimate photochemical states in the atmosphere but also to indicate long-range pollutant transport (for example, in association with the Arctic Haze phenomenon) (Rasmussen and Khalil 1983, Rasmussen et al. 1983).

Aromatic hydrocarbons, benzene and toluene, are also produced primarily from anthropogenic sources and are reactive in the troposphere. Average concentrations of benzene and toluene in the North Atlantic (66± 32 pptv and 17±10 pptv, respectively) are similar to those in the North Pacific (49±10 pptv and 20±12 pptv, respectively) (Nutmagul and Cronn 1985). The apparent agreement in the concentrations found in these ocean areas may be fortuitous given the expected larger sources of hydrocarbons surrounding the North Atlantic basin and the short half-life of benzene and toluene in the atmosphere. The concentrations are lower in the South Pacific (benzene = 10±2 pptv, toluene = 6±2 pptv), but even these low concentrations may be supported by a significant hydrocarbon source in the Southern Hemisphere.

Another ubiquitous class of compound found in the atmosphere are $>C_2$ (nonmethane) hydrocarbons (NMHC). Many of these compounds, e.g., aliphatic hydrocarbons, are primarily anthropogenic in origin and have elevated concentrations in the Northern Hemisphere atmosphere although some individual hydrocarbons are thought to have ocean sources (see Fig. 5-1). The abundance of NMHC are orders of magnitude lower than methane; however, their atmospheric reactivity in terms of carbon conversion can exceed that of methane (Ehhalt et al. 1984). Thus the concentration and variation of NMHC can have a significant impact on atmospheric chemistry.

The urban source of NMHC in the United States is characterized by Sexton and Westberg (1984). They report that the average composition of NMHC is 47-69% paraffins, 21-29% aromatics, and 4-10% olefins (excluding ethylene); typical compositions of urban NMHC (measured between 6 a.m. and 9 a.m.) are in Table 5-2. Data from remote continental sites (Table 5-3) show dramatically lower concentrations. Furthermore, careful analysis of the molecular distribution of hydrocarbons suggests that photochemical processes, as well as dilution and transport, affect the concentration and variability of these hydrocarbon species in remote

Table 5-2. Average (geometric mean) concentrations of selected hydrocarbons (ppbC) in urban centers of the United States.

NMHC	Houston	Philadelphia	Baltimore	Washington	Newark	Boston	Milwaukee
Ethane	25	13	11	12	21	8	9
Acetylene	15	6	11	14	10	9	5
Propane	51	29	14	11	22	9	10
Propene	17	10	7	8	9	4	4
i-Butane	33	21	13	12	22	12	7
n-Butane	64	46	41	37	48	29	28
1 Butene	4	1	1	1	2	<1	1
i-Butene	6	1	4	3	3	<1	2
i-2-Butene	5	1	3	3	3	2	1
i-Pentane	67	42	52	54	55	35	23
n-Pentane	38	27	24	23	28	16	12
2-Methylpentane	20	16	14	10	14	12	7
3-Methylpentane	15	11	9	7	9	8	5
n-Hexane	20	13	9	7	11	9	7
Benzene	18	13	11	8	11	8	4
Toluene	48	29	43	38	47	28	16
Ethylbenzene	15	6	7	9	10	4	3
m- and p-Xylene	38	16	19	26	20	11	8
o-Xylene	18	7	8	11	7	5	3
1, 2, 4-Trimethyl-benzene	21	8	8	7	4	5	5

Source: Sexton and Westberg (1984).

Table 5-3. C_2-C_6 hydrocarbon concentration in the nonurban continental and marine troposphere (ppb).

Compound	Continental				West Germany		Marine		
	United States			Brazil			Pacific		Equatorial Atlantic
	Maine	California	Colorado				East	South	
C_2H_6	1.8	2.8	2.24	1.82	2.33	1.84	2.4	0.40	0.90
C_2H_4	1.0	0.3	0.46	2.77	0.45	0.71	0.12	0.20	0.26
C_2H_2	<0.3	<0.3	0.70	---	0.55	0.64	0.46	---	0.15
C_3H_8	0.7	0.8	1.27	0.36	0.54	0.86	0.80	0.10	0.14
C_3H_6	0.2	<0.2	0.16	0.72	0.08	0.33	0.05	0.20	0.12
$\underline{i}$-C_4H_{10}	0.1	<0.1	0.28	---	0.20	0.28	0.21	---	0.02
$\underline{n}$-C_4H_{10}	0.5	0.1	0.51	0.27	0.40	0.71	0.51	0.05	0.08
$\underline{i}$-C_5H_{12}	0.2	0.1	0.57	---	0.27	0.37	0.24	0.10	0.02
$\underline{n}$-C_5H_{12}	0.1	0.1	0.19	0.14	0.12	0.22	0.42	0.1	0.03
$\underline{n}$-C_6H_{14}	---	---	---	0.07	0.05	0.10	---	0.2	---
C_6H_6	---	---	0.24	0.78	0.10	0.12	---	---	---
$C_6H_5CH_3$	---	---	0.14	0.08	0.11	0.22	---	---	---
$\underline{p}$+$\underline{m}$-C_6H_4 $(CH_3)_2$	---	---	---	0.12	0.02	0.06	---	---	---
$\underline{o}$-C_6H_4 $(CH_3)_2$	---	---	---	0.08	0.02	0.04	---	---	---

Source: Data for Maine and California, from Sexton and Westberg (1984); for Colorado and Brazil, from Greenberg and Zimmerman (1984); for West Germany, from Rudolph and Khedim (1985); for the East Pacific, from Singh and Salas (1982); for the South Pacific, from Bonsang and Lambert (1985); and for the Equatorial Atlantic, from Rudolph and Ehhalt (1981).

areas (Rudolph and Khedim 1985). Hydrocarbon concentrations in nonurban atmospheres have an increasing variability with increasing carbon number. Rudolph and Khedim suggest the increased variability is from an increasing chemical reactivity of longer-chain hydrocarbons. Because of the variable reactivity of hydrocarbons, an ''average'' background composition of NMHC is difficult to characterize.

The concentration of NMHC over the ocean is influenced both by transport processes and ocean emissions (Bonsang and Lambert 1985, Bonsang et al. 1987, Greenberg and Zimmerman 1984, Rudolph and Ehhalt 1981, Rudolph and Khedim 1985, Tille et al. 1985). Concentrations of saturated hydrocarbons are generally lower over oceans than over continents. However, various alkenes--particularly ethylene and propylene--and alkanes ($\underline{i}$-C_4H_{10}, $\underline{i}$-C_5H_{12}, and others) may have marine sources. For example, Bonsang et al. (1987) discuss variations in surface level hydrocarbon concentrations during a transect from the North Atlantic to the South Pacific. They report that variations in saturated hydrocarbon concentrations are related to Rn^{222} fluctuations, which suggests that the transport from continental areas is causing increased saturated hydrocarbon concentrations over the ocean. In comparison, the relatively uniform concentrations of unsaturated hydrocarbons in the Northern and Southern Hemispheres suggests that the unsaturated hydrocarbons may have marine sources.

The vertical distribution of NMHC (and other organics) also needs to be characterized so that transport processes and interactions during the long-range transport of these compounds can be more clearly understood. Rudolph (1988) uses data from extensive aircraft measurements to show that the relatively long-lived species--ethane and propane--are sufficiently stable that ''representative'' cross-sections can describe their concentration distributions from year to year. Generally concentrations of these compounds decrease with increasing altitude and latitude. The concentrations of more reactive species showed greater variability, as expected. The high concentrations of reactive hydrocarbons in the upper trosphere suggest fast vertical mixing and transport processes. Such vertical transport processes, which distribute reactive hydrocarbons to the upper troposphere, may also affect the distribution of radical species and the extent of photochemical processes at high altitudes in the troposphere.

The photooxidation of naturally emitted reactive NMHCs probably represents an important source for carboxylic acids in the atmosphere (Harvey and Lang 1985, Jacob and Wofsy 1988). These acids predominate as vapor and have been measured in both impacted and remote regions at concentrations ranging from less than one to several ppbv's (Andreae et al. 1988, Talbot et al. 1988). Particulate-phase concentrations are typically 1 to 2 orders of magnitude lower than those in the vapor phase. Possible exceptions to this generalization include polar regions where extreme temperatures result in larger fractions of these acids in the condensed phase (Li and Winchester 1986).

Because of their high solubility, carboxylic acids, most importantly formic acid (HCOOH) and acetic acid (CH_3COOH), are major chemical constituents of precipitation and cloud water to which they contribute significant fractions of free acidity (e.g., Keene and Galloway 1988). The similarity of aqueous-phase concentrations in both impacted and remote areas suggests that anthropogenic sources are probably not important over broad geographic regions.

Compared to the number of measurements of C_2-C_5 hydrocarbons, there are fewer reports of gas-phase hydrocarbons >C_6 in continental or marine atmospheres. The major sources of C_6-C_{25} hydrocarbons are related to fuel usage and combustion. Thus, as expected, concentrations of these hydrocarbons are higher in continental/urban areas than they are at remote oceanic sites (Doskey and Andren 1986; Duce and Gagosian 1982; Eichmann et al. 1979, 1980; Sexton and Westberg 1984). Mostly because of analytical difficulties, concentrations of C_6-C_{15} hydrocarbons have not been routinely measured. However, gas-phase hydrocarbons >C_{15} have been reported in continental areas (Table 5-4). Concentrations typically decrease with increasing carbon number, ranging between several ng/m^3 to tens of ng/m^3 for individual hydrocarbon species (Table 5-4). (An order of magnitude smaller concentrations are typical of particle-bound alkanes >C_{15}.) Data covering the ocean's atmosphere are summarized by Duce and Gagosian (1982) and Duce et al. (1983), are informally presented in newsletter reports from the SEAREX program, and are reported in

Table 5-4. Representative concentrations of >C_{10} gas-phase hydrocarbons in continental and marine atmospheres (ng/m^3).

| | Continental | | | Marine | | | | |
| | Belgium | | Texas | North Pacific | | South | Atlantic | |
	Feb	Aug	Sep	Equatorial	Midlatitude	Pacific	North	Equatorial
C_{10}	--	--	--	--	--	--	15	--
C_{11}	--	--	--	--	--	--	9	--
C_{12}	--	--	--	--	--	0.07	5	--
C_{13}	--	--	46	0.23	--	0.10	4	--
C_{14}	--	--	29	0.19	--	0.07	3	--
C_{15}	--	--	15	0.66	1.5	0.11	6	1.6
C_{16}	36	251	12	0.13	1.4	0.04	4	2.2
C_{17}	37	346	11	0.55	0.92	0.08	5	16.0
C_{18}	22	130	10	0.07	0.42	0.03	5	5.7
C_{19}	3	13	4	0.07	0.20	0.04	5	3.4
C_{20}	1	10	3	0.07	0.15	0.04	6	3.0
C_{21}	1	6	1.6	0.07	0.10	0.05	6	3.0
C_{22}	--	3	1.1	0.07	0.08	0.04	5	2.0
C_{23}	--	4	0.5	0.08	0.07	0.03	5	1.8
C_{24}	--	--	--	0.09	0.04	0.03	3	1.1
C_{25}	--	--	--	0.10	0.04	0.02	3	1.1
C_{26}	--	--	--	0.08	0.04	--	2	1.5
C_{27}	--	--	--	0.06	--	--	2	1.7

Sources: Data for Belgium from Broddin et al. (1980); for Texas, North Pacific Midlatitude, and the South Pacific, from Atlas (unpublished data available from the author); and for North Pacific Equatorial, and Atlantic North and Equatorial, from Duce et al. (1983).

several other reports (e.g., Eichmann et al. 1979, 1980). Depending on the remoteness of the ocean area, concentrations of individual alkanes are usually <1 ng/m^3 (see Table 5-4). Also, there is some evidence that hydrocarbon concentrations differ greatly between the North Atlantic and North Pacific atmospheres, which qualitatively can be expected because of the different source strengths and transport times associated with the two oceans.

With a few exceptions, evidence of ocean sources for gas-phase high-molecular-weight hydrocarbons is rare. However, we (Atlas and Schauffler 1989a) recently measured high concentrations of pristane in the North Pacific, a hydrocarbon known to be an essential component of several zooplankton species (copepods). We measured concentrations of pristane >30 ng/m^3 intermittently over several days during late spring and early summer. The normal alkanes showed no such increases and typically remained <1 ng/m^3 for individual alkane components (see Table 5-4). This finding underscores the potential importance of ''patchy'' biological emissions to the atmosphere, which can occur in the ocean over varying time and space scales. Such processes must be considered in evaluating distributions of organics observed in the atmosphere; they may also play a role in near-surface photochemistry.

Polynuclear aromatic hydrocarbons (PAH) are a class of compound also related to fuel combustion and other combustion sources (biomass burning, etc.) Typically, 2- to 3-ring condensed PAH are associated with the gas-phase and 4- to 6-ring PAH are measured on particles. Compared to the information available about other major ''pollutant'' hydrocarbons, there are few data available on PAH in remote atmospheres. PAH are relatively reactive in the atmosphere but low levels of fluorene, phenanthrene (and methyl analogues), fluoranthene, and pyrene have been measured in the North Atlantic and Pacific (Broddin et al. 1980, Daisey et al. 1981, Halkiewicz et al. 1987, Marty et al. 1984, Ohta and Handa 1985, Sicre et al. 1987). Table 5-5 summarizes some of this data for both gaseous and particulate species. Typical concentrations of gas-phase PAH are in the low pg/m^3 concentration range, with the more volatile species found in higher concentration. Some evidence from the North Pacific suggests that increased PAH concentrations are correlated with long-range transport from the Asian continent, but data to support this are being evaluated as this chapter is being written. Interestingly, the most abundant PAH identified in the North Pacific is dibenzofuran, an oxygenated PAH the source of which we do not yet know (Atlas and Schauffler 1989b).

Other compounds that have received considerable attention are the halogenated hydrocarbons that have been used as pesticides or as industrial chemicals. Although only trace concentrations of the compounds are found in the atmosphere (10^4-10^5 molecules/cm^3), they are relatively stable in the environment and can accumulate in organisms. Because of the interest in the long-range transport and oceanic deposition of pesticides and their related compounds, more data are available for concentrations in the atmosphere over oceans than over continents. Chlorobenzenes (pentachlorobenzene [PeCB] and hexachlorobenzene [HCB]), hexachlorcyclohexanes (α-HCH, γ-HCH [lindane], and other isomers), chlordane and related compounds, dieldrin, DDTs, polychlorinated biphenyls (PCB), toxaphene, and compounds related to pentachlorophenol are those most

Table 5-5. Comparison of selected particulate and gas-phase PAHs in the marine remote atmosphere (pg/m^3).

	Barrow Alaska	Mediterranean	Enewetak	North Pacific Particulate	Gas-Phase
Particulate					
Fluorene	--	--	<5	0.1	16
Phenanthrene	17	52	<5	2.2	14
Fluoranthene	28	82	<5	1.3	2.0
Pyrene	29	70	<5	1.1	3.0
Chrysene	5	42	<5	1.8	1.0
Benzofluoranthenes	--	35	<5	2.1	--
Benzo(a)pyrene	10	8	<5	0.4	--
Benzo(a)pyrene	BD	21	<5	1.0	--
Benzo(ghi)perylene	10	11	<5	1.0	--
Indeno(c, d)pyrene	--	6	<5	0.7	--
Total PAH	99	327	--	12	36
Oxygenated PAH					
Fluorenone	--	--	--	9.0	33
Dibenzofuran	--	--	--	--	248
Xanthene	--	--	--	--	9.8

Sources: For Barrow, Alaska, nonhaze data taken in August, Daisey et al. (1981); for the Mediterranean data, Sicre et al. (1987); for Enewetak data, Gagosian et al. (1981); and for the North Pacific gas-phase and particulate data, Atlas and Giam (1989).

frequently identified in remote atmospheres. In tropical and temperate climates these high-molecular-weight compounds are mostly (>90%) measured in the gas phase (Atlas and Giam 1981, 1986a, 1988; Bidleman et al. 1986; Foreman and Bidleman 1987; Oehme and Ottar 1984); however, with high dust loads or colder temperatures, a larger fraction may be associated with particles. Bidleman et al. (1986) have extensively studied the distribution of organics between gas and particle phases. Knowing the physical state of a compound is critical to the understanding of the transport and deposition of organics.

Table 5-6 gives a summary of some available concentration data on chlorinated hydrocarbons. On the average, the distribution and concentrations of these compounds in remote areas of the Northern Hemisphere are fairly uniform. This uniformity reflects the diffuse sources and relatively long atmospheric lifetimes of these compounds. However, near major sources of pesticides, such as India, atmospheric concentrations rise sharply (e.g., Bidleman and Leonard 1982, Tanabe and Tatsukawa 1983). Also, in the Arctic region, differences in pesticide concentrations and compositions are correlated with long-range transport and different air-mass histories (Oehme and Ottar 1984, Oehme and Stray 1982, Ottar et al. 1986). Although there are no data from the South Atlantic, measurements in the South Pacific suggest that concentrations of pesticides in the Southern Hemisphere are lower, on the average, than

Table 5-6. Average concentrations of atmospheric organic compounds at continental and marine sites (ng/m^3).

	HCB	ΣHCH	ΣDDT	Chlordane	Dieldrin	PCB (A1254)	Toxaphene	Reference
Continental								
College Station, TX	0.21	0.93	0.33	1.05	0.083	0.29	1.80	Atlas & Giam 1988
Lillestrom, Sweden	0.16	1.19	--	--	--	--	--	Oehme & Ottar 1984
White Sands, NM	0.13	5.40	--	0.068	--	0.11	0.54	Atlas & Giam 1986b
Columbia, SC	0.29	1.10	0.93[a]	1.30	--	1.50	13.10	Bidleman & Christensen 1979
Denver, CO	0.24	0.30	0.021[a]	0.063	--	0.45	--	Billings & Bidleman 1983
New Bedford, MA	0.18	1.00	--	0.24	--	9.30	--	Billings & Bidleman 1983
Marine/Coastal								
North Inlet, SC	--	--	0.036[b]	0.15	--	0.25	1.70	Bidleman & Christensen 1979
Gulf Coast, TX	0.13	0.44	0.028	0.036	0.017	0.056	0.57	Chang et al. 1985
Barbados	--	--	0.004	0.009	0.005	0.057	<0.10	Bidleman et al. 1981
North Atlantic	0.15	0.39[c]	0.006[a]	0.03	0.02	0.69	--	Atlas & Giam 1981
Western Pacific/ Eastern Indian	--	0.55	0.28	--	--	0.30[d]	--	Tanabe et al. 1982
Enewetak Atoll	0.10	0.26	<0.006	0.013	0.010	0.05	<0.09	Atlas & Giam 1981
American Samoa	0.055	0.032	0.0015	0.001	0.002	0.012	--	Giam & Atlas 1986
Bear Island, Arctic*	0.11	0.24	--	--	--	--	--	Oehme & Ottar 1984
Hopen, Arctic*	0.12	0.30	--	0.001	--	0.038	--	Oehme & Ottar 1984
North Pacific	0.11	0.37	0.004	0.007	0.002	0.032	trace	Oehme & Ottar 1984

*Concentrations estimated from graphs.

[a] p,p' - DDE only.
[b] p,p' - DDT only.
[c] α-HCH only.
[d] Total PCB.

in the Northern Hemisphere. Also, interhemispheric concentrations are notably different for different compounds. For example, HCB is a factor of ~2 lower and HCH a factor of 10-15 lower in the Southern Hemisphere. These differences are consistent with the idea that there is a major source of compounds in the Northern Hemisphere, that HCB is a stable, long-lived species in the atmosphere, and that HCH is more easily removed (by precipitation or other processes).

5.3.2. Particulate Organic Material

A significant fraction of atmospheric aerosol is carbonaceous. Carbonaceous material can come from various sources: from plants, soils, ocean waters, or combustion processes or from secondary processes in the atmosphere that transform existing particulate organic carbon species or convert gases to particulate matter. Several recent reviews discuss the detailed composition of atmospheric organic compounds (Duce et al. 1983, Simoneit 1986, Simoneit and Mazurek 1981); however, I only discuss selected aspects in this chapter.

Carbon in aerosols consists of an organic component as well as an ''elemental'' or ''soot'' component. In urban areas, the soot component can be a major fraction of the total aerosol mass. For example, the average fine aerosol fraction (d < 2.1 μm) in urban areas can be 10-15% ''elemental'' carbon (Groblicki et al. 1981, Wolff and Klimisch 1982). In areas far from combustion sources, soot carbon is less significant. Recent studies by Cachier et al. (1986) characterize the concentration and isotopic composition of particulate carbon (Table 5-7 and Fig. 5-2). By careful analysis of stable carbon isotopes, these scientists easily distinguish carbon of marine origin from that of continental origin. They report that the concentration and isotopic composition of particulate carbon from marine areas are similar from both the Northern and Southern Hemispheres (C_{mean} = 0.07 μg/m^3; ^{13}C = -21 ^{o}/oo). Furthermore, this marine carbon component is primarily associated with particles with diameters over 3 μm. The concentration of aerosol carbon originating from a continent is considerably higher in the Northern Hemisphere (0.45 μg/m^3) than in the Southern Hemisphere (0.06 μg/m^3). This carbon is found primarily on the <1-μm fraction aerosol and has a range of isotopic compositions from -23 ^{o}/oo to -28 ^{o}/oo. This ratio changes depending on the origin of the carbon. They differentiate continental carbon from industrial combustion (prevalent in the Northern Hemisphere) from biomass burning (prevalent in equatorial and Southern Hemisphere regions). Such studies show the power of isotopic analysis in characterizing carbonaceous source materials and in evaluating their long-range transport.

Another powerful tool used to examine long-range transport and to understand the chemical processes involving organic compounds in the atmosphere is the molecular characterization of atmospheric aerosols (Gagosian 1986; Gagosian et al. 1981, 1987; Simoneit 1984, 1986; Simoneit et al. 1983). Studies by Gagosian et al., Simoneit et al., and others (e.g., Kawamura and Kaplan 1986, Mazurek 1985, Mukai and Ambe 1986, Yokouchi et al. 1987) have applied ''molecular markers'' in identifying the sources and transport of organic carbon compounds in the remote

Table 5-7. Particulate carbon concentrations in aerosols from remote
       marine areas.

Location	Number of Samples	Mean ($\mu gC/m^{-3}$ STP)	Analytical Method
		Northern Hemisphere	
Atlantic	7	0.40 ± 0.39	Proton ind. X emission
Northern Atlantic	7	0.76 ± 0.42	Solvent extraction
Irish west coast	6	0.57 ± 0.29	Solvent extraction
Eastern Atlantic	4	0.35 ± 0.15	Dry combustion
Bermuda	8	0.37 ± 0.23	Wet oxidation
Bermuda	8	0.29 ± 0.09	Wet oxidation
Sargasso Sea	4	0.44 ± 0.04	Dry combustion
California	7	0.49 ± 0.26	Solvent extraction
Hawaii	7	0.39 ± 0.03	Wet oxidation
Enewetak Atoll	9	0.82 ± 0.17	Proton ind. X emission
Tropical Pacific	4	0.38 ± 0.19	Solvent extraction
Equatorial Pacific	4	0.21 ± 0.11	Proton ind. X emission
		Southern Hemisphere	
Tropical Atlantic	3	0.23 ± 0.08	Proton ind. X emission
Peru	8	0.16 ± 0.07	Dry oxidation
Eastern tropical Pacific	7	0.22 ± 0.14	Solvent extraction
Equatorial Pacific	8	0.15 ± 0.05	Proton ind. X emission
Tasmania	6	0.53	Solvent extraction
Tasmania	4	0.23 ± 0.07	Proton ind. X emission
Amsterdam Island	1	0.15	Dry oxidation
New Zealand	4	0.13 ± 0.015	Dry oxidation
Samoa	6	0.11 ± 0.03	Dry oxidation
Samoa	9	0.22 ± 0.09	Wet oxidation

Source: Cachier et al. (1986).

*Excluding high values near the coast or contaminated samples according
to the authors.

**Fine fraction only: d < 1.7 μm.

atmosphere (Table 5-8). The basic idea in obtaining a detailed chemical
analysis of certain classes of organic compounds is not to characterize
the total composition of organics in aerosols but to obtain information
on sources that is contained in organic molecular composition. For
example, Gagosian et al. (1987), by analyzing the distributions of fatty
alcohol in aerosol samples from New Zealand, are able to discern air
masses of three distinct origins in the South Pacific. The alkanes,
fatty acids, fatty-acid salts, PAH, and certain ketones are only some of
the classes of compounds that have been examined in detail to help
unravel the source of airborne organic material found in the atmospheres
over remote and semiremote areas.

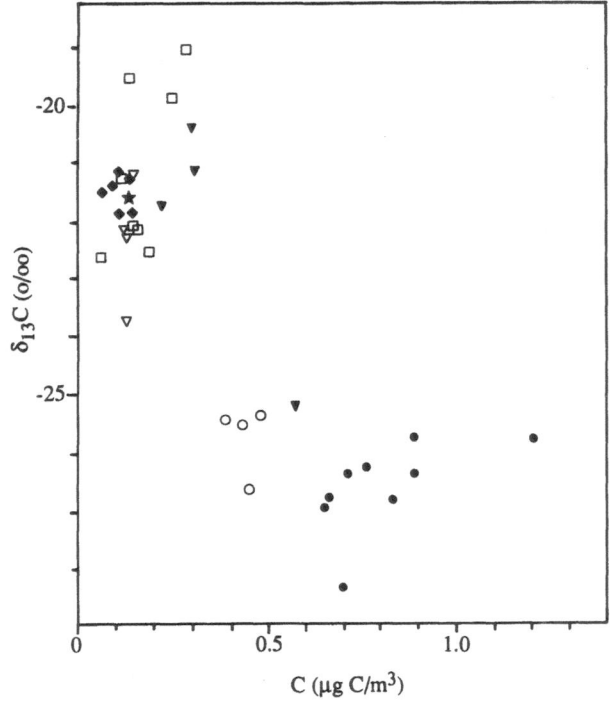

Figure 5-2

Particulate carbon isotopic
composition as a function
of concentration in the
marine atmospheres of New
Zealand (▽), Peru (□),
Amsterdam Island (★), and
Samoa (◆) in the Southern
Hemisphere and Enewetak
Atoll (•), the Sargasso Se
(○), and the Eastern Atlantic Ocean (▼) in the Northern Hemisphere (Cachier et
al. 1986).

Still the total concentration of characterized organic compounds in
remote aerosols does not approach the total organic carbon content. At
Enewetak Atoll, total extractable organic matter ranges up to 3 ng/m³,
of which ~50% is accounted for by alkanes, alcohols, and fatty acids
(Gagosian 1986, Gagosian et al. 1981). This amount is much less than
the average particulate carbon concentration of 820 ng/m³ ± 170 ng/m³
Cachier et al. (1986) report for the same time. Even if much of the
particulate carbon is ''elemental,'' a large fraction of ''continental''
carbon (based on isotopic composition) is unidentified. Carbon-budget
studies by Duce (1978) and the stable-carbon isotope data (Cachier et
al. 1986, Chesselet et al. 1981) suggest a large fraction of small-
particle organic carbon in the remote atmosphere may be formed by gas-to-
particle conversion. However, noone has yet been able to identify the
molecular species involved in such a conversion.

5.4. RATES OF ATMOSPHERIC DEPOSITION

Another major issue we are examining at the NATO workshop is the magnitude of the deposition of organic compounds to the ocean surface and to
remote areas. High-molecular-weight ''pollutant'' compounds are particularly relevant to this topic since the ocean is a major reservoir of
these compounds (Tanabe and Tatsukawa 1986) and atmospheric deposition is
a factor in the accumulation of pollutants by various organisms

Table 5-8. Source marker information.

Hydrocarbons	Fatty Alcohols and Sterols	Fatty Acids
	Terrestrial	
Plant-wax hydrocarbons: Homologous series: $nC_{23}-nC_{35}$ Strong odd/even carbon number predominance	Plant-wax alkanols: Homologous series: $nC_{12}-nC_{36}$ Strong even/odd carbon number predominance β-sitosterol	Plant-wax acids (<10% of wax) Homologous series: $nC_{14}-nC_{36}$ Strong even/odd carbon number predominance Frequently there is no major component
	Marine	
Phytoplankton and zooplankton hydrocarbons: nC_{15}, nC_{17} or nC_{19}, $C_{15:1}$, $C_{17:1}$, $C_{19:1}$, $C_{19:4}$, $C_{19:5}$, $C_{19:6}$, $C_{21:4}$, $C_{21:5}$, $C_{21:6}$ Pristane	Phytoplankton and zooplankton alcohols: generally in range of $C_{12}-C_{22}$, especially nC_{14}, nC_{16}, $C_{20:1}$, $C_{22:1}$ Dinosterol Diatomsterol	Phytoplankton and zooplankton fatty acids: generally in range of $C_{12}-C_{14}$, especially nC_{14}, nC_{16}, nC_{18} and $C_{18:1}$ Polyunsaturated acids are also common: $C_{20:4}$, $C_{20:5}$, $C_{22:5}$, $C_{22:6}$
	Anthropogenic	**Ubiquitous**
Petroleum hydrocarbons: Homologous series: $nC_{15}-nC_{40}$ No odd/even carbon number predominance Unresolved complex mixture common, α-hopane and extended tricyclic terpane series Squalene	nC_{12} and nC_{16}	nC_{16}, $C_{16:1}$, nC_{18}, and $C_{18:1}$

Source: Gagosian (1986).

(Atlas and Giam 1981, 1986b; Atlas et al. 1986). Also, as Gagosian and Peltzer (1986) discuss, the atmospheric deposition of terrestrially derived natural lipids can also have a major impact on marine geochemistry, particularly in deep-sea sediments. The impact of atmospheric deposition is quite clear when estimated atmospheric fluxes of selected synthetic organic compounds to the ocean are compared to riverine fluxes (Table 5-9). For other classes of compounds, such as petroleum hydrocarbons, atmospheric deposition may be less of a factor (Duce and Gagosian 1982).

Several different approaches are used to estimate the magnitude of organic fluxes to the ocean surface (Atlas and Giam 1986b, Atlas et al. 1986, Duce and Gagosian 1982, Zafiriou et al. 1985). Generally, different models are used for different deposition processes. These processes include scavenging of organics on particles and gases by precipitation, dry deposition of particles, and gas deposition or exchange with surface ocean waters. The relative magnitude of each process depends on the physical-chemical properties of each compound, e.g., vapor pressure, solubility, etc. For example, Atlas and Giam (1986b) demonstrate that the relative amount of gas-phase precipitation scavenging for selected synthetic pollutant compounds depends on the compound. They show that compounds, such as lindane, are removed from the atmosphere primarily (>90%) by gas-phase scavenging while less soluble compounds, such as PCB and DDT, are removed by scavenging of the particle-bound phase. Unfortunately, we do not know enough about the chemical properties of many compounds, especially as a function of temperature, to determine how they are removed from the atmosphere. However, some progress is being made in measuring or estimating appropriate Henry's Law constants and other physical properties.

Another major uncertainty in modeling the deposition of organic compounds involves estimating the air-sea exchange rates of gaseous

Table 5-9. Air-sea fluxes (area-weighted averages) and riverine inputs (estimated) of synthetic organic compounds to the North Pacific Ocean (pg/cm²/yr).

Compound	Air-Sea Flux		Riverine Input
	Minimum	Maximum	
PCB (1254)	20	284	10
DDE	0.24	5.3	0.5
DDT	1.3	21	0.5
Chlordane	1.0	23	1
Dieldrin	16	95	1
HCB	4.6	286	1
α-HCH	556	2563	10
Lindane	97	218	10

Source: Atlas and Giam (1986b).

Note: Flux extremes are based on minimum and maximum gas-exchange rates.

compounds. Even with such well-behaved, simple gases as CO_2, accurate modeling of gas-exchange rates in the ocean is complex. Simplified equations that deal with the exchange in terms of Henry's Law Constants and mass-corrected transfer velocities are used to limit the magnitude of gas-exchange rates for high-molecular-weight organics (Atlas and Giam 1986b, Atlas et al. 1982). However, to obtain a gas-exchange flux, the ''real'' degree of the saturation of surface waters must be known. The total concentration of most high-molecular-weight organics in seawater can be affected by association with particulate matter, colloidal material, or dissolved organic compounds. Thus, even in equilibrium with the atmosphere, the total concentration of an organic compound is often higher than predicted, which makes the surface ocean appear supersaturated in comparison to the atmosphere. This uncertainty obviously causes problems in modeling the gas exchange of organic compounds.

Because of these uncertainties in the gas-exchange rates, Giam and I (Atlas and Giam 1986b) have modeled the deposition rates of selected organic compounds based on limits of maximum and zero gas-exchange rates with the sea surface. We also considered the effects of temperature and particle concentration to estimate washout rates and gas-exchange fluxes as a function of latitude in the North Pacific. The resultant profiles show an increased air-sea flux at high latitudes for both insoluble (PCB) and soluble (HCH) compounds (Fig. 5-3). This increase is related to increased solubility and vapor-particle partitioning effects at colder

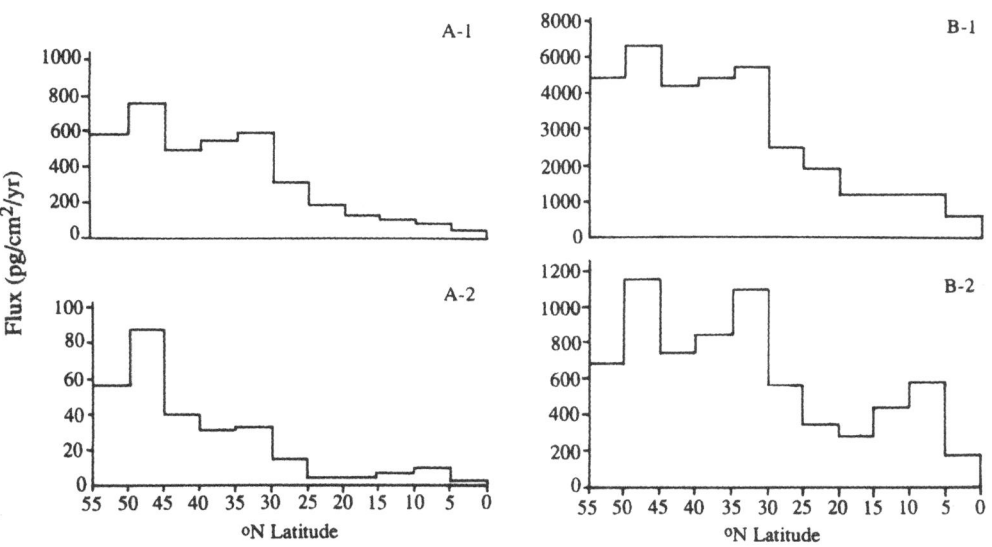

Figure 5-3. Estimated air-sea flux of **A** PCB(1254) and **B** α-HCH in the North Pacific. The top profiles (**A-1** and **B-1**) show the maximum gas exchange and the bottom profiles (**A-2** and **B-2**) show no gas exchange. (Atlas and Giam 1986b).

temperatures. In all cases, the dry deposition of particles is only a minor factor in the total deposition flux. When we include gas exchange in the model, we find the most dramatic change in the calculated fluxes. Total air-sea flux increases by 6-10 times if the ocean absorbs these organic compounds at the maximum rate. These calculations indicate how badly more laboratory and field experiments are needed to determine the actual significance of gas-exchange processes on the deposition of organic compounds to the ocean surface. Still, within the wide limits established by this model, we (Atlas and Giam 1986b) show that calculated atmospheric deposition ranges compare reasonably well to sedimentation rates of synthetic organics determined in deep-sea sediment traps (Knap et al. 1986).

Another test of such models is to compare computed deposition rates to deposition rates measured in the field, which can be done with wet-deposition rates. For example, we (Atlas and Giam 1986b) compare the measured concentrations of synthetic organics in rainfall at Enewetak Atoll to those calculated using appropriate model washout parameters. In most cases, the agreement is good (Fig. 5-4). This agreement (and other model calculations of organic deposition [Leuenberger et al. 1985; Ligocki et al. 1985a, 1985b; Van Noort and Wondergem 1985]) lends confidence to results that relate average atmospheric concentrations to average wet-deposition fluxes. However, such models do not consider the temporal variability of deposition rates and thus may not be applicable

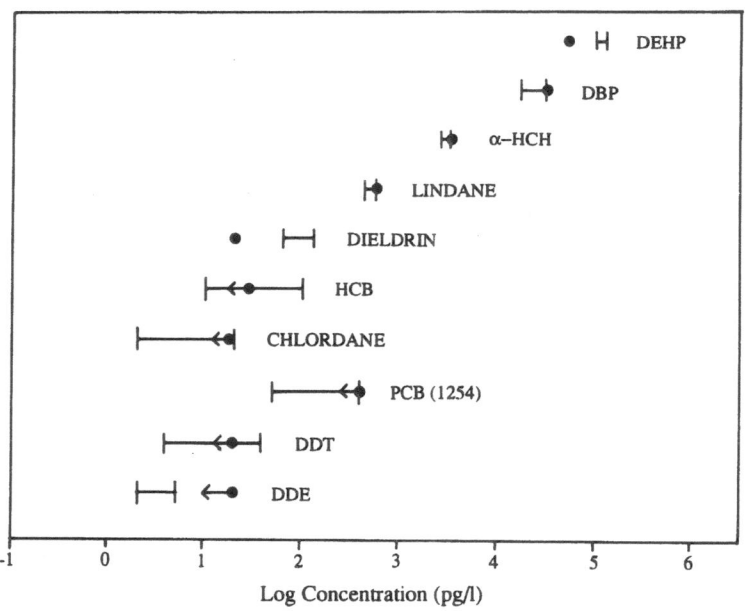

Figure 5-4. Measured concentrations (•) versus predicted concentrations (⊢—⊣) in rainfall at Enewetak Atoll (Atlas and Giam 1986b).

to all oceanic or remote areas. For example, the recent data Knap and Binkley (1988) collected in Bermuda illustrate that a large flux of organic compounds can be associated with only a few storm events. In the case of lindane, Knap and Binkley report that only two rain events during one year account for 37% of the total annual deposition. Furthermore, over the course of 15 months, the concentration of synthetic compounds in rainfall at Bermuda vary by over a factor of three. As Zafiriou et al. (1985) discuss, temporal changes in rain chemistry combined with regional climatological factors can cause large discrepancies between modeled and measured precipitation fluxes of organic compounds at remote sites. Because of the importance of the wet deposition of organics, long-term, continuous measurements of precipitation fluxes must be taken at suitable remote sites so that the magnitude and variability of deposition rates can be characterized.

5.5. TRANSFORMATIONS

Organic compounds in the atmosphere can be transformed in several ways; most are linked to photochemical cycles of OH radical, ozone, and NO_x (Atkinson 1985, Atkinson and Carter 1984, Atkinson and Lloyd 1984). Organic compounds can react to form more stable species; they can be oxidized to more soluble compounds that are removed by precipitation or that can undergo further reaction; or they can be converted from gas to particle phase. In all cases, transformations can alter the behavior of organic material transported across long distances and can limit the lifetime of compounds in the atmosphere. Conversely, the long-range transport of organic material can impact photochemical cycles in remote areas because of the transformation processes occurring there. In the discussion to follow, I focus more on the effect of transformations on organics than on the impact of organics on the photochemistry of remote areas.

It is well beyond the scope of this chapter to discuss kinetics and detailed mechanisms of atmospheric chemical reactions. Excellent reviews are available that provide such information for many compound classes (e.g., Atkinson and Carter 1984, Atkinson and Lloyd 1984, Niki 1982). Much of the information that has been developed concerning chemical reactions and organic transformations is applied to local or regional studies of air pollution, oxidant formation (ozone, peroxy acetylnitrate [PAN]), and acid deposition rather than to problems of remote atmospheres. Studies that consider the longer range transport and effects of organics in remote atmospheres are rare. Most long-range-transport models that I have seen are related to ozone or PAN budgets rather than to transformations and variations of specific organic compounds. (This is probably related to the scarcity of appropriate measurements in the remote troposphere.) Also, transformations of high-molecular-weight compounds, particularly pesticides, are rarely considered although atmospheric reactions of PAH are receiving increasing attention (Atkinson and Aschmann 1987, Atkinson et al. 1987, Gibson et al. 1986, Pitts et al. 1985).

One basic consideration concerning the long-range transport of specific, continentally derived organic compounds is their reactivity with radical species (OH, NO₃) or ozone. Compounds with reaction times << transport times need not be considered in long-range transport models. (Of course, subsequent reactions of oxidized species may certainly affect acid deposition, aldehyde formation, CO, etc.) Reaction rates of various compound classes with OH radical can limit atmospheric half-lives to minutes or hours (Fig. 5-5). Also, the reactivities of terpenes and other olefinic compounds are sufficiently rapid that long-range transport of these compounds should not usually occur (Findlayson-Pitts and Pitts 1986). Because of this high reactivity, I have not included biogenic hydrocarbons (isoprene, terpenes, etc.) in this discussion of long-range transport.

To consider what compounds and reactions may be necessary to include in long-range transport models, Calvert and Madronich (1987)

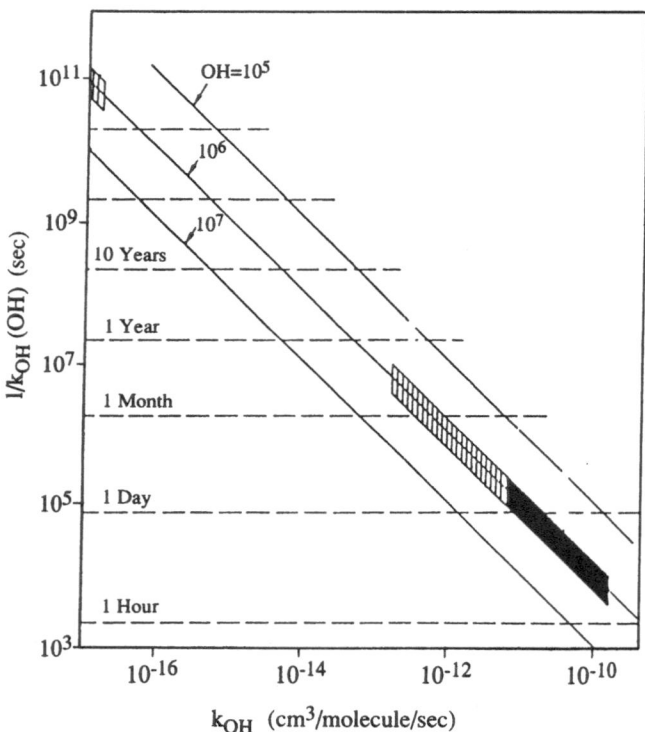

Figure 5-5. Atmospheric lifetimes of gaseous organic compounds as a function of their rate of reaction with the hydroxyl radical (k_{OH}) for OH concentrations of 10^5, 10^6, and 10^7 radicals/cm^3. The wide-striped bar represents haloalkanes; the narrow-striped bar represents alkanes, haloalkenes, alkylnitrates, alcohols, esters, ketones, alkynes, and benzene; the solid bar represents alkenes, terpenes, amines, aromatics, sulfides, mercaptans, and aldehydes. (Unpublished data available from D. Grosjean, DGA, Suite 205, 4526 Telephone Road, Ventura, CA 93003)

model the initial products of oxidation of representative C_2-C_8 hydro-carbons. Table 5-10 shows their calculated product distributions for alkane, alkene, and aromatic hydrocarbon oxidation. The product distri-bution is extremely complex and depends on the presence or absence of NO_x in the initial mixture. Although there are some unique products formed during the oxidation process, only a few of these are sufficiently stable

Table 5-10. Estimated lifetimes and initial products of hydrocarbon oxidation.

Initial Product	Approximate Lifetime (days)
Aldehydes	
Formaldehyde	0.2
Acetaldehyde	0.6
α-Hydroxy-aldehydes	<0.6
γ-Hydroxy-aldehydes	<0.6
α-Dicarbonyl compounds	<0.1
Unsaturated γ-Dicarbonyl compounds	1.8
Aromatic aldehydes	0.8
Methylacrolein	0.4
Ketones	
Acetone	12
Methyl ethyl ketone	3
α-Hydroxy-ketones	3
γ-Hydroxy-ketones	3
Methyl vinyl ketone	0.3
Alcohols	
Methyl alcohol	10.4
Ethyl alcohol	3.9
Dihydric alcohols	1.0
Organic Nitrates	
1-Butyl nitrate	8.3
2-Butyl nitrate	17.3
γ-Hydroxy-alkyl nitrates	4
Phenols	
Phenol	0.4
o-Cresol	0.3
Organic Acids	
Formic acid	24
Acetic acid	19
Organic Hydroperoxides	
Methyl hydroperoxide	0.6
1-Butyl hydroperoxide	1.1

Source: Calvert and Madronich (1987).

Note: Lifetimes were estimated taking [HO] = 10^6, [O_3] = 7.35×10^{11} molecules/cm^{-3}.

126

themselves to be used for long-range-transport studies, e.g., acids, nitrates, and ketones. The acid products, formic and acetic acids, characteristic of alkene oxidation are discussed earlier in this book (see Chapter 4, p. 87). Most other species have not been measured in ambient atmospheres. However, I recently (Atlas 1988) measured $>C_3$ alkyl nitrates in the mid-Pacific atmosphere and was able to correlate these compounds with another continental tracer, ^{222}Rn (Fig. 5-6). These

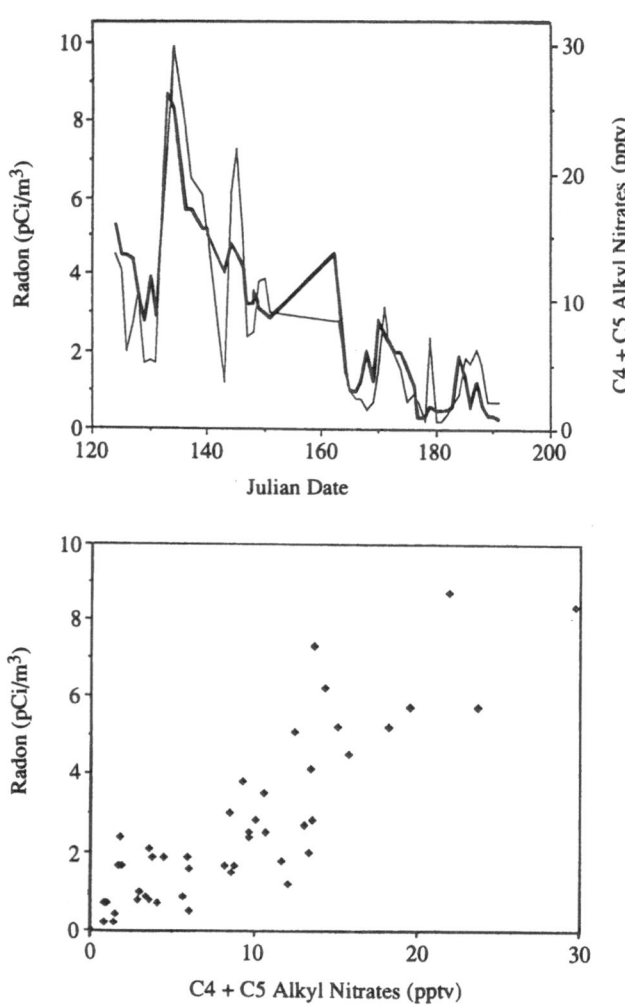

Figure 5-6. A Temporal variations in radon and alkyl nitrates during cruise (**light line** represents radon; **heavy line**, total RONO$_2$) and B covariations of radon and alkyl nitrates. The close correspondence between these chemical tracers suggests that long-range transport is responsible for their variation in the marine atmosphere reactions.

findings suggested that, as predicted, alkyl nitrates are useful indicators of ''polluted'' air masses in remote regions. Clearly, additional
analytical development and field experiments are required to identify and
measure other potential indicators of tropospheric transformations.

Specific transformation products of higher molecular weight PAH are
interesting because of their association with the mutagenicity of atmospheric organic matter (e.g., see discussion in Findlayson-Pitts and Pitts
1986). Laboratory studies demonstrate the formation of nitro-PAH and
hydroxy-nitro-PAH under simulated atmospheric conditions (Atkinson and
Aschmann 1987, Atkinson et al. 1987), and these compounds have been measured in both urban and rural atmospheres (e.g., Nielsen et al. 1984).
Recent experiments by Gibson et al. (1986) also report the in situ
formation of 1-nitropyrene and hydroxy-nitropyrenes in air masses of
continental origin that are transported to Bermuda. Although such observations are qualitatively consistent with theoretical expectations of
atmospheric transformations, simultaneous measurements do not show the
expected losses of another PAH (benz[a]pyrene) in the aged air masses.
Such observations suggest that appropriate measurements are always
required to compare theoretical expectations to reality.

Another aspect to transformation reactions in the atmosphere is
related to the oxidation of naturally occurring particulate organic compounds. Kawamura and Gagosian (1987) discuss one example of these reactions. They have evidence of C_4-C_{14} ω-oxocarboxylic acids and α,ω-
dicarboxylic acids as major organic components of the remote marine aerosol. They propose that the photo-oxidation of unsaturated fatty acids,
either in the atmosphere or in the surface ocean, is a source of these
compounds. The examination of polar organics seems to provide useful
information on the fate and sources of atmospheric organic compounds even
though few studies have tried to characterize the polar fraction of organic particles (Ip et al. 1984, Simoneit 1986, Wauters et al. 1979,
Yokouchi and Ambe 1986, Yokouchi et al. 1987) or elucidate the possible
mechanisms for the formation of polar organic compounds (Kawamura and
Gagosian 1987).

A goal of examining transformation processes in detail is to adequately combine transport, deposition, and chemistry to model the distribution and variation of chemical species in the troposphere and to
assess the impact on remote areas. Indeed, that is the main thrust of
the NATO workshop held at the Bermuda Biological Station. A large amount
of useful information has been gathered for atmospheric organic compounds
and new concepts have been developed; but there is still much to learn
and numerous questions to be answered (Fig. 5-7). For many organic compounds, if not most, source strengths are difficult to obtain and they
vary in time and space. What is the most appropriate way to characterize
these sources? Deposition rates of organic compounds also show large
variability. How large? How is this variability related to gas-particle
partitioning, to atmospheric concentrations, and to meteorological factors? What is the vertical distribution of HMW organics in the atmosphere? Are heterogeneous and aqueous-phase transformations significant
for certain organic species? What is the relationship between atmospheric deposition and biological uptake of synthetic organic compounds?
What is the impact of the long-range transport of organic compounds on

128

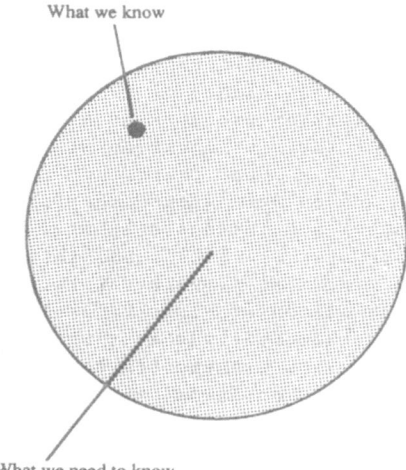

What we know

What we need to know

Figure 5-7. The present knowledge of the long-range transport of organic
compounds.

global budgets and the distribution of ozone and radical species? How
can measurements of appropriate organic compounds in remote areas be used
to probe and elucidate other photochemical processes? These are only a
few of the questions we examined; many others were discussed and evaluated during sessions of our working group at the NATO workshop.

5.6. REFERENCES

Andreae, M. O., R. W. Talbot, T. W. Andreae, and R. C. Harriss. 1988.
    Formic and acetic acids over the central Amazon region, Brazil. 1.
    Dry season. J. Geophys. Res. 93:1616-1624.
Atkinson, R. 1985. Kinetics and mechanisms of the gas-phase reactions of
    the hydroxyl radical with organic compounds under atmospheric con-
    ditions. Chem. Rev. 86:69-201.
Atkinson, R., and S. M. Aschmann. 1987. Kinetics of the gas-phase reac-
    tions of alkylnaphthalenes with O_3, N_2O_5, and OH radicals at 298 ±
    2 K. Atmos. Environ. 21:2323-2326.
Atkinson, R., and W. P. L. Carter. 1984. Kinetics and mechanisms of the
    gas-phase reactions of ozone with organic compounds under atmos-
    pheric conditions. Chem. Rev. 84:437-470.
Atkinson, R., and A. C. Lloyd. 1984. Evaluation of kinetic and mechanis-
    tic data for modeling of photochemical smog. Phys. Chem. Ref. Data
    13:315-444.
Atkinson, R., J. Arey, B. Zielinska, and S. M. Aschmann. 1987. Kinetics
    and products of the gas-phase reactions of OH radicals and N_2O_5 with
    naphthalene and biphenyl. Environ. Sci. Technol. 21:1014-1022.

Atlas, E. L. 1988. Evidence for ≥C_3 alkyl nitrates in rural and remote atmospheres. Nature 331:426-428.

Atlas, E. L., and C. S. Giam. 1981. Global Transport of organic pollutants: Ambient levels of organic contaminants in the remote marine atmosphere. Science 211:163-165.

Atlas, E. L., and C. S. Giam. 1986a. Sampling organic compounds for marine pollution studies. In Strategies and Advanced Techniques for Marine Pollution Studies (C. S. Giam and H. J.-M. Dou, eds.) Berlin: Springer-Verlag, 209-230.

Atlas, E. L., and C. S. Giam. 1986b. Sea-air exchange of high molecular weight synthetic organic compounds. In The Role of Air-Sea Exchange in Geochemical Cycling (P. Buat-Ménard, ed.), NATO Series C, Vol. 185, Dordrecht:Reidel, 295-330.

Atlas, E. L., and C. S. Giam. 1988. Ambient concentrations and precipitation scavenging of atmospheric organic pollutants. Water Air Soil Pollut. 38:19-36.

Atlas, E. L., and C. S. Giam. 1989. Sea-air exchange of high-molecular weight organic compounds--Results from the SEAREX program. In Chemical Oceanography (J. P. Riley and R. A. Duce, eds.) New York: Academic Press, in press.

Atlas, E. L., and S. Schauffler. 1989a. Concentration and variation of trace organic compounds in the North Pacific atmosphere. In Long-range Transport of Pesticides (D. Kurtz, ed.) Chelsea, MI:Lewis Publishers, in press.

Atlas, E., and S. Schauffler. 1989b. Observations of oxygenated hydrocarbons in the marine troposphere. In Measurement of Toxic and Related Air Pollutants, Procs., 1989 EPA/APCA Symp., in press.

Atlas, E. L., R. Foster, and C. S. Giam. 1982. Air-sea exchange of high-molecular weight organic pollutants: Laboratory studies. Environ. Sci. Technol. 16:283-286.

Atlas, E. L., T. Bidleman, and C. S. Giam. 1986. Atmospheric transport of PCBs to the oceans. In PCB and the Environment (J. S. Waid, ed.) Boca Raton, FL:CRC Press, 79-100.

Bidleman, T. F., and E. J. Christensen. 1979. Atmosphere removal processes for high molecular weight organochlorines. J. Geophys. Res. 84:7857-7868.

Bidleman, T. F., and R. Leonard. 1982. Aerial transport of pesticides over the northern Indian Ocean and adjacent seas. Atmos. Environ. 16:1099-1107.

Bidleman, T. F., E. J. Christensen, W. N. Billings, and R. Leonard. 1981. Atmospheric transport of organochlorines in the North Atlantic gyre. Marine Res. 39:443-464.

Bidleman, T. F., W. N. Billings, and W. T. Foreman. 1986. Vapor-particle partitioning of semivolatile organic compounds: Estimates from field collections. Environ. Sci. Technol. 20:1038-1043.

Billings, W. N., and T. F. Bidleman. 1983. High-volume collection of chlorinated hydrocarbons in urban air using three solid adsorbents. Atmos. Environ. 17:383-391.

Bonsang, B., and G. Lambert. 1985. Nonmethane hydrocarbons in an oceanic atmosphere. Atmos. Chemistry 3:257-271.

Bonsang, B., B. C. Nguyen, A. Gaudry, and G. Lambert. 1987. Comments on the residence time of aerosols and SO_2 in the long-range transport over the ocean. Atmos. Chemistry 5:367-370.

Broddin, G., W. Cautreels, and K. Van Cauwenberghe. 1980. On the aliphatic and polyaromatic hydrocarbon levels in urban and background aerosols from Belgium and the Netherlands Atmos. Environ. 14:895-910.

Cachier, H., P. Buat-Ménard, M. Fontugne, and R. Chesselet. 1986. Long-range transport of continentally derived particulate carbon in the marine atmosphere: Evidence from stable carbon isotope studies. Tellus 38B:161-177.

Calvert, J. G., and S. Madronich. 1987. Theoretical study of the initial products of the atmospheric oxidation of hydrocarbons. J. Geophys. Res. 92:2211-2220.

Chang, L. W., E. L. Atlas, and C. S. Giam. 1985. Chromatographic separation and analysis of chlorinated hydrocarbons and phthalate esters from ambient air samples. Int. Environ. Anal. Chemistry 19:145-153.

Chesselet, R. M. Fontugne, P. Buat-Ménard, U. Ezat, and C. E. Lambert. 1981. The origin of particulate organic carbon in the marine atmosphere as indicated by its stable carbon isotopic composition. Geophys. Res. Ltrs. 8:345-348.

Class, T., and K. Ballschmiter. 1986. Chemistry of organic traces in air, V. Determination of halogenated C_1-C_2 hydrocarbons in clean marine air and ambient continental air and rain by high resolution gas chromatography using different stationary phases. Fresenius Z. Anal. Chemie 325:1-7

Class, T., and K. Ballschmiter. 1987. Global baseline pollution studies. Fresenius Z. Anal. Chemie 327:198-207.

Daisey, J. M., R. J. McCaffrey, and R. A. Gallagher. 1981. Polycyclic aromatic hydrocarbons and total extractable particulate organic matter in the Arctic aerosol. Atmos. Environ. 15:1353-1363.

Doskey, P. V., and A. W. Andren. 1986. Particulate and vapor phase n-alkanes in the northern Wisconsin atmosphere. Atmos. Environ. 20: 1735-1744.

Duce, R. A. 1978. Speculations on the budget of particulate and vapor phase nonmethane organic carbon in the global troposphere. Pure Appl. Geophys. 116:244-273.

Duce, R. A., and R. B. Gagosian. 1982. The input of atmospheric $n-C_{10}$ to $n-C_{30}$ alkanes to the ocean. J. Geophys. Res. 87:7192-7200.

Duce, R. A., V. A. Mohnen, P. R. Zimmerman, D. Grosjean, W. Cautreels, R. Chatfield, R. Jaenicke, J. A. Ogren, E. D. Pellizzari, and G. T. Wallace. 1983. Organic material in the global troposphere. Rev. Geophys. Space Physics 21:921-952.

Ehhalt, D. H., J. Rudolph, and U. Schmidt. 1986. On the importance of light hydrocarbons in multiphase atmospheric systems. In Chemistry of Multiphase Atmospheric Systems (W. Jaeschke, ed.) Berlin: Springer-Verlag, 321-350.

Eichmann, R., P. Neuling, G. Ketseridis, J. Hahn, R. Jaenicke, and C. Junge. 1979. n-alkane studies in the troposphere: I. Gas and particulate concentrations in North Atlantic air. Atmos. Environ. 13: 587-599.

Eichmann, R., G. Ketseridis, G. Schebeske, R. Jaenicke, J. Hahn, J. Warnech, and C. Junge. 1980. n-alkane studies in the troposphere: II. Gas and particulate concentrations in Indian Ocean air. Atmos. Environ. 14:695-703.

Farmer, C. T., and T. L. Wade. 1986. Relationship of ambient atmospheric hydrocarbon (C_{12}-C_{32}) concentrations to deposition. Water Air Soil Pollut. 29:439-452.

Findlayson-Pitts, B. J., and J. N. Pitts, Jr. 1986. Atmospheric Chemistry. New York:Wiley, 1098 pp.

Foreman, W. T., and T. F. Bidleman. 1987. An experimental system for investigating of vapor-particle partitioning of trace organic pollutants. Environ. Sci. Technol. 21:869-875.

Gagosian, R. B. 1986. The air-sea exchange of particulate organic matter: The sources and long-range transport of lipids in aerosols. In The Role of Air-Sea Exchange in Geochemical Cycling (P. Buat-Ménard, ed.), NATO ASI Series C, Vol. 185, 409-442.

Gagosian, R. B., and E. T. Peltzer. 1986. The importance of atmospheric input of terrestrial organic material to deep sea sediments. Org. Geochemistry 10:661-669.

Gagosian, R. B., E. T. Peltzer, and O. C. Zafiriou. 1981. Atmospheric transport of continentally derived lipids to the tropical North Pacific. Nature 291:312-314.

Gagosian, R. B., E. T. Peltzer, and J. T. Merrill. 1987. Long-range transport of terrestrially derived lipids in aerosols from the South Pacific. Nature 325:800-803.

Giam, C. S., and E. Atlas. 1986. Strategies and approaches to marine pollution research. In Strategies and Advanced Techniques for Marine Pollution Studies (C. S. Giam, H. J.-M. Dov, eds.) Berlin: Springer Verlag, 33-41.

Gibson, T. L., P. E. Korsong, and G. T. Wolff. 1986. Evidence for the transformation of polycyclic organic matter in the atmosphere. Atmos. Environ. 20:1575-1578.

Greenberg, J. P., and P. R. Zimmerman. 1984. Nonmethane hydrocarbons in remote tropical, continental, and marine atmospheres. J. Geophys. Res. 89:4767-4778.

Groblicki, P. J., G. T. Wolff, and R. J. Countess. 1981. Visibility-reducing species in the ''Denver brown cloud''--I. Relationships between extinction and chemical composition. Atmos. Environ. 15: 2473-2484.

Halkiewicz, J., H. Lamparczyk, J. Grzybowski, and A. Radecki. 1987. On the aliphatic and polycylic aromatic hydrocarbon levels in the southern Baltic Sea atmosphere. Atmos. Environ. 21:2057-2063.

Hargrave, B. T., W. P. Vass, P. E. Erickson, and B. R. Fowler. 1988. Atmospheric transport of organochlorines to the Arctic Ocean. Tellus 40B:480-493.

Harvey, G. R., and R. F. Lang. 1985. Biogenic non-methane organics in and over the remote ocean. Paper presented, IAMAP|IAPSO Joint Assembly, Int. Union of Geodesy and Geophysics, Honolulu, August (available from Dr. Harvey, NOAA, 4301 Rickenbacker Causeway, Miami, FL 33149).

Hidaka, H., S. Tanabe, and R. Tatsukawa. 1983. DDT compounds and PCB isomers and congeners in Weddell seals and their fate in the Antarctic marine ecosystem. Agric. Biol. Chemistry 47:2009-2017.

Hov, O., S. A. Penkett, I. S. A. Isaksen, and A. Semb. 1984. Organic gases in the Norwegian Arctic. Geophys. Res. Ltrs. 11:425-428.

Ip, W. M., R. J. Gordon, and E. C. Ellis. 1984. Characterization of organics in aerosol samples from a Los Angeles receptor site using extraction and liquid chromatography methodology. Sci. Total Environ. 36:203-208.

Jacob, D. J., and S. C. Wofsy. 1988. Photochemistry of biogenic emissions over the Amazon forest. J. Geophys. Res. 93:1477-1486.

Kawamura, K., and R. B. Gagosian. 1987. Implications of omega-oxocarboxylic acids in the remote marine atmosphere for photooxidation of unsaturated fatty acids. Nature 325:330-332.

Kawamura, K., and I. R. Kaplan. 1986. Biogenic and anthropogenic organic compounds in rain and snow samples collected in Southern California. Atmos. Environ. 20:115-124.

Keene, W. C., and J. N. Galloway. 1988. The biogeochemical cycling of formic and acetic acids through the troposphere: An overview of current understanding. Tellus 40b:322-334.

Knap, A. H., and K. S. Binkley. 1988. The occurrence and distribution of trace organic compounds in Bermuda precipitation. Atmos. Environ. 22:1411-1424.

Knap, A. H., K. S. Binkley, and W. G. Deuser. 1986. Synthetic organic chemicals in the deep Sargasso Sea. Nature 319:572-574.

Leuenberger, C., M. P. Ligocki, and J. F. Pankow. 1985. Trace organic compounds in Rain: 4. Identities, concentrations, and scavenging mechanisms for phenols in urban air and rain. Environ. Sci. Technol. 19:1053-1058.

Li, S. M., and J. W. Winchester. 1986. Organic anions in course and fine aerosols at Barrow, Alaska, during AGASP-2, late winter 1986. Paper presented, American Geophys. Union Fall Meeting, San Francisco.

Ligocki, M. P., C. Leuenberger, and J. F. Pankow. 1985a. Trace organic compounds in rain: II. Gas scavenging of neutral organic compounds. Atmos. Environ. 19:1609-1617.

Ligocki, M. P. C. Leuenberger, and J. F. Pankow. 1985b. Trace organic compounds in rain: III. Particle scavenging of neutral organic compounds. Atmos. Environ. 19:1619-1626.

Likens, G. E., E. S. Edgerton, and J. N. Galloway. 1983. The composition and deposition of organic carbon in precipitation. Tellus 35B:16-24.

Marty, J. C., M. J. Tissier, and A. Saliot. 1984. Gaseous and particulate polycyclic aromatic hydrocarbons (PAH) from the marine atmosphere. Atmos. Environ. 18:2183-2190.

Mazurek, M. A. 1985. Geochemical investigations of organic matter contained in ambient aerosols and rainwater particulates. Ph.D. dissert., Univ. of California, Los Angeles, 372 pp.

Mukai, H., and Y. Ambe. 1986. Characterization of a humic acid-like brown substance in airborne particulate matter and tentative identification of its origin. Atmos. Environ. 20:813-819.

Nielsen, T. B. Seitz, and T. Ramdahl. 1984. Occurrence of nitro-PAH in the atmosphere of a rural area. Atmos. Environ. 18:2159-2165.

Niki, H. 1982. Homogeneous gas-phase oxidation processes in the troposphere. In Atmospheric Chemistry (E. D. Goldberg, ed.) New York:
Springer-Verlag, 301-312.

Nutmagul, W., and D. R. Cronn. 1985. Determination of selected atmospheric aromatic hydrocarbons at remote continental and oceanic locations using photo-ionization‖flame-ionization detection. Atmos.
Chemistry 2:415-433.

Oehme, M., and B. Ottar. 1984. The long-range transport of polychlorinated hydrocarbons to the Arctic. Geophys. Res. Ltrs. 11:1133-1136.

Oehme, M., and H. Stray. 1982. Quantitative determination of ultra-traces
of chlorinated compounds in high-volume air samples from the Arctic
using polyurethane foam as collection medium. Fresenius Z. Anal.
Chem. 311:665-673.

Ohta, K., and N. Handa. 1985. Organic components in size-separated aerosols from the western North Pacific. Oceanograph. Soc. Japan 41:
25-32.

Ottar, B., Y. Gotaas, O. Hov, T. Iversen, E. Joranger, M. Oehme, J.
Pacyna, A. Semb, W. Thomas, and V. Vitols. 1986. Air pollutants in
the Arctic. OR Rept. 30/86, Lillestrøm, Norway:Norwegian Institute
for Air Research, n.p.

Penkett, S. A. 1982. Non-methane organics in the remote troposphere. In
Atmospheric Chemistry (E. D. Goldberg, ed.) New York:Springer-
Verlag, 329-355.

Pitts, J. N., J. A. Sweetman, B. Zielinska, R. Atkinson, A. M. Winer, and
W. P. Harger. 1985. Formation of nitroarenes from the reaction of
polycyclic aromatic hydrocarbons with dinitrogen pentaoxide. Environ. Sci. Technol. 19:1115-1121.

Prinn, R. G., P. G. Simmonds, R. A. Rasmussen, R. D. Rosen, F. N. Alyea,
C. A. Cardelino, A. J. Crawford, D. M. Cunnold, P. J. Fraser, and
J. E. Lovelock. 1983. The atmospheric lifetime experiment. I:
Introduction, instrumentation and overview. J. Geophys. Res.
88:8353-8367.

Prinn, R., D. Cunnold, R. Rasmussen, P. Simmonds, F. Alyea, A. Crawford,
P. Fraser, and R. Rosen. 1987. Atmospheric trends in methylchloroform and the global average for the hydroxyl radical. Science 238:
945-950.

Rasmussen, R. A., and M. A. K. Khalil. 1983. Natural and anthropogenic
trace gases in the lower troposphere of the Arctic. Chemosphere
12:371-375.

Rasmussen, R. A., M. A. K. Khalil, and R. J. Fox. 1983. Altitudinal and
temporal variations of hydrocarbons and other gaseous tracers of
Arctic haze. Geophys. Res. Ltrs. 10:144-147.

Rudolph, J. 1988. The two-dimensional distribution of light hydrocarbons:
Results from the STRATOZ III experiment. J. Geophys. Res. 93: 8367-
8377.

Rudolph, J., and D. H. Ehhalt. 1981. Measurements of C_2-C_5 hydrocarbons
over the North Atlantic. J. Geophys. Res. 86:11,959-11,964.

Rudolph, J., and A. Khedim. 1985. Hydrocarbons in the nonurban atmosphere: Analysis, ambient concentrations and impact of the chemistry
of the atmosphere. Intern. Environ. Anal. Chemistry 20:265-282.

Rudolph, J., C. Jebsen, A. Khedim, and F. J. Johnen. 1984. Measurements of the latitudinal distribution of light hydrocarbons and halocarbons over the Atlantic. In Physico-chemical Behavior of Atmospheric Pollutants (B. Versine and C. T. Angeletti, eds.) Dordrecht:Reidel, 492-501.

Sexton, K., and H. Westberg. 1984. Nonmethane hydrocarbon composition of urban and rural atmospheres. Atmos. Environ. 18:1125-1132.

Sicre, M. A., J. C. Marty, A. Saliot, X. Aparicio, J. Grimalt, and J. Albaiges. 1987. Aliphatic and aromatic hydrocarbons in Mediterranean aerosol. Int. Environ. Anal. Chemistry 29:73-94.

Simmonds, P. G., F. N. Alyea, C. A. Cardelino, A. J. Crawford, D. M. Cunnold, B. C. Lane, J. E. Lovelock, R. G. Prinn, and R. A. Rasmussen. 1983. The atmospheric lifetime experiment. 6: Results for carbon tetrachloride based on 3 years data. J. Geophys. Res. 17: 8427-8441.

Simoneit, B. R. T. 1984. Organic matter of the troposphere: III. Characterization and sources of petroleum and pyrogenic residues in aerosols over the western United States. Atmos. Environ. 18:51-67.

Simoneit, B. R. T. 1986. Characterization of organic constituents in aerosols in relation to their origins and transport: A review. Int. Environ. Anal. Chemistry 23:207-237.

Simoneit, B. R. T., and M. A. Mazurek. 1981. Air pollution: The organic components. Critical Rev. Environ. Control 11:219-276.

Simoneit, B. R. T., M. A. Mazurek, and W. E. Reed. 1983. Characterization of organic matter in aerosols over rural sites: Phytosterols. Advances in Organic Geochemistry 1981 355-361, 1983.

Singh, H. B., and L. J. Salas. 1982. Measurement of selected light hydrocarbons over the Pacific Ocean: Latitudinal and seasonal variations. Geophys. Res. Ltrs. 9:842-845.

Singh, H. B., L. J. Salas, and R. E. Stiles. 1983. Selected man-made halogenated chemicals in the air and oceanic environment. J. Geophys. Res. 88:3675-3683.

Subramanian, A., S. Tanabe, H. Hidaka, and R. Tatsukawa. 1986a. Bioaccumulation of organochlorines (PCBs and ρ,ρ'-DDE) in Antarctic Adelie penguins (Pygoscelis adeliae) collected during a breeding season. Environ. Pollut. 40:173-189.

Subramanian, A., S. Tanabe, Y. Fujise, and R. Tatsukawa. 1986b. Organochlorine residues in Dall's and True's porpoises collected from the Northwestern Pacific and adjacent waters. Memoirs, National Inst. Polar Res., Spec. Issue 44:167-173.

Talbot, R. W., K. M. Stein, R. C. Harriss, and W. R. Cofer, III. 1988. Atmospheric geochemistry of formic and acetic acids at a mid-latitude temperate site. J. Geophys. Res. 93:1638-1652.

Tanabe, S., and R. Tatsukawa. 1983. Chlorinated hydrocarbons in the Southern Pacific Ocean, 1983. Memoirs, National Inst. of Polar Res., Special Issue (Procs., BIOMASS Colloquium, 1982) 27:64-76.

Tanabe, S., and R. Tatsukawa. 1986. Distribution, behavior and load of PCBs in the ocean. In PCBs and the Environment (J. S. Waid, ed.) Boca Raton, FL:CRC Press, 143-162.

Tanabe, S., R. Tatsukawa, M. Kawano, and H. Hidaka. 1982. Global distribution and atmospheric transport of chlorinated hydrocarbons: HCH (BCH) isomers and DDT compounds in the western Pacific, eastern Indian, and Antarctic Oceans. Oceanograph. Soc. Japan 38:137-148.

Tanabe, S., H. Tanaka, and R. Tatsukawa. 1984. Polychlorobiphyenyls, sigma-DDT, and hexachlorocyclohexane isomers in the western North Pacific ecosystem. Arch. Environ. Contam. Toxicol. 13:731-738.

Tille, K. J. W., M. Savelsberg, and K. Bachmann. 1985. Vertical distributions of nonmethane hydrocarbons over western Europe: Seasonal cycles of mixing ratios and source strengths. Int. Environ. Anal. Chemistry 21:9-22.

Van Noort, P. C. M., and E. Wondergem. 1985. Scavenging of airborne polycyclic aromatic hydrocarbons by rain. Environ. Sci. Technol. 19: 1044-1048.

Wauters, E., F. Vangaever, P. Sandra, and M. Verzele. 1979. Polar fraction of air particulate matter. Chromatography. 170:133-138.

Wolff, G. T., and R. L. Klimisch, eds. 1982. Particulate Carbon: Atmospheric Life Cycle. New York:Plenum.

Yokouchi, Y., and Y. Ambe. 1986. Characterization of Polar organics in airborne particulate matter. Atmos. Environ. 20:1727-1734.

Yokouchi, Y., T. Ito, and Y. Ambe. 1987. Identification of C_6-C_{15} gamma-lactones in atmospheric aerosols. Chemosphere 16:1143-1147.

Zafiriou, O. C., R. B. Gagosian, E. T. Peltzer, and J. B. Alford. 1985. Air-to-sea fluxes of lipids at Enewetak Atoll. J. Geophys. Res. 90: 2409-2423.

Zell, M., and K. Ballschmiter. 1980. Baseline studies of global pollution: II. Global occurrence of hexachlorobenzene (HCB) and polychlorocamphenes (Toxaphene)(PCC) in biological samples. Fresenius Z. Anal. Chem. 300:387-402.

6. ARCTIC AIR POLLUTION: A CASE STUDY OF CONTINENT-TO-OCEAN-TO-CONTINENT TRANSPORT

Leonard A. Barrie
Atmospheric Environment Service
4905 Dufferin Street
Downsview, Ontario, Canada M3H 5T4

6.1. THE PHENOMENON OF ARCTIC AIR POLLUTION

In the last ten years, the scientific community has recognized that sub-stantial amounts of the air pollutants that cause regional scale problems of acid rain, oxidants, and toxic deposition of heavy metals and organic compounds travel beyond regional boundaries to pollute the atmosphere on an hemispheric scale. Arctic haze is a prime example of this; it is the remnant of industrial emissions, mostly from the Eurasian continent. From a meteorological point of view, Eurasia is favored over North America as a source of Arctic pollution.

This chapter briefly describes the phenomenon of Arctic air pollu-tion, which has already been widely studied, and its nature and origin. For a more comprehensive overview, please consult Barrie (1986a) and Stonehouse (1986).

Arctic air pollution occurring mainly in winter consists of a mix-ture of anthropogenic suspended particulate matter and gases. The average composition of particulate matter during winter/spring is ~ 2 $\mu g/m^3$ of $SO_4^=$, 1 $\mu g/m^3$ of organic compounds, 0.3-0.5 $\mu g/m^3$ of black carbon, a few tenths of a $\mu g/m^3$ of other substances (e.g., metals) and a few $\mu g/m^3$ of water. Important gaseous compounds include SO_2, peroxy-acetylnitrate, high-temperature-combustion products (CO_2, CO), industri-al and commercial chlorofluorocarbons, and pesticides, such as chlordane.

First let us consider Arctic air pollution as represented by one of its major particulate constituents, sulfate. Sulfate concentrations in the Arctic show strong seasonal variations that persist from year to year throughout the region (Fig. 6-1). Concentrations peak from January to April and are minimal between June and August. These seasonal varia-tions are caused partially by the removal of pollutants through precipi-tation (Fig. 6-2) and partially by their south-to-north transport from Eurasia. The seasonal behavior of other particulate constituents roughly parallels that of $SO_4^=$ (Fig. 6-3).

In 1986 I published estimations of the average vertical distribution of suspended particulate matter in the Arctic atmosphere (Fig. 6-4) (Barrie 1986a). These estimations are based on data from 23 aircraft flights made mainly during March and April, part of the polluted winter season. Although on any given day there are large variations from the mean profile, on the average most of the pollution is concentrated in the troposphere below 5 km with the highest concentrations in the lowest 2 km.

A. H. Knap (ed.), The Long-Range Atmospheric Transport of Natural and Contaminant Substances, 137–148.
© 1990 Kluwer Academic Publishers.

138

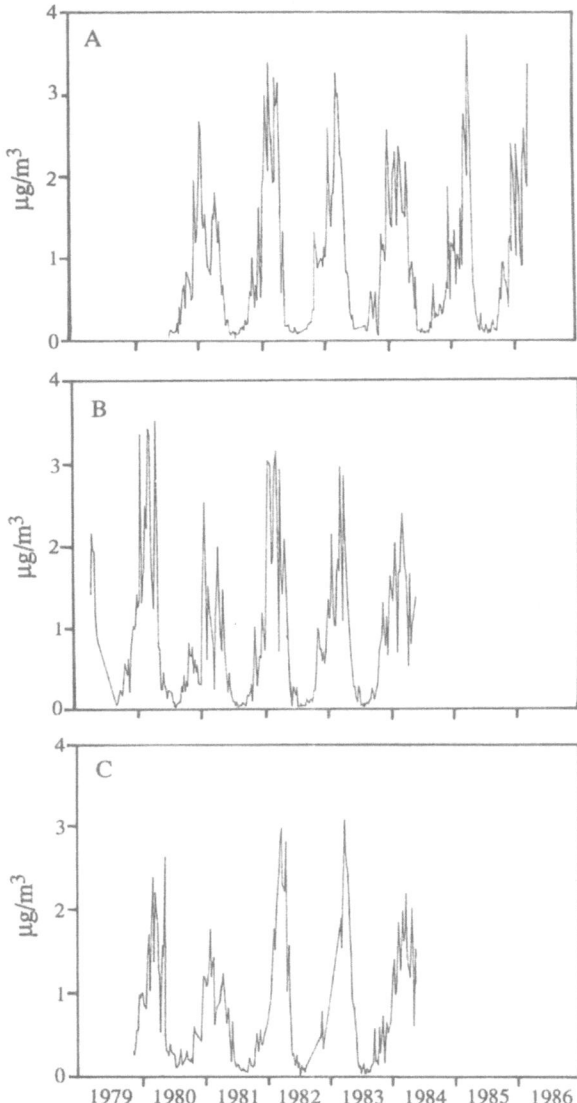

Figure 6-1. The temporal variations of weekly averaged, ground-level
sulfate concentrations at (A) Alert, (B) Mould Bay, and (C) Igloolik
in the high Canadian Arctic in the early 1980s. These are typical
of concentrations found throughout the Arctic.

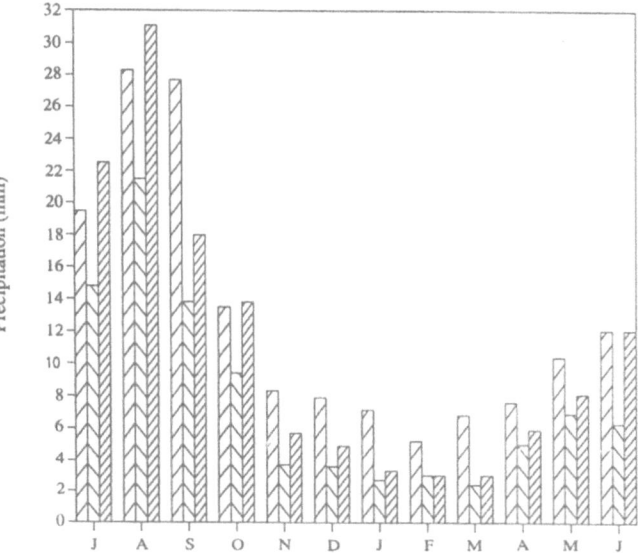

Figure 6-2

The seasonal varia-
tions in monthly
precipitation amounts
at Alert (▨), Mould
Bay (▨), and Reso-
lute (▨) in the
Canadian high Arctic.

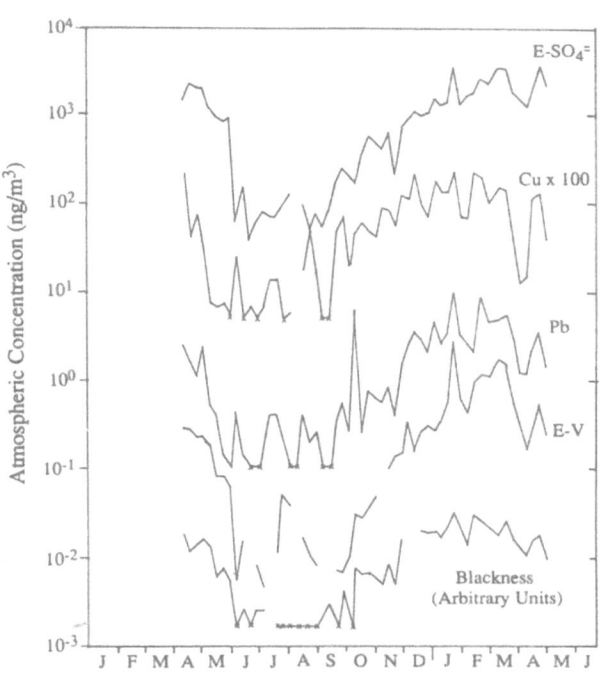

Figure 6-3

Temporal variations
of atmospheric con-
centrations compared
to those of $SO_4^=$ at
Mould Bay in the
Canadian high Arctic;
black-carbon concen-
trations are given in
arbitrary units (Bar-
rie et al. 1981).

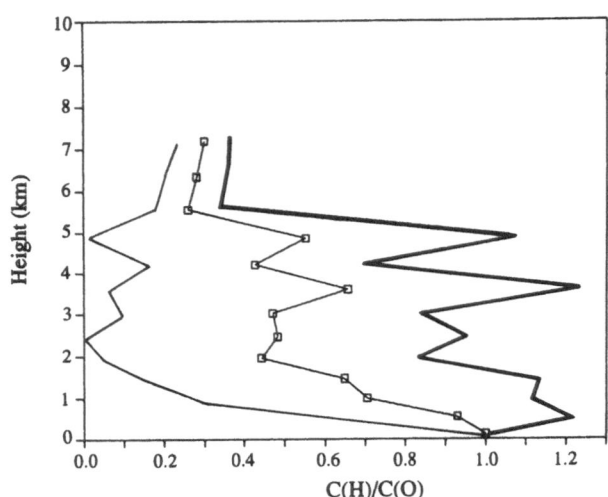

Figure 6-4

The estimated mean ver-
tical profile of the
concentration of anthro-
pogenic aerosol mass in
the high Arctic during
March and April based
mainly on aircraft ob-
servations of aerosol
light scattering. The
light line (far left)
represents the mean
minus 1 standard devia-
tion; the center line
represents the mean; and
the heavy line (right)
represents the mean plus
1 standard deviation.
(Barrie 1986a)

Figure 6-5 shows the spatial distribution of Arctic air pollution
for January to April of 1980 as represented by the spatial distribution
of the average concentration of ground level $SO_4^=$. Arrows showing the
predominant pathway of air in and out of the Arctic during the winter
months are superimposed on this diagram. These arrows clearly show a
plume from anthropogenic sources in the Eurasian continent and a well-
mixed Arctic reservoir. The Siberian high-pressure cell over the eastern
Soviet Union during winter and the associated clockwise flow around it
means that south-to-north transport from Eurasia to the Arctic is
favored. In contrast, the prevailing winds over the eastern North
American industrial source region are from the west/southwest, carrying
emissions eastward over a stormy high-precipitation area of the North
Atlantic. In summer, the Siberian high disappears leaving a weak south-
to-north transport link.

Lack of precipitation, a relatively inert and aerodynamically smooth
underlying surface, and poor vertical mixing because of the strong vertical stability are unique characteristics of the Arctic air mass that
result in pollutants having longer residence times in the atmosphere over
the Arctic than in any other air mass on the globe. In other words, the
potential for polluting the Arctic air mass is highest.

Because of the long residence times, air pollutants are transported
from Eurasia to the Arctic and then out of the Arctic into North America
and into the climatologically persistent Aleutian and Greenland low-
pressure cells. This transport into North America is clear from our
research in central Canada; concentrations of $SO_4^=$ and SO_2 in air from
the north reach a peak in the winter months (Barrie 1986a).

That Arctic air pollution has been with us for much of this century
is evident in historical records of glacial ice composition on Greenland

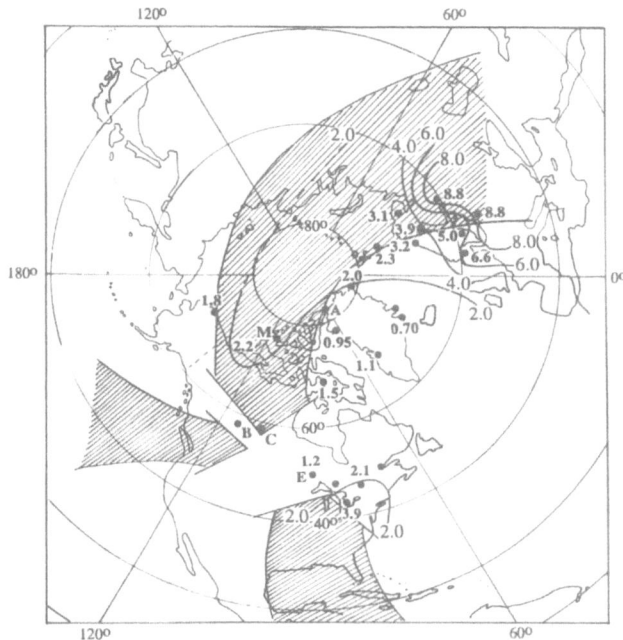

Figure 6-5. The three main sources of air for the North American airshed
(i.e., the Arctic, Pacific, and Caribbean) superimposed on the spatial distribution of arithmetic mean $SO_4^=$ air concentrations ($\mu g/m^3$)
for January to April (1980) when the Arctic air mass is most polluted (Barrie 1986b).

and northern Ellesmere Island, Canada. Figure 6-6 shows an increase in
concentration in this century for $SO_4^=$ of about 1 $\mu eq/l$ to current levels
of 1.7 $\mu eq/l$ and for NO_3^- of 1.3 $\mu eq/l$ to 2 $\mu eq/l$ currently. Figure 6-7
shows that emissions remained roughly constant in the first part of the
century and ice-core conductivity remained roughly constant, except for a
peak in 1912-1913 associated with the eruption of Mt. Katmai in the
Aleutian Islands in 1912. Then between 1956 and 1977, as SO_2 emissions
in Europe doubled, ice-core conductivity underwent a marked increase of
75%. Current levels of wintertime acidity are 12-14 $\mu eq/l$ compared to
the 7-9 $\mu eq/l$ before 1956.

The effects of Arctic air pollution that are of concern can be
classed into three groups: 1) the reduction of visibility, 2) the climatic effects, and 3) the effects associated with the deposition of
anthropogenic, acidic, or other potentially toxic substances. A reduction of visibility from the clean-air limit of several hundred kilometers
to 30-50 km can be expected from the highest levels of particulate pollution observed. However, visibilities of 10 km are often reported in the
winter Arctic even in the absence of blowing snow (Leaitch et al. 1984).

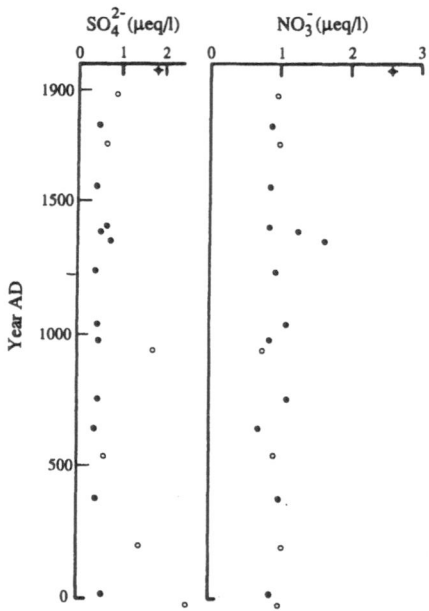

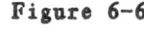

Figure 6-6

An historical record of $SO_4^=$ and NO_3^- from Greenland ice at Dye 3 in which the open circles (o) denote intervals with volcanic impurities; the solid circles (●), periods without. Each point represents a multiyear average of from 2 yr to 37 yr; the sample from this century is emphasized by a cross. (Herron 1982)

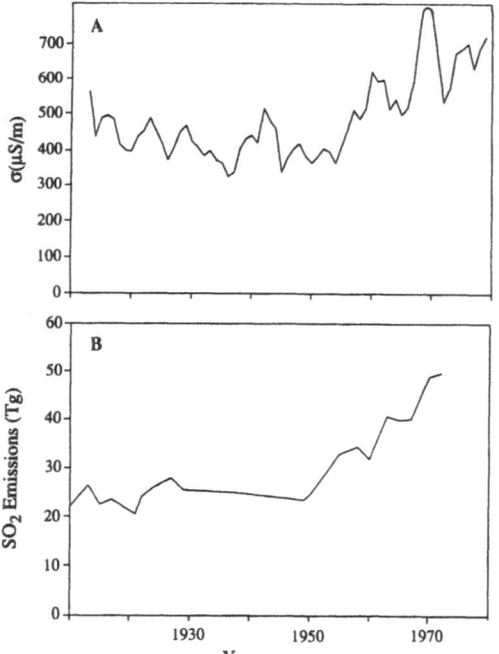

Figure 6-7

Comparison of the historical record of annual maximum conductivity (a 3-yr running mean) in A melted ice from Agassiz glacier, Ellesmere Island, and B SO_2 emissions (10^6 Mt/yr) from Europe (including the western Soviet Union). The conductivity is highly correlated with the acidity. (Barrie et al. 1985)

Most likely this is caused by the presence of ice crystals in the atmosphere, which often accompany haze aerosols (Hoff 1988). The physical-
chemical link between aerosols and ice crystals needs to be investigated
on a more quantitative basis. If there is indeed a link, as current
evidence suggests (Otake 1984), then reduced visibilities are indirectly
caused by anthropogenic aerosols.

Climatic effects of widespread Arctic pollution are expected to
occur from black-carbon particles and radiatively active gases, such as
CO_2, CH_4, and perfluorocarbons, influencing the atmospheric energy budget
and perturbing atmospheric circulation patterns. Black-carbon aerosols
act--either while suspended in air above the surface or while deposited
in the snow pack--to increase the amount of solar energy trapped by the
atmosphere. The net result of energy perturbations caused by pollutants
is the potential modification of local climate and possibly the hemispheric climate. Atmospheric climate modelers are in the process of
assessing the magnitude of these impacts. The most comprehensive assessment done to date (Blanchet 1987) indicates that present aerosol pollution increases the temperature of the Arctic troposphere during March and
April by $1°$ to $2°$ C.

As evident from glacial records mentioned above, snow falling in the
Arctic is slightly acidic (pH ranges from 4.9 to 5.2). To my knowledge,
analyses of the effects of acidic snow on delicate ecosystems in the
Arctic have not been published.

Potentially toxic metals and synthetic organics when deposited in
the Arctic may seriously threaten plant and wildlife. Because many of
these substances bioaccumulate, they are often most concentrated in
organs of fish and animals at the top of the food chain, such as polar
bears, whales, and man. Few budgets of organics and trace elements have
been calculated to assess the importance of the atmosphere as a source.
One notable exception is the budget estimate Rahn (1981) has done for
the Arctic Ocean for Al, V, Mn, Cd, Pb, $SO_4^=$, and NO_3^-. He concludes
that Pb is the only element whose direct atmospheric source equals or
exceeds riverine or oceanic sources. This does not preclude an atmospheric source for the latter. There is an update of Rahn's estimates of
metals deposition in Chapter 9 (page 177).

6.2. SULFUR TRANSPORT INTO THE ARCTIC

My colleagues and I (Barrie et al. 1989) did a recent study of the transport of sulfur into the Arctic from July 1979 to June 1980. Using a
trajectory transport model, we have calculated the flux of sulfur into
the Arctic as a function of month, longitude, and altitude. We were also
looking for the fraction of anthropogenic sulfur emissions released from
midlatitudes that enters the Arctic. Figure 6-8 is a map of the model
domain, the gridded SO_2 emissions, the Arctic boundary, and the sample
trajectories. The details of the model calculations are in our original
study. What we have done is to divide the Arctic's circular boundary
into eighteen $20°$-longitude segments and three vertical layers (0-1.2 km,
1.2-2.4 km, and 2.4-3.6 km). We have calculated the sulfur flux as the
product of the meridional velocity and the model-estimated sulfur

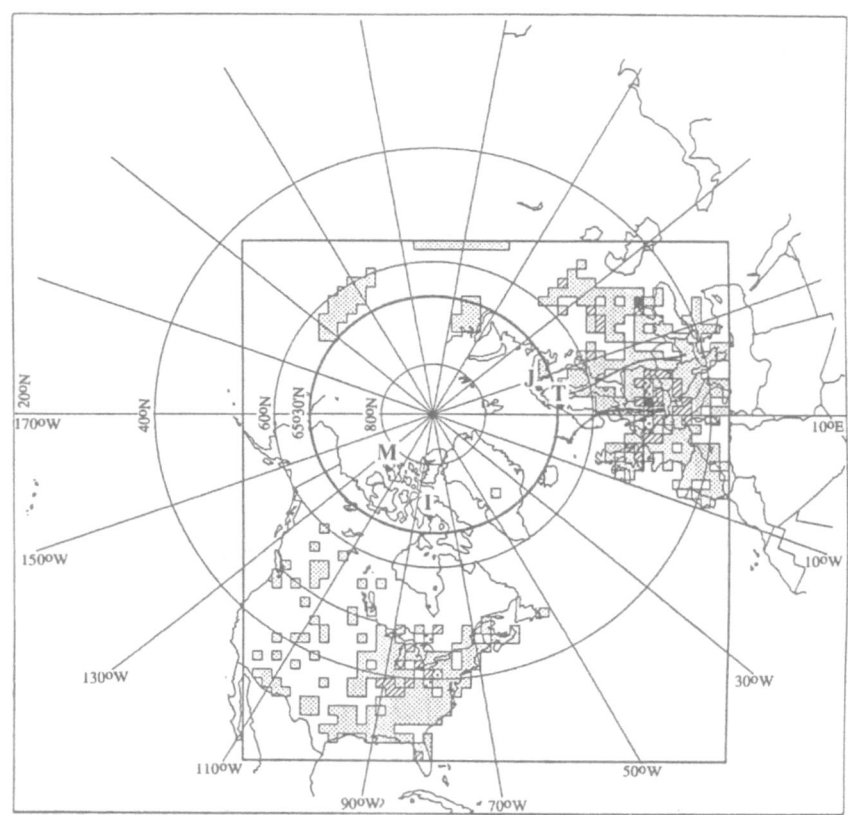

Figure 6-8. The model domain (**large rectangle**) showing gridded sulfur
    emissions from North America and Europe to the Arctic circle for
    which Barrie et al. (1989) calculated sulfur fluxes. We report
    ground-level sulfur concentrations for Jergul (**J**) and Tustervaten
    (**T**), Norway. Examples of two 5-day back trajectories from the
    Arctic circle are also shown. ⊞ = > 3,000 SO_2 Ktonnes/yr, ■ =
    2,500-3,000, ◪ = 2,000-2,500, ▨ = 1,600-2,000, ⊡ = 1,000-1,600,
    ▨ = 500-1,000, ⊞ = 25-500, ☐ = < 25. (Barrie et al. 1989)

concentration for each layer in each segment on a 12-hour basis.
The concentration is obtained from a box trajectory model using three-
dimensional back trajectories and gridded fields of SO_2 emissions, mixing depths, and precipitation. Pollutant removal en route to the Arctic
circle is parameterized using scavenging ratios and dry-deposition
velocities. Parameterizations are adjusted to yield good agreement
between model predictions and observations at Jergul (Fig. 6-9) and
at Tustervaten (not shown).

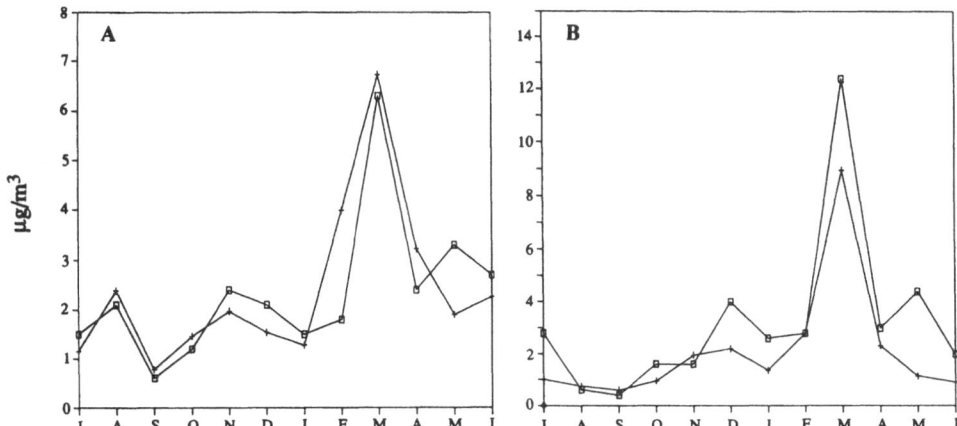

Figure 6-9. Average atmospheric concentrations of **A** SO$_4^=$ and **B** SO$_2$
observed at Jergul (+) from July, 1979, to June, 1980, compared to
the model estimates for the same period ([]) calculated for the
model domain and with the back trajectories shown in Figure 6-8
(Barrie et al. 1989).

As a function of longitude, the annual flux of sulfur (Fig. 6-10)
into the Arctic is concentrated between 30° W and 100° E.   The sulfur
fluxes from the Soviet Union, eastern Europe, and western Europe are also
shown.   These results quantitatively confirm the qualitative picture
drawn above showing that Eurasia is the most probable source of anthropogenic pollution found in the Arctic.   Similarly, the monthly variation of
the net sulfur flux (Fig. 6-11) is consistent with the seasonal variability in the south-to-north transport.   The input of sulfur to the Arctic
between October and May is about 2.5 times greater than it is during the
rest of the year.   Figure 6-11 clearly shows that most of the sulfur
entering the Arctic circle does so in the lower troposphere.

In the Table 6-1, the relative contributions of the various source
regions are compared to total anthropogenic sulfur emissions and to total
sulfur input to the Arctic.   The right-hand column gives the fraction of
a source region's emissions that enters the Arctic; in other words, the
percentage of the region's emissions that contribute to the Arctic air
pollution.   On an annual basis, 4.8% of the anthropogenic sulfur emissions of Eurasia and North America enter the Arctic.   Eurasia contributes
much more than North America (94% versus 6%).

146

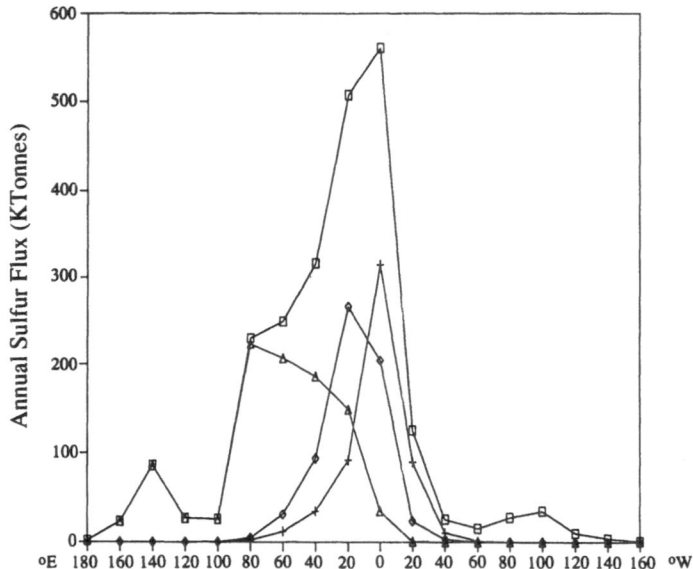

Figure 6-10. The longitudinal distribution of sulfur flux into the
Arctic atmosphere between 0-3.5 km, July 1979 to June 1980 as a
function of source region; [] = total flux, + = flux from western
Europe, ⟨⟩ = flux from eastern Europe, and △ = flux from the Soviet
Union (Barrie et al. 1989).

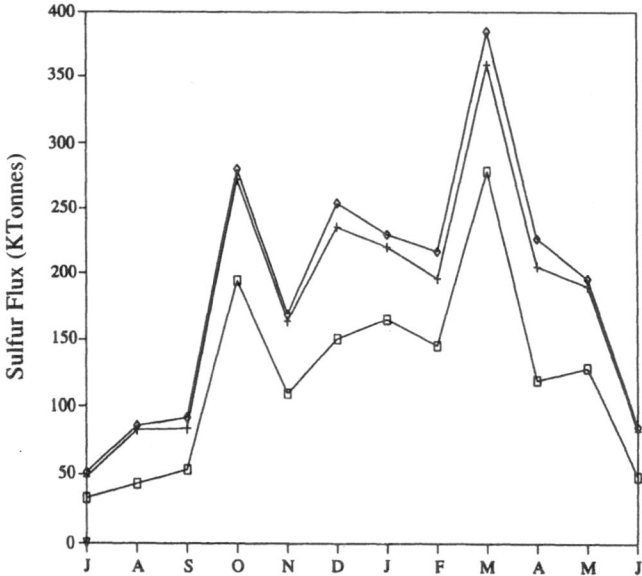

Figure 6-11. Temporal variations of the monthly flux of sulfur into the
Arctic from July 1979 to June 1980 as a function of altitude ([] =
0-1 km, + = 0-2 km, and ⟨⟩ = 0-3.5 km) (Barrie et al. 1989).

Table 6-1. Annual SO_2 emissions from sources in the Northern Hemisphere affecting the Arctic and the sulfur flux into the Arctic during July 1979 to June 1980.

Source Region	SO_2 Emissions (Mt S/yr)	(% of total)	Sulfur Flux to Arctic (Mt S/yr)	(% of total)	Percentage of Emissions Entering the Arctic
Western Europe	10	21	0.56	25	5.5
Eastern Europe	11	23	0.63	27	5.9
Soviet Union	11	24	0.96	42	8.5
North America	15	32	0.12	6	0.8
Total	47	100	2.27	100	

6.3. KNOWLEDGE GAPS AND FUTURE RESEARCH

Although there is now a solid knowledge base on the input of anthropogenic sulfur to the Arctic, little is known about that of other substances. Studies similar to the one described in the previous section need to be done for metals, nitrogen oxides, and potentially toxic organic compounds.

Once pollutants are in the Arctic what happens to them? They can be deposited by wet and dry removal or they can flow out of the sides and top of the Arctic domain. Few reliable precipitation-chemistry data are available covering the polar region during the low-precipitation period between November and May because of sampling difficulties associated with light snowfall and blowing snow conditions. Methodology development is needed. Although estimates of total deposition could be obtained from available glaciers, this has not been done for the vast majority of the Arctic region because of the limited number of glaciers. Snow-pack sampling must be viewed with caution as a viable alternative because the Arctic is a cold desert where the soil, which is not buried in winter, often contaminates snow-pack samples.

6.4. REFERENCES

Barrie, L. A. 1986a. Arctic air pollution: An overview of current knowledge. Atmos. Environ. 20:643-663.

Barrie, L. A. 1986b. Background pollution in the Arctic air mass and its relevance to North American acid rain studies. Water Air Soil Pollut. 30:765-777.

Barrie, L. A., R. M. Hoff, and S. M Daggupaty. 1981. The influence of midlatitudinal pollution sources on haze in the Canadian Arctic. Atmos. Environ. 15:1407-1519.

Barrie, L. A., D. Fisher, and R. M. Koerner. 1985. Twentieth century trends in Arctic air pollution revealed by conductivity and acidity observations in snow and ice in the Canadian high Arctic. Atmos. Environ. 19:2055-2063.

Barrie, L. A., M. P. Olson, and K. K. Oikawa. 1989. The flux of anthropic sulphur into Arctic from midlatitudes in 1979/1980. Atmos. Environ. 23, in press.

Blanchet, J. P. 1987. The effects of Arctic aerosols on a general circulation model simulation: A sensitivity study. In Aerosols and Climate (P. V. Hobbs and M. P. McCormick, eds.) Hampton, VA:Deepak Publishing (Sci. Technol. Corp.), 373-383.

Herron, M. M. 1982. Impurity sources of F^-, Cl^-, NO_3^- and $SO_4^=$ in Greenland and Antarctic precipitation. J. Geophys. Res. 87:3052-3060.

Hoff, R. M. 1988. Vertical structure of Arctic haze observed by Lidar. Appl. Meteorol. 27:125-139.

Leaitch, W. R., R. M. Hoff, S. Melnichuk, and A. Hogan. 1984. Some physical and chemical properties of the Arctic winter aerosol in northeastern Canada. Climate Appl. Meteorol. 23:916-928.

Otake, T. 1984. Ice crystal nucleation on microscopic aerosols. Procs. 11th Int. Conf. Atmos. Aerosols, Condensation and Nuclei, Budapest: Hungarian Meteorolog. Service, Vol. II, 44-48.

Rahn, K. A. 1981. The Arctic air-sampling network in 1980. Atmos. Environ. 15:1349-1352.

Stonehouse, B. 1986. Arctic Air Pollution. Cambridge, England:Cambridge University Press, 327 pp.

7. CHERNOBYL: A SUMMARY CASE STUDY WITH EMPHASIS ON THE TRANSPORT AND DEPOSITION TO SCANDINAVIA

Anton Eliassen
The Norwegian Meteorological Institute
P. O. Box 43, Blindern N-0313
Oslo 3, Norway

On April 26 at 0123 local time, reactor No. 4 of the nuclear power station at Chernobyl (51° 17' N, 30° 15' E) exploded and caught fire. The reactor had been in operation since December, 1983. Under normal operation, the reactor had a thermal effect of 3200 MW and delivered 1000 MW of electric power. The reactor was a graphite-moderated channel-boiler type. Before the accident, there were twelve such reactors in operation in the Soviet Union.

The Soviet Union has reported estimates of the emissions from the accident. In the actual explosion and during the first 24 hours of the accident, about 0.80×10^{18} Bq were released, excluding rare gases. Table 7-1 gives a summary of the total release of the various radionuclides. Figure 7-1 shows the daily release of all nuclides together with similar data for ^{137}Cs calculated from data contained in the Soviet report to the IAEA.

On April 27, the plume height from the reactor fire exceeded 1200 m. On subsequent days, however, the plume height did not exceed 200–400 m (Izrael et al. 1987). On the night of the explosion, a ground-based temperature inversion extended up to 400–500 m. Below the inversion, winds were weak and from the ESE. Above the inversion, the wind was also ESE but stronger, up to about 12 m/sec.

Table 7-1. Estimated total release of the various radionuclides from the Chernobyl accident.

Nuclide	$T_{1/2}$		10^{18} Bq	Nuclide	$T_{1/2}$		10^{18} Bq
^{89}Sr	50.5	day	0.094	^{137}Cs	30.2	yr	0.037
^{90}Sr	29	yr	0.0081	^{140}Ba	12.7	day	0.28
^{95}Zr	64.0	day	0.16	^{141}Ce	32.5	day	0.13
^{99}Mo	2.75	day	0.16	^{144}Ce	285	day	0.088
^{103}Ru	39.4	day	0.14	^{239}Np	2.36	day	0.97
^{106}Ru	372	day	0.059	^{238}Pu	87.7	yr	3.0×10^{-5}
^{131}I	8.04	day	0.67	^{239}Pu	24110	yr	2.6×10^{-5}
^{132}Te	3.26	day	0.45	^{240}Pu	6560	yr	3.7×10^{-5}
^{134}Cs	2.06	yr	0.019	^{242}Cm	163	day	7.8×10^{-4}

Source: Persson et al. (1987).

A. H. Knap (ed.), The Long-Range Atmospheric Transport of Natural and Contaminant Substances, 149–162.
© 1990 Kluwer Academic Publishers.

150

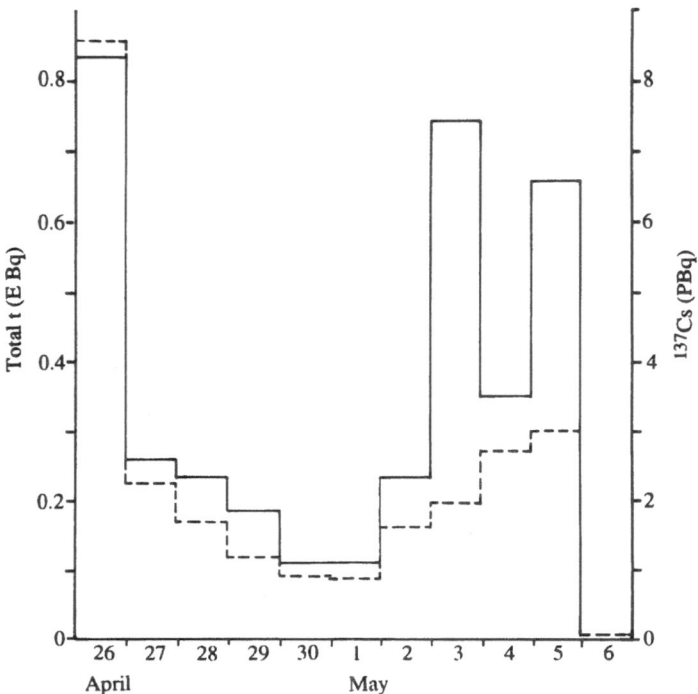

Figure 7-1. Daily radioactive releases from the Chernobyl accident; the
dashed line represents total activity (excluding noble gases), the
solid line, the release of ^{137}Cs (EBq = 10^{18} Bq, PBq - 10^{15})
(Persson et al. 1987)

At the time of the accident, the weather situation in Europe was
characterized by a high-pressure area with its center over the northwest
Soviet Union. This produced warm-air advection over Byelorussia and the
Ukraine towards the Baltic and the Nordic countries. Weather maps from
April 25-28 are given in Figure 7-2. These maps show that a small low-
pressure area with associated precipitation formed over South Scandinavia
on April 28 on the front between the warm advected air and the colder air
in the northwest.
    The initial release was transported towards the Nordic countries.
Wind shear gradually affected the plume in such a way that the lower part
was transported towards Sweden and the higher parts towards Finland.
There are indications that a low-level jet transported some of the material towards Denmark. The arrival time both in Finland and Sweden was
about 00 GMT on April 28.
    Trajectories of the initial release have been calculated by quite a
few scientific groups. Samples are shown in Figures 7-3 to 7-8.
Although the trajectories exhibit some differences, it seems clear that
the transport towards Sweden and Norway occurred at a relatively low

level, say below 1000 m high. The 925-HPa trajectories seem to explain the observed deposition reasonably well. The three-dimensional trajectories are not necessarily better than two-dimensional ones.

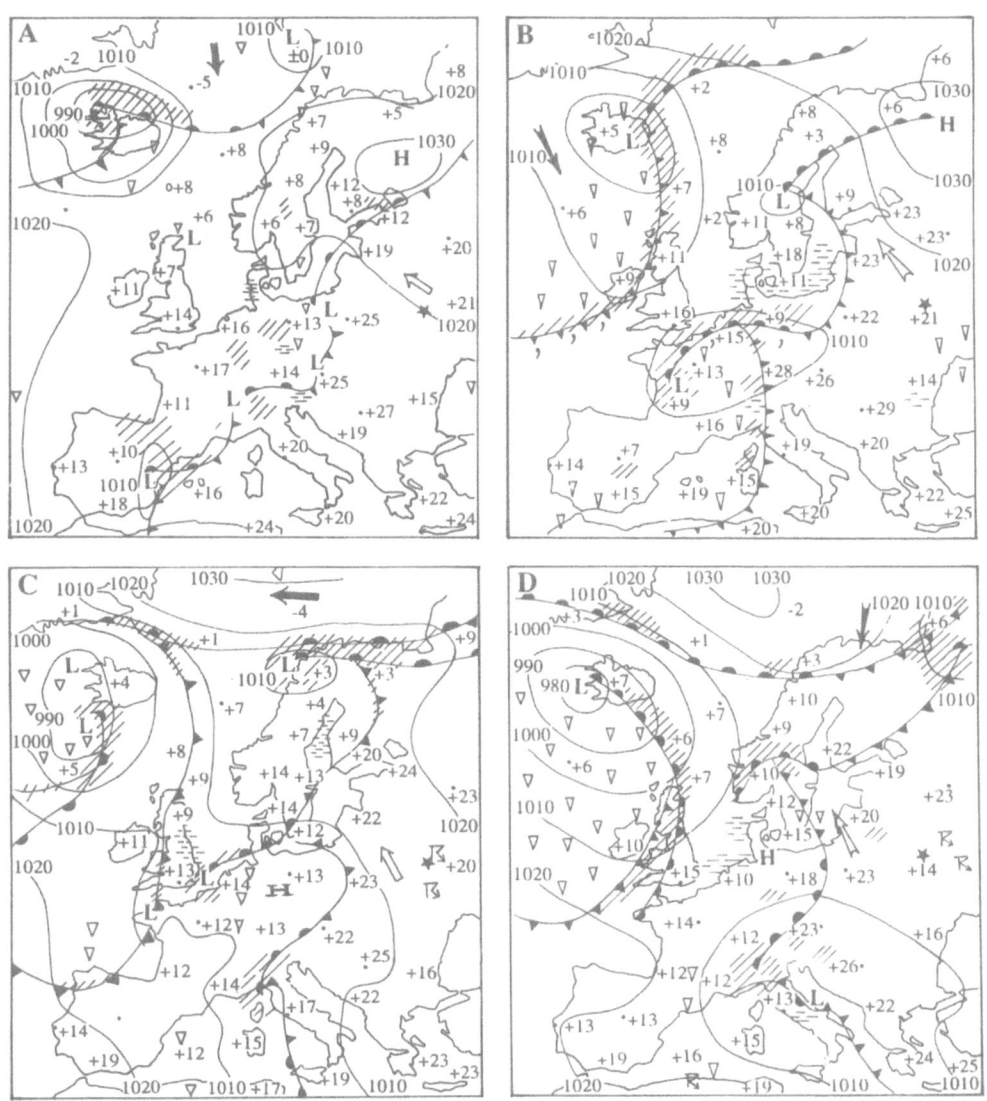

Figure 7-2. Simplified surface weather maps for 12 GMT from April 25 through April 28, 1986 (Persson et al. 1987).

152

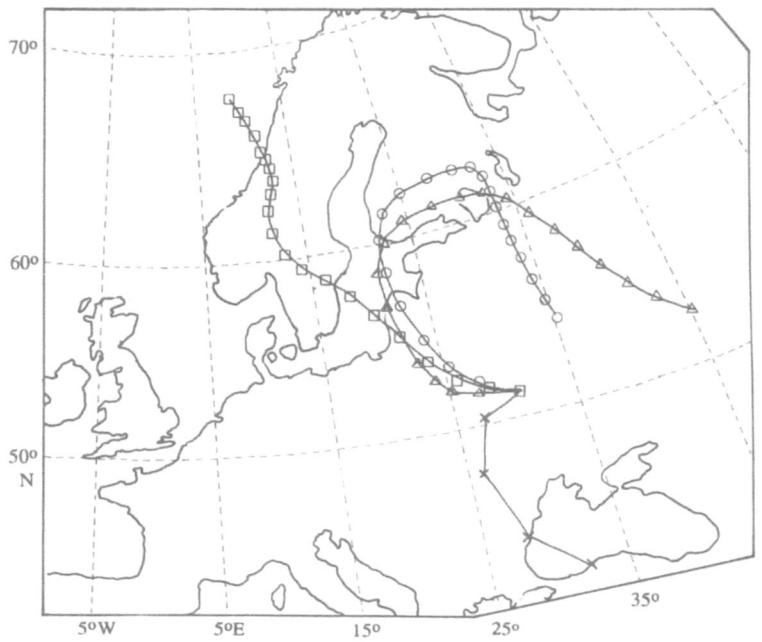

Figure 7-3. Trajectories at 500 mb (+), 700 mb (△), 850 mb (○),
and 925 mb (□) plotted from Chernobyl at 00 GMT on April 26 using
data from the Atmospheric Environment Service (4905 Dufferin Street,
Downsview, Ontario, Canada M3H 5T4).

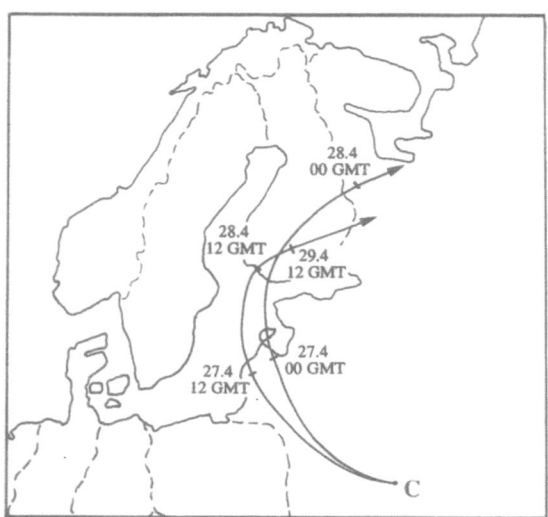

Figure 7-4

Two trajectories (850 HPa)
plotted from 00 GMT and 12
GMT on April 27, 1986 (27.4),
using data from the Finnish
Meteorological Institute
(Sahaajankatu 22E, SF-00810,
Helsinki, Finland).

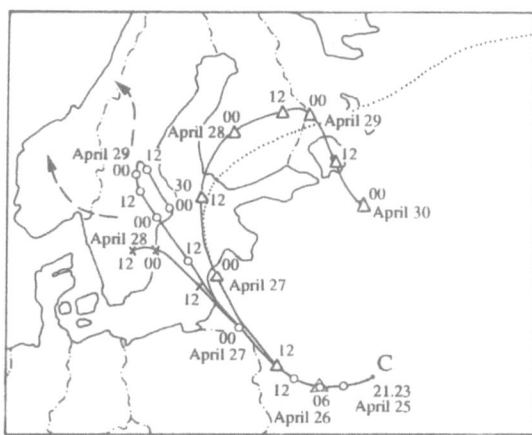

Figure 7-5

Trajectories at 300 m
(×), 750 m (○), 1500
m (△), and 3000 m (·····)
originating at the time of
the Chernobyl explosion
(2123 GMT on April 25)
calculated with data from
the Swedish Meteorological
and Hydrological Institute.
A 3-dimensional trajectory
was used from the origin to
0600 GMT on April 26.
(Persson et al. 1987)

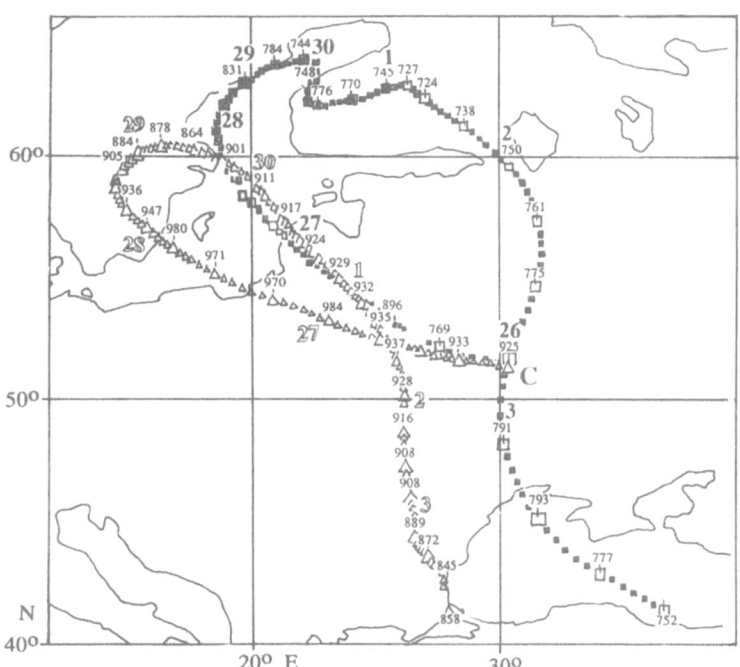

Figure 7-6. Using data from ECMWF, two 3-dimensional trajectories were
plotted originating at Chernobyl at 00 GMT at 925 HPa and 850 HPa on
April 26 and followed until 00 GMT on May 4. The major position
marks are at 6-hr increments.

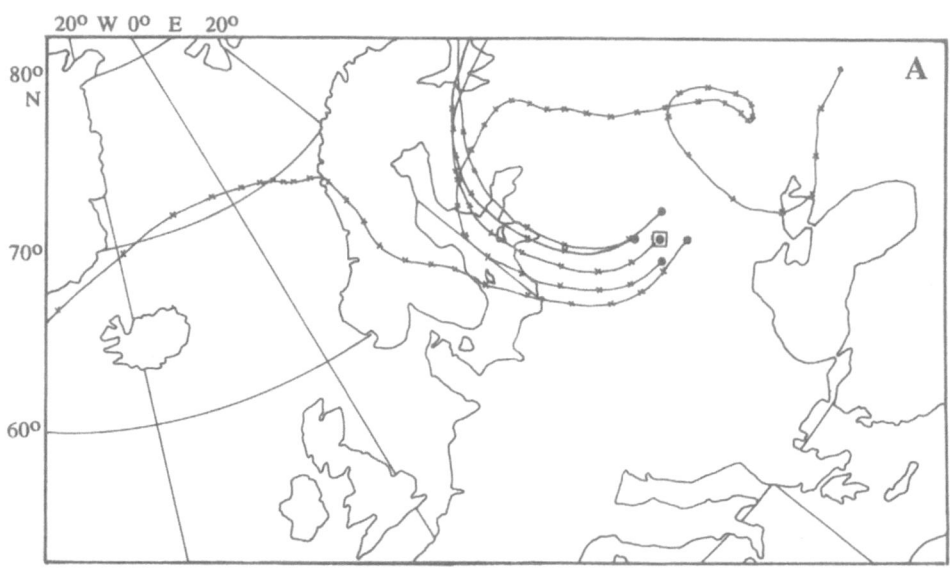

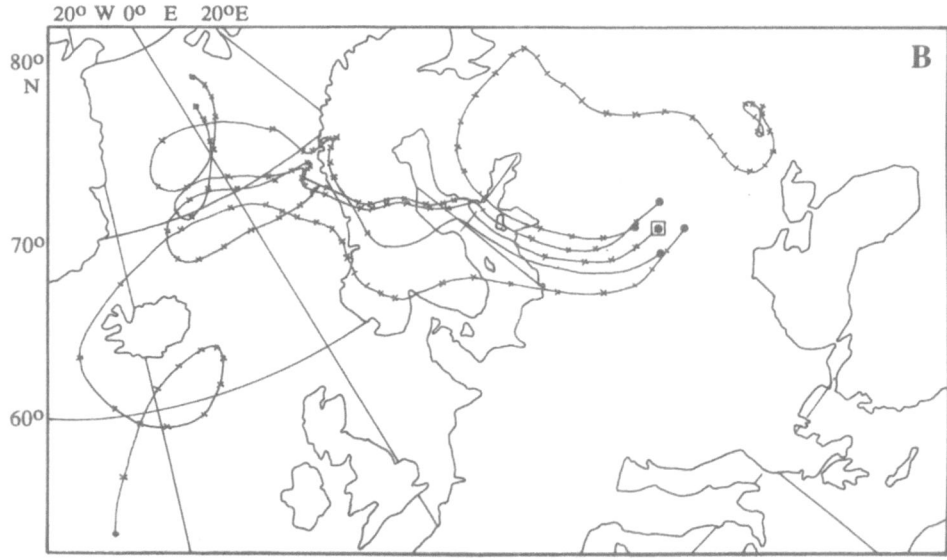

Figure 7-7. Five trajectories at **A** 850 HPa and **B** 925 HPa calculated using data from the Norwegian Meteorological Institute (Box 320, Blindern, Oslo 3, Norway N-0314) that originate from Chernobyl and four points 150 km away in each grid direction from 00 GMT, April 26, to 00 GMT, May 5; major position marks at 6-hr increments.

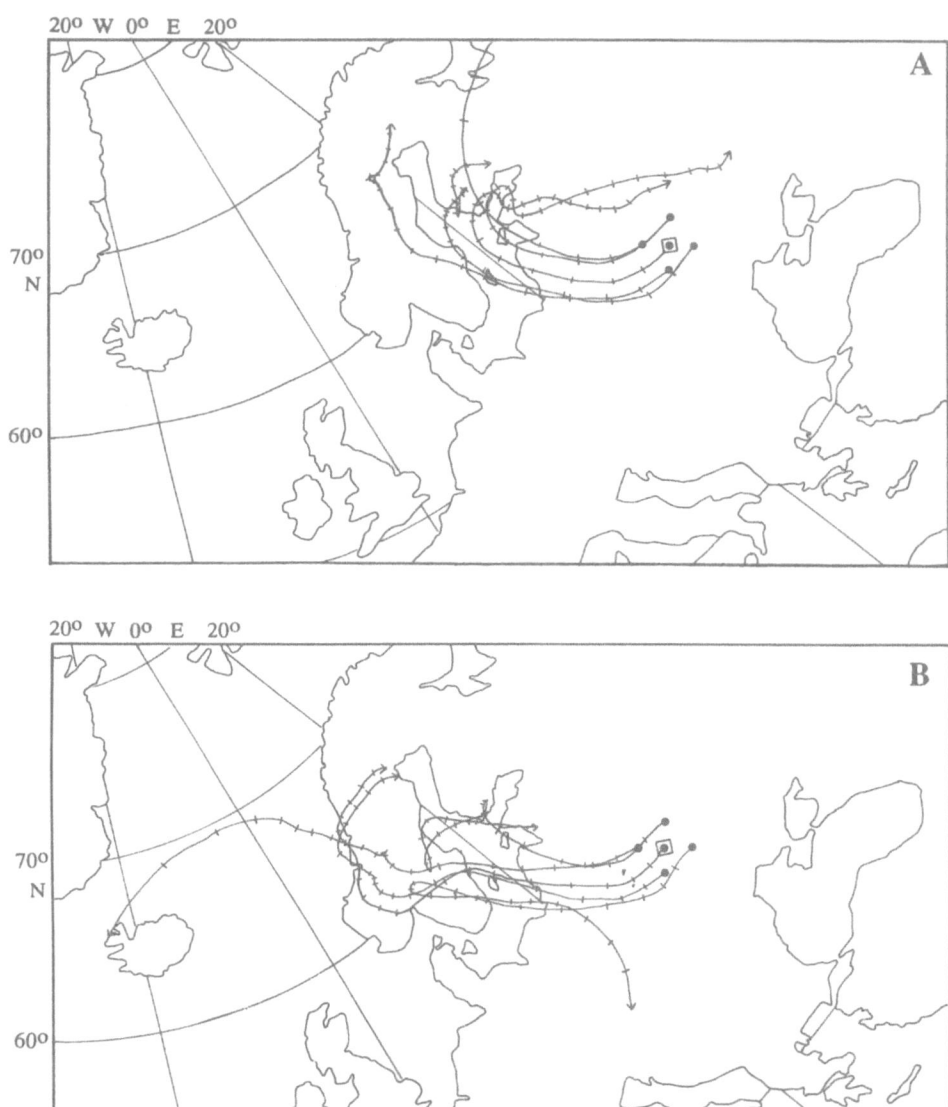

Figure 7-8. Five trajectories at **A** 850 HPa and **B** 925 HPa calculated using data from the Norwegian Meteorological Institute (Box 320, Blindern, Oslo 3, Norway N-0314) that originate from Chernobyl and four points 150 km away in each grid direction from 06 GMT, April 26, to 00 GMT, May 2; major position marks at 6-hr increments.

The deposition of ^{137}Cs in Finland, Sweden, and Norway was strongly correlated with observed precipitation amounts, showing that the deposition at this distance from the source was mainly wet deposition. Figure 7-9 shows the deposition pattern of ^{137}Cs over Norway. The concentration of radioactive material in precipitation varied considerably. Figure 7-10 shows the estimated concentration of ^{137}Cs in precipitation over Sweden. Approximately 10% of the released ^{137}Cs seems to have been deposited over Sweden where the initially released material first encountered precipitation (Persson et al. 1987).

After initially being dispersed towards the Nordic countries, the wind shifted, gradually transporting the material over large parts of Europe. Figure 7-11 shows the estimated day-by-day outlines of the Chernobyl plume (Smith 1987).

The radioactive plume entered southern England in the early hours of May 2. This material had been transported more than 4,000 km from the source. Convective rain caught up with the plume over the United Kingdom, causing significant deposition of radioactive material. The deposition pattern over the United Kingdom is shown in Figure 7-12.

Parts of the plume were transported to higher altitudes and were caught by stronger and gradually predominantly westerly winds. The plume crossed the Asian continent and the Pacific Ocean, arriving at the west coast of North America about 13 days after the initial release (Fig. 7-13).

7.1. CONCLUSIONS

1. In case of such an accident, the deposition close to the source (within 100 km) will always be large due to the effective dry deposition of radioactive particles that are large enough to have a significant gravitational settling velocity.

2. In general the dry-deposition pattern will depend much on the particle spectrum, which can only be very roughly estimated during an accident.

3. Deposition of ^{137}Cs far from the source (more than 100 km) seems to occur mainly as wet deposition.

4. The deposition is greatly enhanced when the radioactive plume encounters precipitation. This can produce local deposition maxima several thousand kilometers from the source.

5. Diagnostic estimates of the dispersion pattern from the Chernobyl accident made by different scientific groups show differences, probably due to differences in wind-field analyses. Predictive models for dispersion and deposition of radioactive material, in conjunction with numerical weather prediction routines, are now being developed. Considerable uncertainties are to be expected in predictions of radioactive deposition.

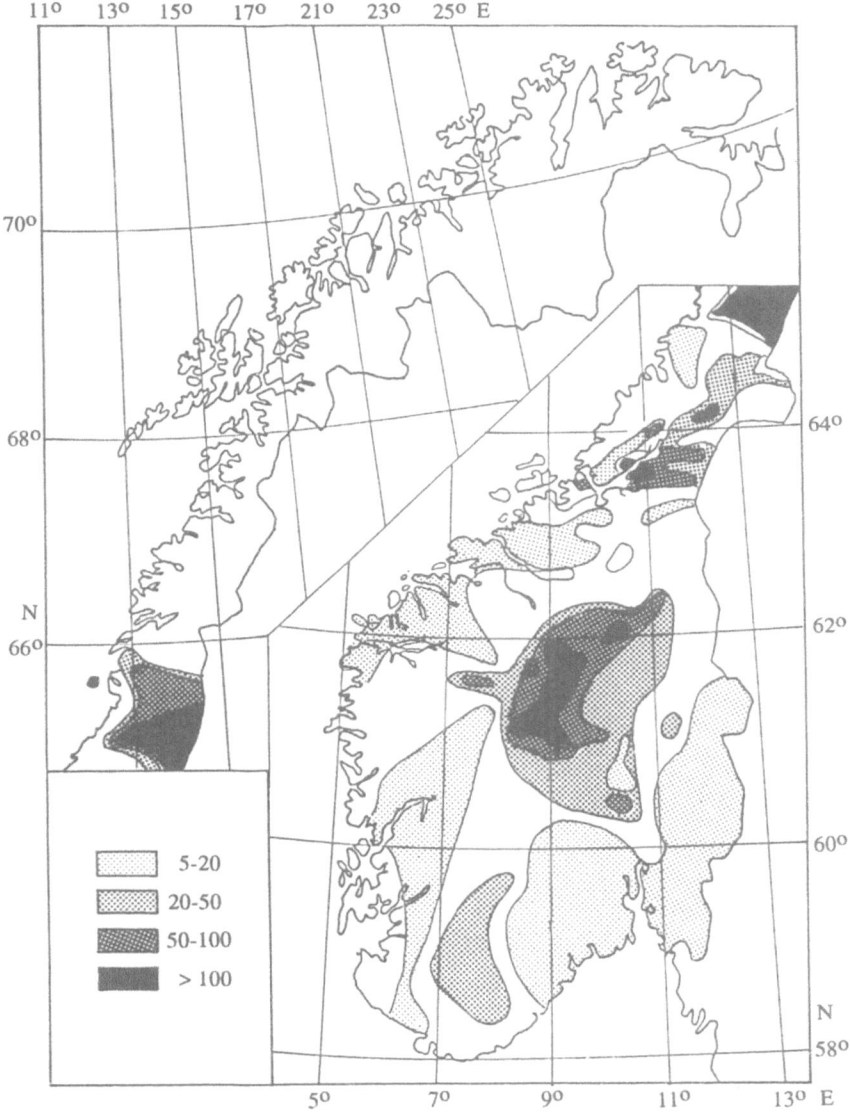

Figure 7-9. Total deposition (kBq/m^2) of ^{137}Cs and ^{134}Cs over Norway plotted using data from the Statens Institute for Stralehygiene.

158

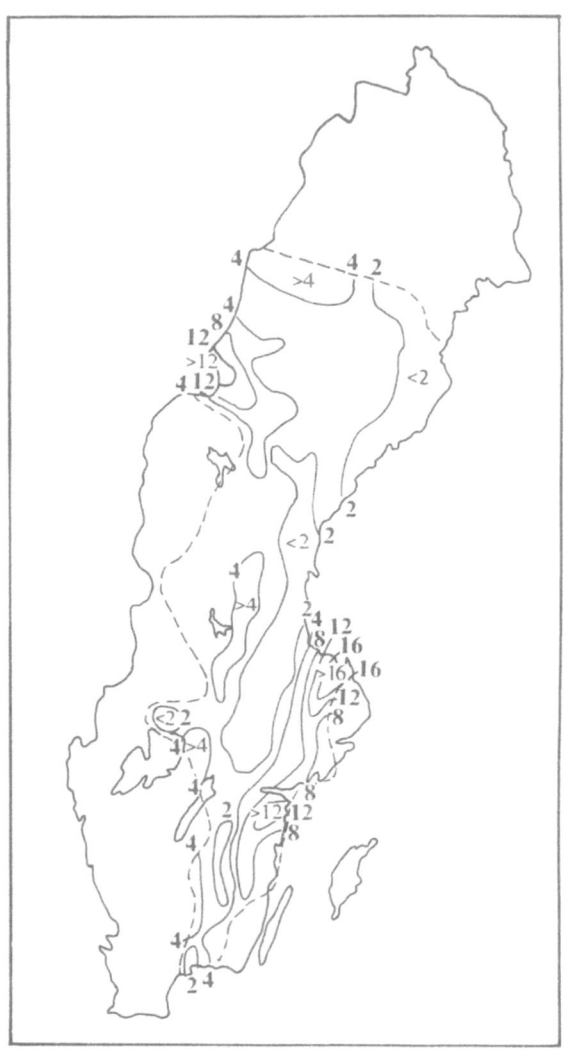

Figure 7-10. Estimated concentrations of ^{137}Cs in precipitation over
Sweden (Persson et al. 1987).

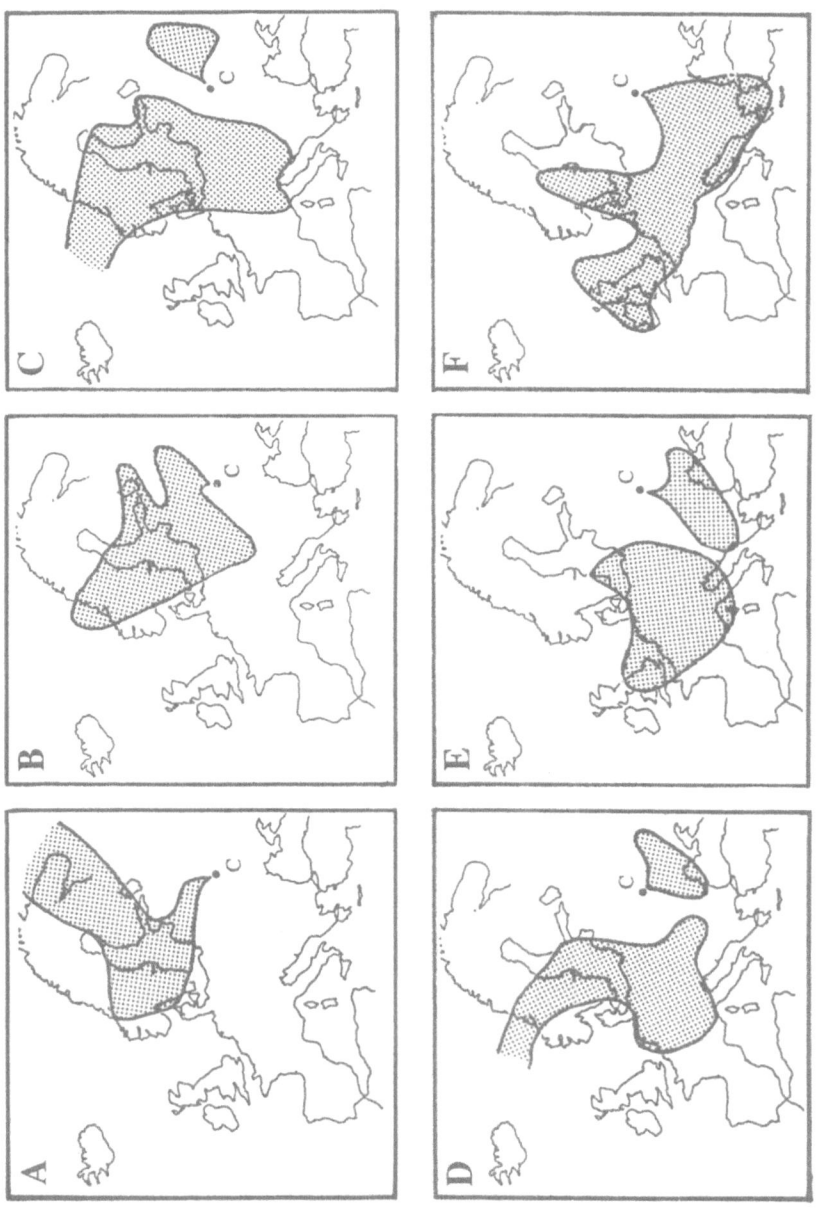

Figure 7-11. Day-by-day outlines (A April 28, B April 29, C April 30, D May 1, E May 2, F May 3) of the Chernobyl plume inferred from radiological data and 925 HPa trajectories (Smith 1987).

Figure 7-12. Estimated total (wet and dry) deposition (1000 Bq/m^2) of ^{137}Cs over the United Kingdom from the passage of the Chernobyl plume.

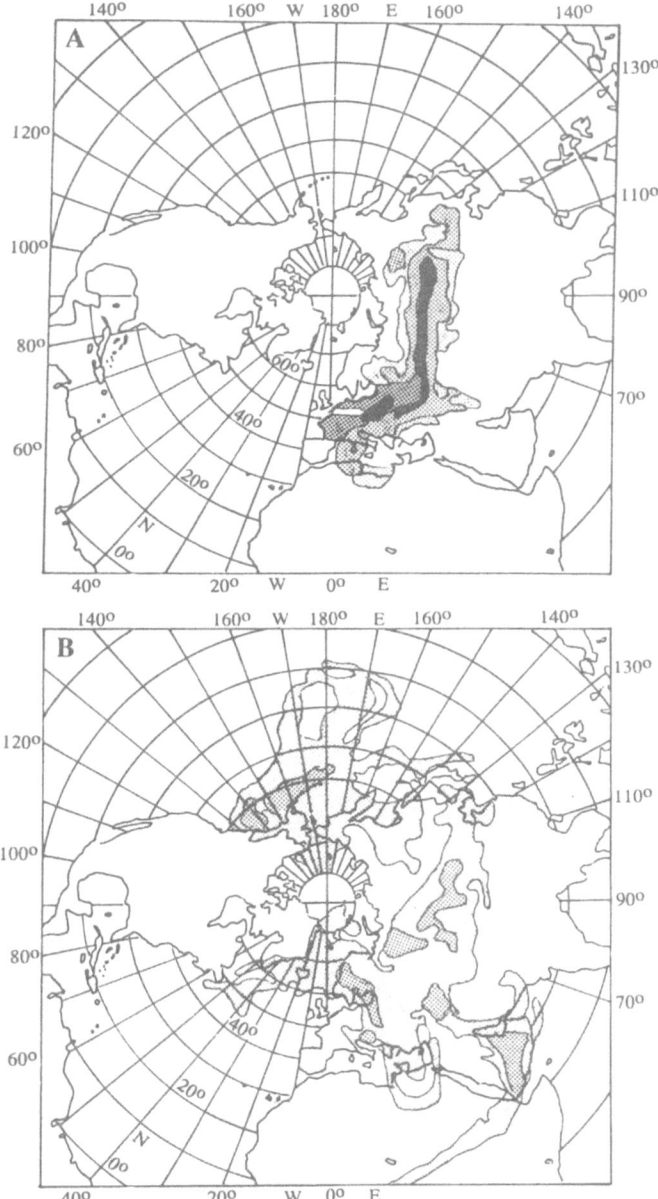

Figure 7-13. Calculated positions of the Chernobyl plume across the Northern Hemisphere on A May 1, 1986, after 5 days, and B May 9, 1986, after 13 days (Pudykiewicz 1988).

7.2. REFERENCES

Izrael, Y. A., V. N. Petrov, D. A. Severov. 1987. Radioactive fallout simulation in the close-in area of the Chernobyl nuclear power plant. Soviet Meteorol. Hydrol. (Special English translation), 17 pp.

Persson, C., H. Rodhe, and L.-E. De Geer. 1987. The Chernobyl accident-- A meteorological analysis of how radionucleides reached and were deposited in Sweden. Ambio 16:20-31.

Pudykiewicz, J. 1988. Numerical simulation of the transport of radioactive cloud from the Chernobyl nuclear accident. Tellus 40B:241-259.

Smith, F. B. 1987. The environmental consequences of the Chernobyl nuclear reactor accident. Provisional Rept., Bracknell, England: Meteorological Office, 41 pp.

8. GLOBAL SOURCE STRENGTH AND LONG-RANGE ATMOSPHERIC TRANSPORT OF TRACE ELEMENTS EMITTED BY VOLCANIC ACTIVITY

Patrick Buat-Ménard
Centre des Faibles Radioactivités
Laboratoire Mixte CNRS-CEA
F-91198 Gif-sur-Yvette, CEDEX, FRANCE

8.1. INTRODUCTION

It has long been recognized that volcanic eruptions significantly affect the earth's environment and particularly the chemistry of the atmosphere (Devine et al. 1984, Kennett and Thunell 1975, Lamb 1970, Rampino and Self 1982). Depending on the type and the strength of an eruption, the release of volcanic gases and particles can affect the atmosphere on a local or regional scale or on a global scale. The more violent eruptions can occur every several decades or not for thousands of years and they can lead to important physical and chemical changes in the stratosphere. Large amounts of gas (mainly sulphur dioxide and halogenated acids) and dust (generally silicate ash) are released and may remain in the stratosphere for several weeks or for several years (Cadle et al. 1976).

Although still controversial, it is generally accepted that the stratospheric aerosol layer is formed mostly by the conversion of SO_2 into sulfuric acid droplets, which have a much longer residence time than ash particles. This reduces atmospheric transmissivity and consequently may affect surface-air temperatures. For example, the mean-temperature negative anomalies of about negative 1° C recorded in Europe and the United States in the spring and summer of 1816 are attributed to the effects of the 1815 Tambora eruption (Rampino and Self 1982).

The role of volcanoes in atmospheric chemistry goes far beyond the role of the major but rare volcanic eruptions. Moderate events as well as noneruptive activity are now believed to influence the global atmospheric budgets of several trace substances, both gaseous and particulate. Volcanoes are likely to be a major natural source for sulfur as well as for the volatile heavy metals and metalloids. Volcanoes enrich volatile elements on aerosols by rather large factors in roughly the same way that high-temperature industrial activity does. It is, therefore, essential to assess the magnitude of volcanic sources as well as their long-range transport pathway through the troposphere if we want accurately to assess the effects of anthropogenic perturbations on several atmospheric biogeochemical cycles (Cu, Zn, As, Se, Hg, Pb).

Because the composition of volcanic plumes varies with time as well as from one volcano to another, there are large uncertainties in the current estimates of volcanic source strengths. However, in the last few years, because of the development of better sampling and analytical techniques, the available data base has increased considerably, enabling more accurate order-of-magnitude estimates to be obtained. In this chapter I

A. H. Knap (ed.), The Long-Range Atmospheric Transport of Natural and Contaminant Substances, 163–175.
© 1990 Kluwer Academic Publishers.

briefly review the estimates for S and heavy metals and discuss possible approaches for assessing the long-range transport and deposition of substances emitted by volcanoes.

8.2. THE EFFECT OF VOLCANOES ON THE ATMOSPHERIC SULFUR BUDGET

In 1972, Kellogg et al. calculated the global volcanic sulfur emission from an estimate of the mean volume of emitted lavas and ashes, their density, and their SO_2 content. They estimated that the emission rate is 0.75 Tg S/yr and considered this to be negligible compared to the global sulfur emission (~100 Tg S/yr). However, this amount is now considered to be a gross underestimate since it neglects the SO_2 released during noneruptive periods. With the development of new methods for directly measuring the sulfur-emission rate of volcanoes, such as remote-sensing correlation spectrometry (Stoiber and Jepsen 1973), more reliable estimates were available at the time of our NATO ARW in Bermuda. The best evaluation by extrapolating direct measurements is by Berresheim and Jaeschke (1983) for H_2S, SO_2, and SO_4^{2-}. Their assessment is based on all the measurements available in the literature and a suitable classification of the different types of volcanic activity, both eruptive and noneruptive, as well as the frequency of each activity class (Table 8-1). They estimated that the global volcanic-sulfur release is 12 Tg S/yr, with the values for SO_2, H_2S, and SO_4^{2-} being 1.5 Tg/yr, 1 Tg/yr, and

Table 8-1. Volcanic activity classes and characteristic SO_2 emission rates.

Category	SO_2 (10^6 g/day)	F*	Number of Volcanoes
Preeruptive	$500-10^3$		
Intraeruptive	$500-10^3$		
Posteruptive	$50-500$		365
Extraeruptive	$0-10$		100
Eruptive Activity			
Icelandic	10^3	---	
Hawaiian	10^3	10.46	
Strombolian	$300-10^3$	760.00	
Vulcanian	10^4	9.40	
Vesuvian	7×10^4	4.13	
Plinian	$1-2 \times 10^5$	0.53	
Pelean	$1-2 \times 10^5$	0.12	
Krakatoan	?	---	

Source: Berresheim and Jaeschke (1983).

*F = Frequency of activity classes a year with the average number of erupting volcanoes being 55 a year; values calculated for 1961-1979.

10 Tg/yr, respectively. Their most interesting finding is that 90% of
the SO_2 is emitted during noneruptive activities, such as permanent fuma-
rolic activity, and injected directly into the troposphere.

Using a completely different approach, Lambert et al. (1988) estimated that the global volcanic output of SO_2 is on the order of 50 Tg/yr,
which is somewhat higher than the estimate of Berresheim and Jaeschke
(1983). Lambert et al. normalized the SO_2 flux to the global volcanic
^{210}Po flux from simultaneous measurements of SO_2 and ^{210}Po in volcanic
gases and aerosols.

It seems from these estimates that volcanoes contribute at least 10%
of the present global atmospheric sulfur budget. Since about half of
this budget is due to human activities, it follows that the volcanic
source strength is comparable in magnitude to other major natural
sources, such as biospheric emissions from the sea and land. In the
troposphere of the Southern Hemisphere, natural sources of S probably
dominate anthropogenic sources (Chapter 4, p. 87). Since the influence
of natural biospheric sources is primarily confined below the planetary
boundary layer and since the summit of many volcanoes is above this
layer, a significant fraction of the sulfur burden of the free troposphere of the Southern Hemisphere must come from volcanoes. How the
global volcanic source strength of S has varied in the past is extremely
difficult to assess. Records of volcanic ash in deep-sea sediments suggest that volcanism has dramatically increased during the last two million years (Kennett and Thunell 1975). Unfortunately, today such a
record from sediments is almost useless for S, which is most soluble in
seawater, because the ratio of S to ash in volcanic plumes varies considerably from one type of eruption to another (Rampino and Self 1982).
As I discuss in the following section, past changes in the volcanic sulfur budget caused by major volcanic eruptions can, however, be assessed
from polar ice-core analyses.

8.3. THE VOLCANIC SULFUR FALL-OUT IN POLAR SNOW AND ICE

Polar ice cores have yielded much information on past volcanism, not,
however, because of the deposition of ash particles. Ash particles are
generally too large or too few to reach these remote environments in
amounts that would significantly raise the continental dust background.
Indeed, elemental and mineralogical analyses have shown that most insoluble microparticles in polar ice are aluminosilicates originating from
arid areas (De Angelis et al. 1987). However, volcanic acids deposited
shortly after large volcanic eruptions have been found in Greenland and
Antarctic ice by Delmas and Boutron (1980), Delmas et al. (1985), Hammer
(1977), Hammer et al. (1980), and Legrand and Delmas (1987). These studies have allowed one to reconsider the history of explosive activity
worldwide as well as the history of changes in the global sulfur budget.
These researchers have used methods based on electrical conductivity and
acidity as well as sulfate measurements in layers generally of less than
one year's accumulation. Their results indicate acidity peaks lasting
for one or two years superimposed over background values. Thanks to
accurate dating of the snow layers, most of these observed peaks can be

related to historically recorded volcanic events. Figure 8-1 gives a
220-year continuous record of volcanic H_2SO_4 in the Antarctic ice sheet.
The longest event recorded probably spans more than four years (Peak 4)
and corresponds to the sum of the effects of the eruptions of Krakatoa
(1883) and Tarawera (1886).

Because of the spatial variations of the deposition flux of any
given event, a global fall-out of H_2SO_4 cannot be extrapolated from the
concentration peaks. Also, volcanic signals are better defined in the
central areas of the Antarctic than in the more coastal areas. This
might be because of a variable fall-out of volcanic debris from the stratosphere; Legrand and Delmas (1987) have shown, however, that, at least
for the Agung and Tambora eruptions, such spatial variations are modu-
lated mainly by the snow-accumulation rate.

By comparing data from the Greenland and Antarctic ice sheets, the
influence of a single volcanic eruption can be evaluated. Large volcanic
eruptions at low latitudes that disperse debris into both hemispheres are
of global concern. By assuming a linear relationship between the deposited flux and the snow-accumulation rate, Legrand and Delmas (1987)
quantitatively compared the Tambora and Agung global fallouts. Although
they reported a balanced distribution for Tambora, there was probably a
higher input to the southern stratosphere following the Agung event,
which agreed with the stratospheric sulfate measurements (Castleman et
al. 1974). As pointed out by Legrand and Delmas (1987), these differences are probably related to the altitudes at which the debris was
injected into the stratosphere. Other more moderate eruptions, generally
at middle and high latitudes, have been recorded only in Greenland or in
Antarctica (Delmas et al. 1985, Hammer et al. 1980) and appear to have
perturbed the troposphere only on a hemispheric basis.

The total fallout from eruptions of global concern can be calculated
from polar ice-sheet values by using the global fallout patterns observed
in the 1960s for nuclear-test debris (Hammer et al. 1980). Over the last
200 years, these values have ranged from 20-50 Tg S per eruption, the
highest value of 150 Tg S being for the Tambora eruption (1815). This
indicates that a major volcanic eruption can significantly affect the
global sulfur budget, especially in the Southern Hemisphere, for as long
as a year or two with resulting increases in precipitation acidity.

Although SO_2 may be the major precursor for sulfuric acid droplets
in the stratosphere (Cadle et al. 1979), Rasmussen et al. (1982) reported
that other sulfur compounds, COS and CS_2, adsorbed on ash particles during major volcanic eruptions and eventually released into the upper
atmosphere can also be important contributors. However, Belviso (1986)
and Khalil and Rasmussen (1984) estimated that the direct release of COS
and CS_2 from volcanoes is much smaller than the production of these compounds from terrestrial and marine biospheres.

Although H_2SO_4 is the most frequent contributor to the acidity peaks
recorded in the ice cores, halogenated acids are also significant. Based
on the chemical analyses of pumice erupted from the volcano, Sigurdsson
and Carey (1987) estimated that the Tambora eruption emitted more HCl and
HF than H_2SO_4. These authors also pointed out that the HCl emission is
about 100 times the current nonvolcanic stratospheric chlorine budget.

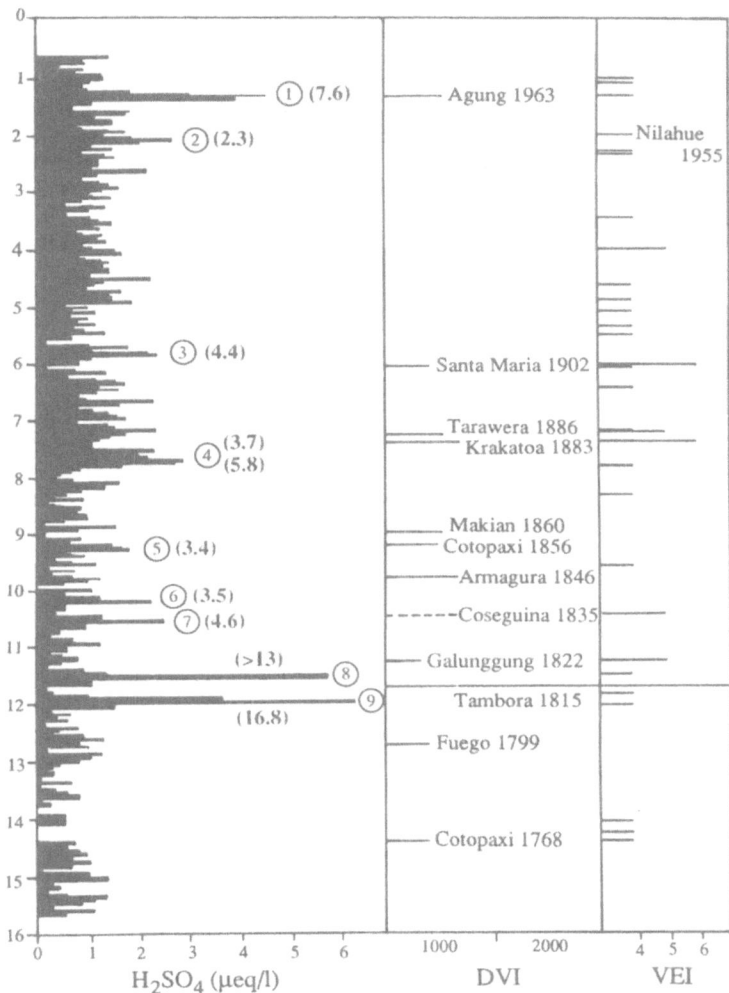

Figure 8-1. Profile of H_2SO_4 concentrations from Dome C (74°40'S, 125°10'E) from the past 220 years compared to the records of explosive volcanic eruptions south of 20°N; **VEI** is the volcanic explosivity index and **DVI** is the dust-veil index. The major H_2SO_4 spikes, **1-9**, are obviously of volcanic origin; the accompanying values (kg/km²) are for concentrations deposited at Dome C (adapted from Legrand and Delmas 1987).

It is, therefore, possible that great volcanic eruptions significantly affect the ozone layer.

Finally, the ice-core data support the stratospheric evidence (Cadle et al. 1976) that, on a global scale, ash particles are transported less efficiently than sulfuric acid droplets. In any event, the impact of volcanism on the chemistry of the global atmosphere is likely to be a sporadic phenomenon. Although Antarctic ice-core measurements of SO_4 covering the entire last climatic cycle indicate major long-term changes in the H_2SO_4 content of the Antarctic ice sheet (Legrand et al. 1988), Legrand et al. reported that these changes are primarily from H_2SO_4 emanating from marine biogenic sources.

8.4. VOLCANIC OUTPUT OF TRACE ELEMENTS INTO THE ATMOSPHERE

Over the last 15 years, there have been several studies of the chemical composition of volcanic plumes in the troposphere based on ground-based and aircraft measurements (Bergametti et al. 1984, Buat-Ménard and Arnold 1978, Patterson and Settle 1987, Varekamp and Buseck 1981, Zoller 1984, Zoller et al. 1983). At the temperature of the ambient atmosphere, most of the elements studied are attached to aerosol particles, such as ash particles and S- and Cl-rich particles, except for Hg which is mostly in the gas-phase. Their general findings were that airborne particles from volcanoes have a chemical composition markedly different from that of the parent magma. Volatile elements on such aerosols are enriched by several orders of magnitude and are found on the smallest size fraction (<1 μm), which has the greatest ability to undergo long-range transport. These enriched elements include several potentially toxic heavy metals and metalloids (such as Zn, Cu, Au, Pb, As, Cd, Sb, Se) emitted by both low- and high-temperature anthropogenic activities.

It is, however, extremely difficult to ascertain how much emissions from volcanoes actually contribute to the global tropospheric and stratospheric burden of such elements. Not only do we not know the magnitude of total particulate emissions but also the composition of the emissions varies considerably from one volcano to another and even during different eruptive periods of the same volcano. For example, the relative abundances to S of enriched elements, such as Cd, Zn, Se, As and Hg, as measured in plumes from several volcanoes span over two orders of magnitude (Table 8-2). Large variations have also been reported for various stages of eruptions of Mt. Etna (Buat-Ménard and Arnold 1978) and Mt. Augustine (Lepel et al. 1978).

Such changes in the composition of the emissions are certainly related to the physical and chemical processes occurring in the magma. Lambert et al. (1985) designed an outgassing model to explain the variability of Mt. Etna gaseous emissions of the volatile radionuclides $210Pb$, $210Bi$, and $210Po$. This model assumes that the emission of volatile materials depends not only on their partitioning between condensed and gaseous phases but also on the turnover of superficial magma bringing new nondegassed materials to the surface. Much higher enrichments are expected when emissions originate from deep nondegassed magma, such as during cataclysmic volcanic eruptions.

Table 8-2.   Element-to-sulfur weight ratio for various volcanic plumes.

	Cd	Zn	Se	As	Hg
United States					
Mt. St. Helens	5.6×10^{-6}	10×10^{-4}	1×10^{-5}	80×10^{-5}	9×10^{-5}
Costa Rica					
Poas	3.2×10^{-6}	0.5×10^{-4}	1×10^{-5}	0.3×10^{-5}	10×10^{-5}
Arenal	270×10^{-6}	2.4×10^{-4}	7.6×10^{-5}	4×10^{-5}	30×10^{-5}
Mexico					
Calima	23×10^{-6}	0.34×10^{-4}	2.1×10^{-5}	1×10^{-5}	---
El Chichon	---	0.25×10^{-4}	0.5×10^{-5}	25×10^{-6}	22×10^{-5}
Italy					
Mt. Etna	11×10^{-6}	9×10^{-4}	40×10^{-5}	7×10^{-5}	5×10^{-5}
Antarctica					
Mt. Erebus	120×10^{-6}	7×10^{-4}	1×10^{-5}	20×10^{-5}	2×10^{-5}
Indonesia					
Mt. Merapi	60×10^{-6}	15×10^{-4}	20×10^{-5}	70×10^{-5}	12×10^{-5}

Sources: For the United States, Costa Rica, and Mexico, Zoller (1984); for Italy, Buat-Ménard and Arnold (1978); for Indonesia, Buat-Ménard (unpublished data available from Dr. Buat-Ménard, CNRS-CEA, F-91198, Gif-sur-Yvette, Cedex, France).

The geology of a volcanic system is another variable that must be considered. For example, Zoller et al. (1983) reported emissions highly enriched in iridium from the January 1983 eruption of Kilauea although they found no Ir enrichment in six other active volcanoes, possibly because Hawaiian volcanoes are more likely to be fed by magma from very deep within the earth, as deep even as the mantle. They hypothesize that this type of emissions accounts for the Ir anomaly at the Cretaceous-Tertiary boundary layer, which is usually explained as the result of a cataclysmic meteoritic impact.

Palais (1987) and Phelan (1983) studied two distinct volcanic groups in their attempts to explain why the composition of emissions varies so much from one volcano to another. The first group consisted of volcanoes, such as Heimaey, Etna, Arenal, Poas, and Calima, in which the parent magma is low in silica and total alkali; i.e., basaltic or andesitic. The second group, in which the parent magma is high in silica and total alkalis; i.e., trachyandesitic, dacitic, or phonolitic, included such volcanoes as El Chichon, Mt. St. Helens, and Mt. Erebus. When the composition of the emissions was normalized to that of Se, these researchers found that the abundances of As, In, Sb, and In relative to Se in the plumes from the volcanoes in the first group are generally higher by at least an order of magnitude. Such differences may allow one to track the origin of volcanic aerosols through the troposphere.

An accurate estimate of global volcanic emission rates of trace elements to the atmosphere cannot be made from the available data base. Up to the time of our meeting in Bermuda, most published estimates were from element-to-S ratios and global volcanic sulfur emissions. However, in spite of the enormous uncertainties, some general geochemical conclusions could be drawn.

Globally, anthropogenic sources overwhelm volcanic sources for many enriched elements (Zoller 1984). For example, Patterson and Settle (1987) estimated that industrial inputs of Pb are on the order of 3×10^{11} g/yr although the volcanic source strength is only approximately 2×10^9 g/yr. Although, the industrial inputs of Cd, Zn, and As probably also dominate over inputs from volcanoes, a volcanic source of Se and Cu appears significant, especially in the Southern Hemisphere.

By comparing volcanic emissions to other natural sources, Zoller (1984) assessed the impact of volcanism on natural cycles in the atmosphere. For Pb, volcanic emissions are of the same order as the natural soil-dust inputs. For such elements as As (Walsh et al. 1979) and Se (Mosher and Duce 1987), volcanic emissions are comparable in magnitude to the low-temperature release from biospheric sources. Unfortunately, because of the uncertainties in the current estimates of emissions from natural sources, the geochemical impact of each individual source cannot be assessed. However, there are at least two trace elements (Bi and ^{210}Po) for which volcanic emissions appear to be the dominant global source. The global volcanic source strength of Bi is on the order of 1 to $5 \ 10^9$ g/yr—10 times more than emissions from other natural or anthropogenic sources (Lambert et al. 1988, Lee et al. 1985, Patterson and Settle 1987). Lee et al. (1986) also stated that the present-day Bi cycle in the world's oceans is probably controlled by aeolian inputs of volcanic origin.

^{210}Po, the last radioactive nuclide of the $^{238}U/^{226}Ra$ series, is generally present in volcanic plumes in concentrations several orders of magnitudes higher than in the usual atmosphere (Bennett et al. 1982, Lambert et al. 1985). Lambert et al. (1982), using a global atmospheric budget of ^{222}Rn and its decay products, demonstrated that volcanoes represent about 50% of the total global source of ^{210}Po in the atmosphere with an uncertainty of less than a factor of two. Further, this isotope can also be assumed to be entirely volatilized from outflowing lava since fresh lava samples do not contain ^{210}Po (Lambert et al. 1985). Consequently, Lambert et al. (1988) estimated the global volcanic source strengths of other trace elements by comparing the volatility of an element to that of ^{210}Po, with a knowledge of concentrations in both air and fresh lava. For instance, the global source strength of lead of 2.5×10^9 g/yr can be derived from the volatility of Pb, which is similar to that of ^{210}Pb, which in turn is known to be 1.5% that of ^{210}Po. This value for the global source strength of lead agrees with other estimates based on normalization to sulfur. By extrapolating the results obtained for Mt. Etna, these researchers evaluated global volcanic emissions for nine elements (Table 8-3) with an uncertainty estimated to be a factor of 3. Their flux calculation was based on the following equation:

$$\phi_x = \phi_{Pb} \times \frac{\varepsilon_x \ C_x}{\varepsilon_{Pb} \ C_{Pb}}$$

Where ϕ is the flux of a metal, C is the average concentration in basalts, and ε its volatility relative to that of ^{210}Po. Although more data are needed from other volcanoes, the values in Table 8-3 are probably the best estimates for global volcanic emissions of the trace elements available at the time of our NATO ARW.

It is worth considering whether polar ice and snow data can be used, as they have been for sulfur, to provide an historical record of global volcanism for trace elements. Although the volcanic gases found in many ice layers attest to the presence of trace elements of volcanic origin, it is extremely difficult to tell how much volcanic emissions actually contributed to the elemental concentrations found in the ice sheets. Most previous studies have been hampered by inadequate sampling and analytical techniques (Wolff and Peel 1985). Moreover, the levels of many trace elements found in present-day polar snows are probably the result of global pollution. When ancient, pre-industrial ice is used, it is almost impossible to distinguish between elements from different natural sources. Up to now this differentiation has only been achieved for Pb (Boutron and Patterson 1986). Boutron and Patterson showed that about half the lead content of Antarctic ice from the Holocene period is probably from volcanoes and the other half from soil dust. However, the enhanced concentrations of lead found for the last glacial period are not from increased volcanic inputs but from enhanced soil-dust transport.

If high-quality data are generated for other trace elements, more information on paleovariations of global volcanic emissions can be obtained. Special attention should be given to measuring concentrations of Bi since this element seems to be, according to present data, emitted

Table 8-3. Global volcanic emissions of metals.

Element	Relative Volatility (ε)* $(\times 10^{-4})$	Average Concentration in Basalts (ppm)	Flux** $(10^9$ g/yr)
Pb	100	5	2.5
Bi	2000	0.15	1.5
Cd	1000	0.2	1
Cu	30	100	15
Zn	10	100	5
Al	0.2	87600	88
Mg	0.2	45000	45
Na	2	19400	194
K	6	8300	250

Source: Lambert et al. (1988).

*ε= 1 for ^{210}Po; see text.

**Assuming a global flux of ^{210}Po = 5 $\times$ 10^6 Ci/yr, a flux of ^{210}Pb = 5 $\times$ 10^2 Ci/yr, and a specific activity of ^{210}Pb of 0.4 dpm per mg of lead.

into the atmosphere primarily from volcanoes. Also, as pointed out above, platinum metals, such as iridium, should be investigated. The concentration levels of heavy metals during episodes of high acidity, which are related to the global fallout of stratospheric sulfur from volcanoes, should also be determined.

Systematic studies of the trace-element content of volcanic plumes should be undertaken in the near future. Samples need to be collected during both eruptive and noneruptive periods. Such data, coupled with the application of long-range-transport models, would certainly allow better predictions of the global volcanic fallout over oceans and continents. An alternative approach would be to use unambiguous tracers of volcanic aerosols.

In the last section of this chapter, I summarize some results obtained in our laboratory that are based on the measurements of the $^{210}Po/^{210}Pb$ ratio in ambient aerosol samples collected over the Mediterranean Sea.

8.5. $^{210}Po/^{210}Pb$ AS A TRACER OF THE LONG-RANGE TRANSPORT OF VOLCANIC AEROSOLS

If all volcanic contributions are ignored, the $^{210}Po/^{210}Pb$ activity ratio in the atmosphere is generally close to 0.1. And, since such a ratio can be as high as 100 in volcanic plumes, it would follow that measured $^{210}Po/^{210}Pb$ ratios higher than the background value reflect the presence of volcanic aerosols.

During several cruises in the Mediterranean Sea, Dulac et al. (1987) found values for this ratio well above 0.1 in several samples collected >100 km from Mt. Etna and Mt. Stromboli. The backward air-mass trajectories that my colleagues and I (1987) analyzed strongly suggest that the sampled air masses are ''contaminated'' by inputs from these volcanoes. By normalizing to excess ^{210}Po and using earlier data on the elemental composition of the volcanic plumes (Bergametti et al. 1984, Buat-Ménard and Arnold 1978), we predicted, for a given sample, the atmospheric concentrations of Pb, Zn, Cu and Cd from volcanic activity. We assumed that no elemental fractionation occurs during long-range atmospheric transport. Such an hypothesis is supported by measurements of the mass-size distributions of the concentrations of the tracers and the metals of interest associated predominantly with the smallest particle-size fraction. Our comparisons between the predicted and the measured concentrations showed that the volcanic contribution is always negligible for Pb and Zn but can be dominant for Cu and Cd. For this atmospheric environment, the available climatological information based on an 18-year data base, forward trajectories from Mt. Etna (Martin et al. 1984), and the mean S and ^{210}Po output from this volcano (Martin et al. 1986) all suggested that the volcanic component of the atmospheric delivery of Cu and Cd is significant over the Central and Western Mediterranean. Such a tracer approach should be applied, whenever possible, to all presently active volcanoes around the world together with air-mass trajectory models or remote-sensing data on volcanic plumes.

8.6. SUMMARY

In this paper, I have reviewed the available data on the emission rates of sulfur and selected trace elements emitted to the atmosphere by volcanic activities. Unfortunately, the observed variability of the composition of volcanic plumes from one volcano to another, and during the various eruptive stages of a given volcano, does not yet allow for the global volcanic source strength to be accurately assessed. At best, order-of-magnitude estimates indicate that on a yearly basis the emission rates due to noneruptive volcanic activity are likely dominant in the troposphere. However, sporadic, but intense, volcanic eruptions that affect the stratosphere likely overwhelm more or less steady volcanic inputs during 1- to 2-year periods. Polar ice-core records suggest that such events may significantly influence the global S budget. The combined use of tracers, such as ^{210}Po or Bi and air-mass trajectory models, or remote sensing of volcanic plumes may be useful tools for assessing the long-range transport and deposition of elements of volcanic origin as well as the relative importance of volcanic emissions compared to other natural and anthropogenic sources.

8.7. REFERENCES

Belviso, S. 1986. Apports volcanique et océanique d'oxysulfure de carbone a l'atmosphere. Ph.D. dissert., Univ. of Paris, 131 pp.

Bennett, J. T., S. Krishnaswami, K. K. Turekian, W. G. Melson, and C. A. Hopson. 1982. The uranium and thorium decay series nuclides in Mt. St. Helen's effusives. Earth Plantary Sci. Ltrs. 60:61-69.

Bergametti, G., D. Martin, J. Carbonnelle, R. Faivre-Pierret, and R. Vie Le Sage. 1984. A mesoscale study of the elemental composition of aerosols emitted from Mt. Etna volcano. Bull. Volcanol., 47-4: 1107-1116.

Berresheim, H., and W. Jaeschke. 1983. The contribution of volcanoes to the global atmospheric sulfur budget. J. Geophys. Res. 88:3732-3740.

Boutron, C. F., and C. C. Patterson. 1986. Lead concentration change in Antarctic ice during the Wisconsin/Holocene transition. Nature 323:222-225.

Buat-Ménard, P., and M. Arnold. 1978. The heavy metal chemistry of atmospheric particulate matter emitted by Mt. Etna volcano. Geophys. Res. Ltrs. 5:245-248.

Buat-Ménard, P., D. Martin, M. Pennisi, N. Risler, B. Ardouin, M. Arnold, M. F. LeCloarec, and G. Lambert. 1987. ^{210}Po/^{210}Pb as a tracer of volcanic aerosols over the Mediterranean Sea. Paper presented at 6th Int. Symp., Commission. Atmos. Chemistry and Global Pollut. of Global Atmos. Chemistry, Trent University, Peterborough, Ontario, Canada, August.

Cadle, R. D., C.S. Kiang, and J.-F. Louis. 1976. The global scale dispersion of the eruption clouds from major volcanic eruptions. J. Geophys. Res. 81:3125-3132.

Cadle, R. D., A. L. Lazrus, B. J. Huebert, L. E. Heidt, W. I. Rose, Jr.,
    D. C. Woods, R. L. Chuan, R. E. Stoiber, D. B. Smith, and R. A.
    Zielinski. 1979. Atmospheric implications of studies of Central
    American volcanic eruption clouds. J. Geophys. Res. 84:6961-6968.
Castleman Jr., A. W., H. R. Munkelwitz, and B. Manowitz. 1974. Isotopic
    studies of the sulfur component of the stratospheric aerosol layer.
    Tellus 26:222-234.
De Angelis, M., N. I. Barkov, and V. N. Petrov. 1987. Aerosol concentra-
    tions over the last climatic cyle (160 Kyr) from an Antarctic ice
    core. Nature 325:318-321.
Delmas, R., and C. Boutron. 1980. Are the past variations of the
    stratospheric sulfate burden recorded in central Antarctic snow and
    ice layers? J. Geophys. Res. 85:5645-5649.
Delmas, R. J., M. Legrand, A. J. Aristarain, and F. Zanalini. 1985.
    Volcanic deposits in Antarctic snow and ice. J. Geophys. Res. 90:
    12901-12920.
Devine, R. J., H. Sigurdsson, A. N. Davis, and S. Self. 1984. Estimates
    of sulfur and chlorine yield to the atmosphere from volcanic erup-
    tions and potential climatic effects. J. Geophys. Res. 89:6309-6325.
Dulac, F., P. Buat-Ménard, M. Arnold, V. Ezat, and D. Martin. 1987.
    Atmospheric input of trace metals to the western Mediterranean sea:
    1. Factors controlling the variability of their atmospheric concen-
    trations. J. Geophys. Res. 92:8437-8453.
Hammer, C. V. 1977. Past volcanism revealed by Greenland ice sheet impu-
    rites. Nature 270:482-496.
Hammer, C. V., H. B. Clausen, and W. Dansgaard. 1980. Greenland ice
    sheet evidence of past glacial volcanism and its climatic impact.
    Nature 288:230-235.
Kellogg, W. W., R. D. Cadle, E. R. Allen, A. L. Lazrus, and E. A.
    Martell. 1972. The sulfur cycle. Science 175:587-596.
Kennett, J. P., and R. C. Thunell. 1975. Global increase in quaternary
    volcanism. Science 187:497-503.
Khalil, M. A. K., and R. A. Rasmussen. 1984. Global sources, lifetimes,
    and mass balances of carbonyl sulfide (OCS) and carbon disulfide
    (CS_2) in the earth's atmosphere. Atmos. Environ. 18:1805-1813.
Lamb, H. H. 1970. Volcanic dust in the atmosphere with a chronology and
    assessment of its meteorological significance. Phil. Trans. Royal
    Soc. London 266:425-533.
Lambert, G., B. Ardouin, and G. Polian. 1982. Volcanic output of long-
    lived radon daughters. J. Geophys. Res. 87:11,103-11,108.
Lambert, G., M.-F. Le Cloarec, B. Ardouin, and J. C. Le Roulley. 1985.
    Volcanic emission of radionuclides and magma dynamics. Earth
    Planet. Sci. Ltrs. 76:185-192.
Lambert, G., M.-F. Le Cloarec, and M. Pennisi. 1988. Volcanic output of
    SO_2 and trace metals: A new approach. Geochim. Cosmochim. Acta
    52:39-42.
Lee, D. S., J. M. Edmond, and K. W. Bruland. 1986. Bismuth in the Atlan-
    tic and North Pacific: A natural analogue to plutonium and lead?
    Earth Planetary Sci. Ltrs. 76:254-262.
Legrand, M., and R. J. Delmas. 1987. A 220-year continuous record of
    volcanic H_2SO_4 in the Antarctic ice sheet. Nature 327:671-676.

Legrand, M., R. C. Lorius, N. I. Barkov, and V. N. Petrov. 1988. Vostok
    (Antarctica) ice core: Atmospheric chemistry changes over the last
    climatic cycle (160,000 yr). Atmos. Environ. 22:317-331.
Lepel, E. A., K. M. Stefansson, and W. H. Zoller. 1978. The enrichment
    of volatile elements in the atmosphere by volcanic activity: Augus-
    tine volcano, 1976. J. Geophys. Res. 83:6213-6220.
Martin, D., D. Cheymal, M. Imbard, and B. Strauss. 1984. Climatology of
    forward trajectories of Mt. Etna plume over an 18-year period.
    Bull. Volcanol. 47-6:1115-1123.
Martin, D., B. Ardouin, G. Bergametti, J. Carbonnelle, R. Faivre-
    Pierret, G. Lambert, M.-F. Le Cloarec, and G. Sennequier. 1986.
    Geochemistry of sulfur in Mt. Etna plume. J. Geophys. Res.
    91:12,249-12,254.
Mosher, B. W., and R. A. Duce. 1987. A global atmospheric selenium bud-
    get. J. Geophys. Res. 92:13,289-13,298.
Palais, J. M. 1987. Trace elements in volcanic emissions: Previous work
    and potential for future investigations. Paper presented, 6th
    Int. Symp., Commmission Atmos. Chemistry and Global Pollut. of
    Global Atmos. Chemistry, Trent University, Peterborough, Ontario,
    Canada, August.
Patterson, C. C., and D. M. Settle. 1987. Magnitude of lead flux to the
    atmosphere from volcanoes. Geochim. Cosmochim. Acta. 51:675-81.
Phelan, J. M. 1983. Volcanoes as a source of volatile trace elements in
    the atmosphere. Ph.D. dissert., Univ. of Maryland, College
    Park, 241 pp.
Rampino, M. R., and S. Self. 1982. Historic eruptions of Tambora (1815),
    Krakatau (1883), and Agung (1963): Their stratospheric aerosols and
    climatic impact. Quart. Res. 18:127-143.
Rasmussen, R. A., A. K. Khalil, R. W. Dalluge, S. A. Penkett, and B.
    Jones. 1982. Carbonyl sulfide and carbon disulfide from the erup-
    tions of Mt. St. Helens. Science 215:665-667.
Sigurdsson, H., and S. Carey. 1987. Tambora 1815 eruption: An event with
    global atmospheric impact. Maritime 3-4:8-10.
Stoiber, R. E., and A. Jepsen. 1973. Sulfur dioxide contributions to the
    atmosphere by volcanoes. Science 182:577-578.
Varekamp, J. C., and P. R. Buseck. 1981. Mercury emissions from Mount
    St. Helens during September 1980. Nature 293:555-556.
Walsh, P. R., R. A. Duce, and J. L. Fasching. 1979. Considerations of
    the enrichment sources and fluxes of arsenic in the troposphere.
    J. Geophys. Res. 84:1719-1726.
Wolff, E. W., and D. A. Peel. 1985. The record of global pollution in
    polar snow and ice. Nature 313:535-540.
Zoller, W. H. 1984. Anthropogenic pertubation of metal fluxes into the
    atmosphere. In Changing Metal Cycles and Human Health (J. O.
    Nriagu, ed.) Berlin:Springer Verlag, 27-41.
Zoller, W. H., J. R. Parrington, and J. M. Phelan-Kotra. 1983. Iridium
    enrichment in airborne particles from Kilauea volcano: January
    1983. Science 222:1118-1121.

9. THE LONG-RANGE TRANSPORT OF TRACE ELEMENTS: FOUR CASE STUDIES

Timothy D. Jickells, Rapporteur
School of Environmental Sciences
University of East Anglia
Norwich, United Kingdom

Richard Arimoto
Center for Atmospheric
  Chemistry Studies
Graduate School of Oceanography
The University of Rhode Island
Narragansett, RI   02882-1197

Leon Mart
Institut fuer Chemie der Kern-
  forschungsanlage Juelich GmbH
P. O. Box 1913
D-5170 Juelich
Federal Republic of Germany

Leonard A. Barrie
Atmospheric Environment Service
4905 Dufferin Street
Downsview, Ontario
Canada M3H 5T4

William T. Sturges
Department of Chemistry
Institute of Aerosol Science
University of Essex
Colchester
Essex CO4 3SQ, United Kingdom

Thomas M. Church
College of Marine Studies
University of Delaware
Newark, DE   19711

William H. Zoller
Department of Chemistry, BG 10
University of Washington
Seattle, WA   98195

Frank Dehairs
Faculteit Wetenschappen
Vrije Universiteit Brussels
Pleinlaan 2
B-1050 Brussels, Belgium

Francois Dulac
Centre des Faibles Radioactivités
Laboratoire Mixte CNRS-CEA,
Avenue de la Terrasse
91198 Gif-sur-Yvette, CEDEX
France

9.1. INTRODUCTION

A general introduction to the current state of knowledge of the long-range transport of trace elements is given in Chapter 2 (p. 37).   In this chapter we present four case studies to illustrate our level of understanding of the processes and to draw attention to the gaps in the current knowledge.   Each case has been treated differently to reflect the complexity of the case and the available data base.   The approaches we used are exploratory and our results, preliminary.   However, these

A. H. Knap (ed.), The Long-Range Atmospheric Transport of Natural and Contaminant Substances, 177–196.

results do illustrate ways in which we might improve our understanding of
the processes involved in long-range transport. The themes common to
each case are explored in the final section.

9.2. THE LONG-RANGE TRANSPORT OF TRACE ELEMENTS TO THE ARCTIC

The case of transport of European pollutant emissions into the Arctic is
unique among those considered here and of special interest because the
Arctic Ocean is ice-covered. There are several qualitative pieces of
evidence that pollution in the Arctic originates mainly from the Eurasian
continent. These include meteorological analyses that point to Eurasia
as the predominant landmass upwind of the Arctic Ocean from November to
April (Raatz and Shaw 1984, Rahn 1982); aerosol trace-element data
(Lowenthal and Rahn 1985, Rahn 1981); lead-isotope data (Sturges and
Barrie 1989); and chemical modeling studies (Barrie et al. 1989).
    In Chapter 6 (p. 137), Barrie discusses the phenomenon of Arctic air
pollution and presents a quantitative model estimate of the fraction of
midlatitudinal anthropogenic sulfur that reaches the Arctic. A detailed
estimate of the input of aerosol trace elements to the Arctic comparable
to that for sulfur has not been made. However, Lowenthal and Rahn (1985)
and Rahn and Lowenthal (1984) used trace-element signatures to suggest
that pollutant aerosol in the Arctic is of European origin.
    Pacyna (1984) reported his estimates of the annual emissions from
Europe of several important anthropogenic trace elements. These estimates provided a data base from which the transport of trace elements to
the Arctic could be estimated. Our assessment was based on an analogy to
the sulfur-transport model described in Chapter 6 (p. 137). Our major
assumption was that the fraction of anthropogenic trace elements released
in Europe annually that entered the Arctic (Fx) was proportional to that
of sulfur (F SO_4). We calculated the annual influx of trace elements to
the Arctic by setting Fx to be half of F SO_4, which, according to Barrie's model, is 6.7%. The results are presented in Table 9-1.
    The assumption that trace elements are transported only half as
efficiently as sulfur, although uncertain, was justified because, with
the exception of Pb, the trace elements we considered are distributed
amongst particles larger than $SO_4^=$ and therefore have more mass deposited
locally than $SO_4^=$. It was justified for Pb because most lead is emitted
from automotive exhausts near ground level in contrast to sulfur, which
is generally emitted at much greater heights from chimneys and stacks.
However, this may not be the case for lead emitted in the far north of
the USSR (NILU 1984). Studies by Rahn et al. (1982) further support the
differential efficiency of sulfur and metal transport close to their
sources; they found that sulfate-to-vanadium ratios double during the
first ten days of transport and thereafter stabilize.
    To answer the question, ''What fraction of trace elements entering
the Arctic are deposited there?'' we could only make rough estimates for
a few elements (Table 9-1). We used two approaches for this assessment.
The first approach (Approach A, Table 9-1) was to estimate the net deposition from Mart's 1983 measurements of trace-element concentrations in
Arctic snowfall and the climatological precipitation amounts (P) in the

Table 9-1. Estimates of anthropogenic trace-element inputs and deposition to the Arctic.

Element	European Emissions (10^6/yr)	Input to the Arctic (10^6/yr)	Deposition to the Arctic Approach A[1] (10^6/yr)	%[2]	Approach B[1] (10^6/yr)	%[2]
As	6,500	218				
Be	50	2				
Cd	2,700	91	6.7	7	47.8	53
Co	2,000	67	<14	<21		
Cr	18,900	635				
Cu	15,500	521	133	26		
Mn[3]	17,700	595			375	63
Mo	850	29				
Ni	16,000	538	277	52		
Pb	123,000	4,133	258	6	1,331	32
Sb	380	13				
Se	420	14				
V[3]	34,500	1,159			102	9
Zn	80,000	2,688				
Zr	1,700	57				
S[4]	32×10^6	2.1×10^6	0.25×10^6	12		

[1]Approach A and B are described in text.

[2]% is percentage of input.

[3]Noncrustal.

[4]Estimate based on $SO_4^=$ concentration in winter snowfall of 18 μeq/l and winter precipitation of 41 mm water equivalent and on negligible anthropogenic $SO_4^=$ deposition in the summer half of the year.

Arctic from Chapter 6 (p. 137). For the warm periods of June to October and the cold periods from November to May, precipitation was estimated to be 78 mm and 41 mm (1 mm = 1 l water/m^2), respectively. The corresponding concentrations of Cd, Pb, Cu, Ni, and Co in snow for the summer period were 0.4 ng/kg, 13 ng/kg, 13 ng/kg, 22 ng/kg, and <5 ng/kg; for the more polluted winter period, the concentrations were 7 ng/kg, 275 ng/kg, 130 ng/kg, 280 ng/kg, and <7 ng/kg, respectively. However, these concentrations were based on only a few snowfall records.

The second approach (Approach B, Table 9-1) we used was simply an extrapolation of Rahn's (1981) estimated deposition rates for noncrustal V, noncrustal Mn, Cd, and Pb for the Arctic Ocean (area 1.2 × 10^{13} m^2) to yield an estimate for the whole Arctic cap above 66.5° N (area 2.1 × 10^{13} m^2). Rahn's mean values (from his Table 4) are based on three different approaches: (1) Applying a total ground-based deposition velocity to observed aerosol concentrations; (2) Using the snow-composition figures from North Slope, Alaska, during winter from Weiss et al. (1978) with a precipitation amount (P) of 104 mm; and (3) Using snow-composition figures from southern Greenland from Boutron (1979) and Herron et al. (1977) with a P of 350 mm. Number (1) was highly uncertain because the total deposition velocity is not well known. Number (2) was an upper estimate of annual deposition because Weiss et al. used the chemical composition of winter snow, which is not representative of summer precipitation (Mart 1983). Number (3) was not very representative of the lower tropospheric Arctic because the southern Greenland ice cap is at an altitude of 3 km and receives more precipitation than other areas of the Arctic. Furthermore, the concentration reported for the lead, and possibly other trace elements, may have been high because of contamination since the samples were taken before modern ultra-clean sampling procedures were in use (e.g., Boutron and Batifol 1985, Mart 1983).

Although the results shown in Table 9-1 (using the estimates of Barrie and Mart) were highly uncertain, they did suggest that between 6% and 60% of the trace elements entering the Arctic from Europe were deposited in the Arctic. Obviously more data need to be collected in the Arctic but the information available to us at the time of the NATO ARW in Bermuda did suggest that substantial amounts of trace elements were transported to and deposited in the Arctic. The impact of such deposition needs to be assessed.

9.3. THE LONG-RANGE TRANSPORT OF TRACE ELEMENTS TO THE PACIFIC OCEAN

Studies at several places in the Pacific Ocean provide some of the most dramatic evidence supporting the long-range atmospheric transport of natural and anthropogenic trace elements to the remote areas of the world's oceans. The long-range transport of natural and anthropogenic trace elements from Asia to the mid-Pacific was first documented during experiments at Enewetak Atoll in the Marshall Islands ($11°N$, $162°E$) as part of the SEAREX program (Arimoto et al. 1985; Duce et al. 1980, 1983; Settle and Patterson 1982) and independently at the Moana Loa Observatory in Hawaii (Parrington et al. 1983, Shaw 1980). In addition to these state-of-the-art chemical studies, more than ten years of data on the physical properties of aerosols (e.g., condensation nuclei concentrations and particle sizes and numbers) have been collected at the U.S. National Oceanic and Atmospheric Administration's Geophysical Monitoring for Climatic Change Stations at Moana Loa and American Samoa (Bodhaine 1983). This long-term record indicates that, even though the remote Pacific is very clean compared to continental regions, environmental contaminants and mineral aerosols are transported into the area, primarily from Asia.

Results from the samples at stations in the Pacific have shown that volcanic emissions not only affect the concentrations of trace elements in the atmosphere but are also important for SO_2 and ash (Zoller et al. 1983). The chemical characteristics of volcanic emissions vary greatly depending on the type of volcano and the source of the magma. Volcanic hot spots, such as the Hawaiian Islands, have the mantle as their source of magma, and the chemistry of their emissions are enriched in elements, such as Ir and Os. Margin volcanism is much more common than hot-spot volcanism, and margin volcanoes (e.g., Mt. St. Helens) are supplied with magma from the plate-mantle interface. Explosive eruptions associated with margin volcanisms produce a characteristic suite of trace elements, including As, In, Sb, Cd, Pb, Se, S, and the halogens. These emissions mainly affect local cycling of trace elements but at certain times can have regional and even global effects (see Chapter 8, p. 163).

Assessing the impact of long-range transport requires information on the natural and anthropogenic sources of trace elements although both are poorly known. Results from the Pacific have shown that the oceans themselves are a globally important source for a variety of trace elements (see Chapter 2, p. 37). More information on the strength of the oceanic source is required to evaluate the recycling of trace elements in sea spray--a process that evidently accounts for a major fraction of the dry-deposition fluxes of lead (Settle and Patterson 1982) and other trace elements (Arimoto et al. 1985) to the Pacific Ocean. Oceans are also a significant source of trace elements involved in biological cycles or with geochemically significant vapor phases. These elements include Se (Mosher and Duce 1983, 1987; Mosher et al. 1987), Hg (Fitzgerald et al. 1984, Kim and Fitzgerald 1986), and I (Duce et al. 1983, Rasmussen et al. 1982).

Although major discoveries concerning the sources, transport, and fluxes of trace elements have been made in the past ten years, extrapolating from the limited data available to derive figures for the entire Pacific basin was not warranted at the time of our meeting. In fact, one of the clearest lessons we have learned from the studies in the Pacific is that the patterns of trace-element transport and deposition are highly heterogeneous and vary on time scales ranging from days to months. Uematsu et al. (1983, 1985) estimated that the air/sea exchange of mineral aerosol in five latitudinal bands from 0° to 50° N for the region between 150°E and 130°W is ~20×10^6g/yr. Their data indicate a strong zonal variability in transport and fluxes with differences between zones exceeding a factor of ten. Similar patterns of zonal transport and deposition are known to occur for anthropogenic lead (Settle and Patterson 1982), but the geographical coverage of trace-metal sampling has been extremely limited. Furthermore, Uematsu et al. (1985) noted that the annual fluxes of mineral dust are highly episodic and dominated by transient sporadic events.

In addition to the transport of gases and particles from Asia, there is also the potential for the transport from the western coasts of North and South America to the mid-Pacific. In fact, recent studies indicate that nitrogen oxides and anthropogenic lead are transported thousands of kilometers westward from North and South America to the tropical North Pacific (Galasyn et al. 1987, Settle and Patterson 1982 respectively). Gagosian et al. (1987) have reported some effects of a variety of organic substances emitted from Australia on the atmosphere of the South Pacific and Arimoto's preliminary results (unpublished data available from Dr. R. Arimoto, Center for Atmospheric Chemistry Studies, The University of Rhode Island, Narragansett, RI 02882-1197) from New Zealand have suggested a clear continental signature in his trace-element and radon data. Because the available information concerning the composition and quantities of chemical emissions from source regions in the Pacific basin countries is so limited, we could not evaluate trace-element transport and deposition through mass-balance accounting as was done for other regions. What was clear, however, was that the rapid industrial development of China, Korea, and other countries bordering the Pacific is altering both the character and the quantities of emissions. Equally important, the fluxes of sulfur and various trace elements to the Pacific are likely to increase significantly in the next 10-20 years (see Chapter 4, p. 87).

9.4. THE LONG-RANGE TRANSPORT OF TRACE ELEMENTS TO THE MEDITERRANEAN SEA

In the context of the long-range transport of trace elements, the Mediterranean Sea is rather a small area and the deposition of contaminants is enhanced by their transport from local continental coastal source areas (Maring et al. 1987). Nevertheless, the region is particularly well studied and of interest here for several reasons:

1. Several sources of atmospheric trace elements have already been identified (Arnold 1985, Dulac et al. 1987), including local and Atlantic sea-salt production, African desert dust, volcanoes (Mounts Etna and Stomboli), and various anthropogenic activities. These multiple sources cause the daily atmospheric concentrations of trace elements to vary considerably (Bergametti 1987).

2. There is a temporally and geographically complex meteorology (Miller et al. 1987) resulting in seasonal patterns in atmospheric concentrations and deposition of trace elements (Bergametti 1987).

3. Atmospheric inputs play a major role in the biogeochemical cycles of several elements (Boyle et al. 1985, Dehairs et al. 1987, Elderfield and Greaves 1982, Hydes 1985). Deposition to the Mediterranean of radionuclides from atmospheric weapon tests over the Pacific (GESAMP 1985) and from the Chernobyl accident (Fowler et al. 1987) have been reported.

Current atmospheric data sets are available for trace elements in the western Mediterranean area (Arnold 1985; Bergametti 1987; Bergametti et al. 1989; Chester et al. 1984; Dulac 1986; Dulac et al. 1987, 1989; Seghaier 1984).

9.4.1. Evidence of the Long-range Transport of Trace Elements

The trend in the mass-median diameter of Al (derived predominantly from the Saharan desert to the south) over the western Mediterranean basin strongly suggests a decrease in particle size from en-route gravitational settling of the coarse aerosol particle fraction. The mean mass-median diameter of Al in Corsica, 1,000 km away from any African sources, is

2.3 μm (Dulac et al. 1989), which is similar to that reported at Enewetak Atoll, 5,000 km away from any continental sources (Arimoto et al. 1985). Therefore, the mineral dust remaining in the fine fraction is probably transported far further north than the larger particles. By contrast, metals of predominantly anthropogenic origin (e.g., Pb and Cd) are transported primarily on small particles; elements originating from both natural and man-made sources show bimodal distributions (Bergametti 1987, Dulac 1986).

Enrichment factors indicate that such elements as Pb and Cd are associated with air pollution, including the fraction associated with the largest particles (diameter > 7 μm); factor analysis corroborates their continental origin (Dulac et al. 1985). The $^{210}Po/^{210}Pb$ ratios indicate that the volcanic sources are occasionallly important for atmospheric trace elements in the Mediterranean (see Chapter 8, p. 163). Moreover, the trace-element-to-sulfur ratios for volcanic emissions have shown that the volcanic source of Cu and Cd may occasionally be dominant (see Chapter 8, p. 163). Bergametti (1987) and Dulac (1986) have used enrichment factors to calculate the following man-made contributions to the atmospheric load of S (95%), V (90%), and Mn (60%). Dulac et al. (1987) also analyzed the Br/Na and Pb/Na ratios and reported that 45% of the Br derives from long-range transport from urban areas although the work of Sturges and Harrison (1986) suggested that ratios involving Br must be treated cautiously.

9.4.2. Source Regions

Because regional emissions can have unique chemical signatures, elemental ratios may be used to distinguish atmospherically transported contaminants from coastal sources. Dulac et al. (1987) used inventories of European man-made emissions from Pacyna (1984) to distinguish between urban and industrial sources (Pb/Cd > 150 and < 5, respectively) and Bergametti (1987) found some inter-elemental ratios to be unambiguous tracers of African sources.

Dulac et al. (1987) also used 3-dimensional air-mass trajectories to identify source regions for atmospheric particles. The vertical dimension is particularly useful in determining when the air-mass traveled in the boundary-layer where pollutants are predominantly emitted. Further, for several cases of transport, Maring et al. (1987) assessed the influence of the different source regions on Pb inputs, using stable lead isotope analyses and 3-dimensional air-mass trajectories. Miller et al. (1987), using climatologies of the airflow patterns in the western Mediterranean, reported that the long-range westerly flow is dominant about 45% of the time. Air is transported from the north and from the south 30% and 15% of the time, respectively; transport from the east (where there is volcanic activity) is negligible. The frequency of short-range transport (limited to countries bordering the sea within a 4-day back-trajectory) is about 15%. Surprisingly these frequency patterns do not vary much with the seasons but there are some large differences between years.

Dulac et al. (1987) used airflow climatologies together with the mean atmospheric concentrations in air masses coming from these different

sectors to evaluate the contributions of the different source regions on
a yearly time scale. They reported that, because of the dominant westerly flow, the Iberian Peninsula and southwestern France contribute 55-
60% of the atmospheric particles over the Mediterranean basin. However,
more recent studies by Bergametti (1987) suggested the contribution of Al
associated with flow from the south may have been underestimated because
of lofting of warm African air masses to considerable height. He estimated that up to 70% of the Al deposited comes from Africa.

9.4.3. Deposition

Bergametti (1987) also reported on the influence of rainfall during
transport and the influence of daily rainfall rates on scavenging ratios
(Slinn 1983) for the Mediterranean. Trace elements of crustal origin
(Al, Ti) are apparently scavenged more efficiently than those from other
sources, probably because of their larger particle size (see also Chapter 2, p. 37). Both Bergametti (1987) and Dulac (1986) reported that
sporadic events (rains or African dust storms) dominate the inputs of
trace elements to the Mediterranean Sea. Dulac et al. (1989) estimated
that wet deposition accounts for two-thirds of the total Al deposited in
the western Mediterranean. Dulac (1986) reported that more than 15% of
the Al in rain may sometimes be dissolved although for Cd this proportion
is close to 100%.
   For 1985, the mean total deposition of Pb on Corsica is 29 mg/m^2
(Bergametti 1987). When we extrapolated Bergametti's figure to the
entire northwestern Mediterranean (about 3×10^{11} m^2), we obtained an
atmospheric input of about 9×10^9 g/yr, which represented about 7% of the
total European emissions of Pb (123×10^9 g/yr, including the USSR; Pacyna
1984). Flux estimates suggested that atmospheric inputs overwhelm fluvial Pb inputs.

9.5.   THE TRANSPORT OF TRACE ELEMENTS ACROSS THE NORTH ATLANTIC OCEAN

From the analyses of enrichment factors, stable isotopes, and air-mass
trajectories (e.g., Church et al. 1984, Duce et al. 1976, Veron 1988), it
is clear that the atmosphere above the North Atlantic is contaminated by
emissions from North America and Europe. This contamination has significantly perturbed the oceanic biogeochemical cycles of several trace elements in the North Atlantic (Boyle et al. 1986, Jickells et al. 1987,
Shen 1986, Veron et al. 1987).
   Based on analyses of aerosols and rainwater data stratified by
transport direction, North America is clearly a substantial source of the
trace elements in the northwest Atlantic (Duce et al. 1976; unpublished
data available from Dr. T. M. Church, College of Marine Studies, University of Delaware, Newark, DE 19711). Unfortunately, no accurate estimates of either the atmospheric emissions from North America or the
export off the North American East Coast were available when we met at
the Bermuda Biological Station. However, Whelpdale et al. (1988) have
estimated the export of sulfur as 5×10^{12} g S/yr using aerosol data from
a relatively large network of stations, climatological wind data, and

Table 9-2. Aerosol trace-element data from Lewes, Delaware (ng/m^3).

Element	Aerosol Concentrations			
	A	B	C	D
Al	377	432		
Fe	281	149		
Mn	6.8	4.7		
Cu	--	4.5		
Ni	--	5.4		
Pb	--	63		
V	6.2	12		
Zn	19.4	25		
Se	1.2	1.9		
As	0.7	1.4		
SO_2			4.5	6.7
non-sea-salt SO_4			2.1	1.3

Sources: A: Unpublished 1985 data available from Dr. William Zoller, Department of Chemistry, BG-10, University of Washington, Seattle, WA 98195; B: Unpublished 1983 data available from Dr. Wolff; C: Wolff et al. (1986); D: Hastie et al. (1988).

atmospheric transport models. Although comparable data sets for trace elements were not available, a series of measurements believed to be reliable had been made at Lewes, Delaware (Table 9-2). Aerosol and gas-phase sulfur measurements were also taken at the Delaware site.

   To obtain a crude estimate of trace-element export off the East Coast, we multiplied the trace-element/sulfur concentration ratios in the Delaware samples (Table 9-2) by the total estimated sulfur-export rate from North America. The results of our calculations are presented in Table 9-3.

Table 9-3. Estimated export of trace elements off the North American East Coast (10^{12} g/yr).

Element	Estimated Export	Estimated Deposition West of Bermuda
Al	0.2	0.14
Fe	0.1	0.065
Mn	3×10^{-3}	3.5×10^{-3}
Cu	2×10^{-3}	1.8×10^{-3}
Ni	3×10^{-3}	1.8×10^{-3}
Pb	0.03	0.007
V	5×10^{-3}	1.3×10^{-3}
Zn	12×10^{-3}	14×10^{-3}

Although crude, the calculations in Table 9-3 probably reflected the trace-element export from North America within an order of magnitude. In these emission estimates, we used the sulfur flux (SO_2 + non-sea-salt SO_4) without any assumed fractionation of sulfur and other trace elements as discussed earlier for the Arctic. We believed that this was reasonable because such fractionation probably only occurs during the early stages of transport (Rahn et al. 1982) and hence before the aerosols reach the coast. Church et al. (1984) have compared the concentrations of trace metals and sulfate in rainfall on Bermuda and at Lewes. When we assumed that Bermuda rainfall was derived from an air mass with an initial composition similar to that for masses at Lewes, a comparison of sulfate and trace-element concentrations deposited at the two sites provided a crude test of this sulfur/trace-element-transport analogy.

For the predominantly crustal elements usually associated with relatively large aerosol particles (Fe, Mn, V), there was a more rapid decline in concentration than for sulfate. For cadmium and copper, the ratios of average concentrations at Lewes and Bermuda were very similar although for nickel, lead, and zinc, they suggested a modest preferential removal (factor of two) with respect to sulfate. These comparisons suggested that using sulfur transport as an analogue for trace elements might result in only a modest overestimate of fluxes. Of course, the validity of an analogy between sulfur and trace-element transport in this context was uncertain because of the differences in locations and types of sources for the different constituents.

Veron (unpublished data available from Dr. Veron, Centre des Faibles Radioactivites, Laboratoire Mixte CNRS-CEA, Avenue de la Terrasse, 91198 Gif-sur-Yvette, CEDEX, France) estimated that lead emissions in North America for 1985 are 150×10^9 g/yr. Using this estimate for Pb emissions, we calculated about 30% of this lead was transported eastwards from North America. Whelpdale et al. (1988) estimated that about 30% of the sulfur emitted in North America is transported eastwards, which would indicate that our transport estimates were plausible.

Using the average wet-deposition rates of trace elements at Lewes, Delaware, from Church et al. (1984) and the total deposition rates at Bermuda from Jickells et al. (1987), we made crude estimates of trace-element deposition rates to the area of the North Atlantic west of Bermuda (area 2.5×10^{12} m^2; Jickells et al. 1982) (Table 9-3).

Values from the larger data base available on sulfur transport emphasize the variability of rainwater contamination over the North Atlantic. Because Bermuda lies somewhat to the south of the main plume of contamination emanating from North America, deposition may be underestimated. Elliot and Reed's (1984) estimates of the precipitation rates over the North Atlantic suggested that Bermuda may receive more rainfall than the surrounding ocean. However, Jickells et al. (1982) suggested that no such ''island effect'' is evident. Galloway and Whelpdale (1987) used a rainfall figure 50% lower than Jickells et al. (1987) used. If the rainfall on Bermuda is higher than over the surrounding oceans, then the estimated deposition figures of Jickells et al. (1987) are high.

These uncertainties (the differential removal with respect to sulfate, the location of Bermuda in the United States' ''plume,'' and the rainfall rates) tended to cancel one another out and led us to believe that the crude budgetary approach attempted here provided an order of magnitude estimate of the export of these trace elements over the central North Atlantic Ocean. Taken literally, the estimates in Table 9-3 implied that from as much as three quarters (Cu) to as little as a negligible fraction (Zn, Mn) of the trace elements emitted from North America were exported beyond the western Atlantic. For sulfur and nitrogen, Galloway and Whelpdale (1987) estimated 35-65% and 0-33%, respectively, are similarly exported, which would suggest that our trace-element estimates were indeed plausible.

Over the northwestern North Atlantic, the dominant westerly airflow leads to the export of contaminants from North America to Europe. Over the northeastern Atlantic, contaminants are exported from Europe westward as is clearly seen for sulfur (Whelpdale et al. 1988). Veron (1988) estimated the origin of atmospheric lead over and the deposition of atmospheric lead to the northeast Atlantic from aerosol lead concentrations, climatological transport patterns, and lead-isotope data. Based on values extrapolated from one area to the whole northeast Atlantic (20-60°N, east of 20°W, area 5×10^{12} m^2), the total lead input was 15×10^9 from the east and 3×10^9 g/yr from the west. Based on the estimated emissions for Europe from Pacyna (1984) and for North America from Veron (1988), 10-15% and <5% of European and North American emissions, respectively, were deposited in this area. For comparison, we estimated that <1% of European lead inputs were deposited in the Arctic and 7% in the Mediterranean.

Although these calculations provided plausible first-order estimates of the export of some trace elements to the North Atlantic, we had several reservations. First, for our estimates we extrapolated the available data sets beyond the areas and times of measurements. Atmospheric deposition rates of lead vary over the ocean; they are higher at the United States East Coast (Church et al. 1984) than they are in the northeast Atlantic (Veron 1988) and higher in the northeast Atlantic than they are on Bermuda (Jickells et al. 1987). Church (unpublished data available from Dr. T. M. Church, College of Marine Studies, University of Delaware, Newark, DE 19711) believes that deposition rates are probably lower on the west coast of Ireland than on Bermuda. Such differences are reflected in the water-column concentrations of lead (Veron 1988). Boyle et al. (1986) suggested that the atmospheric deposition of North American lead emissions may be primarily confined to a limited area beneath the prevailing westerly winds, as Whelpdale et al. (1988) suggested for sulfur. This spatial inhomogeneity may, in part, explain the complex water-column profiles from the western North Atlantic. These observations all emphasize the difficulties of extrapolating data bases, no matter how good the bases are in themselves, to large oceanic areas.

The results of these various budgets are summarized in Figure 9-1, a map of the North Atlantic showing our estimates of the transport of lead. We selected lead because the quality and scale of the data base was so much better than for any of the other elements.

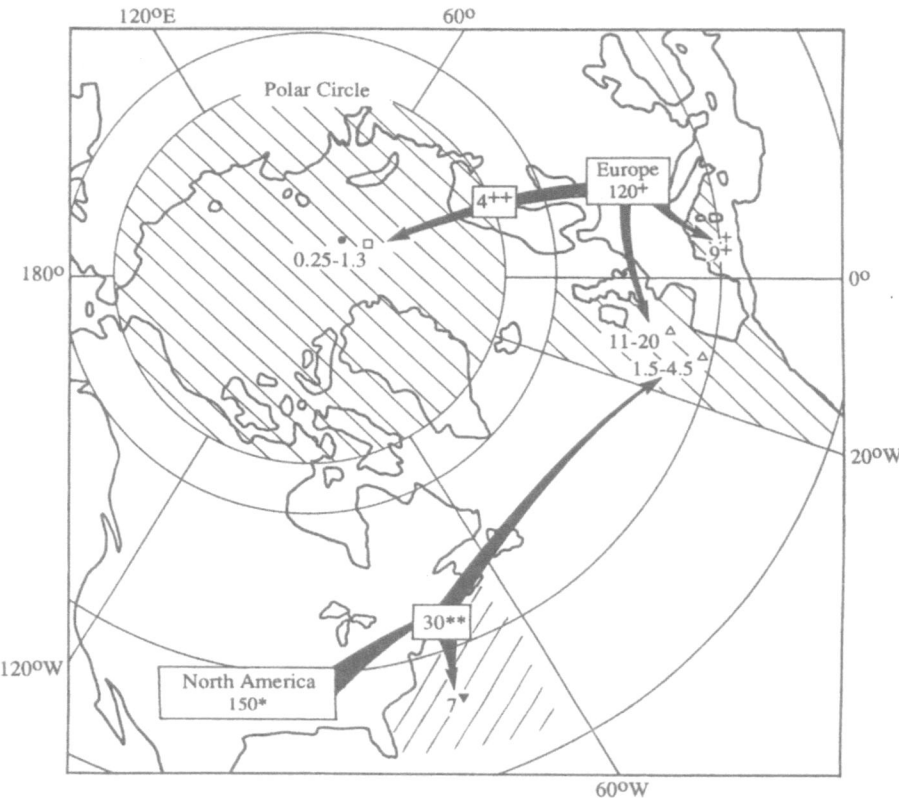

Figure 9-1. Tentative partial atmospheric budget for lead over the Atlantic and Arctic Oceans (10^9 g/yr). * = total 1985 emissions from North America (unpublished data available from Dr. A. Veron, Centre des Faibles Radioactivites, Laboratoire Mixte CNRS–CEA, Avenue de la Terrasse, 91190 Gif-sur-Yvette, France);** = estimated export from the North American east coast; ▼= deposition to part of the northwest Atlantic Ocean (Jickells et al. 1987); + = total European emissions (Pacyna 1984, 1986); ++= estimated export from Europe to the Arctic; •= deposition to the Arctic Ocean; □= deposition to the Arctic Ocean (Rahn 1981); ‡= deposition to the northwest Mediterranean Sea; △= deposition to the northeast Atlantic Ocean (Veron 1988).

9.6. CONCLUSIONS

The case studies presented in this chapter provided clear evidence of the long-range transport of trace elements from continent to ocean and suggested that there was also transport from continent to continent. Two of the case studies (those discussing the Mediterranean and the Pacific) demonstrated the power of multiple tracers, most notably metals and stable nuclides or radionuclides, when used to reveal emission sources. The other two cases (those discussing the Arctic and the North Atlantic) were based on budget calculations with analogies to sulfur-transport models. Because of the much larger data bases from which sulfur transport could be estimated, the possibility of using sulfur-transport models to estimate trace elemental transport was explored. However, because of the differences in emission sources and differences in the atmospheric chemistry of sulfur compared to other trace elements, the validity of this approach required testing. Although a preliminary attempt at such validation was made in this chapter, the data might exist for a more extensive effort; we recommend that the sulfur/trace-element analogy be critically examined.

A common feature of the studies we presented was that the data bases used to estimate long-range transport were fragmentary and, therefore, inadequate to make anything but rough estimates. This was true for the data on sources, atmospheric concentrations, and deposition alike. This unfortunate situation could only be remedied by extensive measurement campaigns. Although past data-gathering efforts have been limited by the problems of sampling and analysis of the low concentrations encountered, this situation was already improving at the time of our NATO workshop in Bermuda; several groups had the capacity to make the required measurements. A second problem was that the atmosphere is highly variable in space and time; this severely limited our confidence in extrapolating existing data bases to larger regions and to times other than those sampled.

Our attempts to understand the effects of atmospheric deposition were hindered by an inadequate knowledge of the physico-chemical speciation of trace elements in aerosols and rain. We needed to know the fractions of metals that are soluble, the extent to which acidity affects solubility, and the chemical forms of the various trace elements in the atmosphere and in precipitation. Again research underway at the time of our meeting was moving in this direction, but our level of understanding then did not allow us to evaluate fully the impact of atmospheric deposition on the oceans. In particular, our understanding of deposition processes was inadequate to allow the modeling of atmospheric transport and deposition. A major component of this uncertainty stemmed from an inadequate knowledge of the particle-size distributions of aerosols, for which more research was needed. In addition our knowledge of precipitation amounts over the oceans was rudimentary.

A promising approach to studying the long-range transport of trace metals was based on using a combination of emission estimates and atmospheric transport models. This approach cannot replace field measurements of deposition, but it can complement them. Examples of such modeling of the transport of acids are discussed in this book (see Chapter 4, p. 87;

Chapter 6, p. 137; Chapter 11, p. 231). However, these models required more information on vertical distributions of trace elements in the atmosphere.

Pacyna (1986) has reviewed and compiled inventories of natural and anthropogenic sources of trace elements to the atmosphere. However, this and many other such compilations rely to varying extents on earlier compilations (Lantzy and Mackenzie 1979, Nriagu 1979). In addition, for a few elements, (Sn, Byrd and Andreae 1982; Se, Mosher and Duce 1987; B, Fogg and Duce 1985; Hg, Fitzgerald 1986), more recent atmospheric emission inventories were available. For volcanic emissions, estimates of global emissions were available (see Chapter 8, p. 163).

Comparisons of the compilation of anthropogenic emission estimates of Pacyna (1986) to those of Lantzy and Mackenzie (1979) revealed significant differences even though both are primarily from 1975. This introduced an important uncertainty into calculations of interference factors (see Chapter 2, p. 37). Although the reasons for these differences varied, we expected that emissions had changed significantly since the mid-1970s as improved emission control and product-use patterns were introduced in developed countries. At the same time, industrial activity and power consumption had increased in developing countries (see Chapter 4, p. 87). A notable example of these changes was the reduction in the amount of lead added to petrol in the United States since the late 1970s (Boyle et al. 1986). Thus, both the patterns and magnitudes of global emissions continue to change significantly with time. A striking feature of the emission data of Pacyna (1986) was that it is limited to rather few elements despite clear evidence that other elements are enriched on aerosol particles (Wiersma and Davidson 1986). Obviously our knowledge of global-emission inventories was inadequate to define accurately the scale of global interference in atmospheric transport for any but a very limited range of elements.

Although global-emission inventories were useful for indicating the potential ''global averaged'' scale of man's impact on trace-element concentrations in the atmosphere, such inventories cannot be used to model long-range transport from specific regions. Such modeling requires more specific knowledge on source strengths within regions. At the time of the Bermuda meeting, this information was becoming available for Europe (Pacyna 1986) and was being used in modeling efforts (e.g., Krell and Roeckner 1988; Chapter 6, p. 137).

The coverage of regional emission inventories must be extended if the long-range transport of trace elements is to be modeled adequately. In addition information on the particle-size distribution and chemical speciation of trace elements at emission and during transport and deposition must be ascertained. To understand the transport of natural emissions, a parallel effort on natural sources must be undertaken.

9.7. RECOMMENDATIONS

As noted above, information on the magnitudes, locations, and physico-chemical speciation of trace-element emissions must be improved to further our understanding of the long-range transport of trace elements through the atmosphere. Better information on the particle-size distributions of trace elements in emissions and the troposphere at remote locations in particular is required. A better understanding of the deposition to and recycling of trace elements from the sea surface including its parameterizations in transport models is needed. Such models require a knowledge of precipitation processes over the ocean that was not available at the time of our NATO workshop in Bermuda.

Although the analogies of sulfur and trace-element transport offered promise, this approach required validation. In particular, the sulfur-transport models assumed certain vertical distributions of aerosols in the atmosphere. At the time of our meeting, there was little information on the vertical profiles of trace elements in the atmosphere.

Several techniques provided convincing evidence of long-range transport; however, to use these techniques quantitatively, more research was required. For example, records of deposition, such as the data from ice or sediment cores, could have been used quantitatively if the processes of deposition in these environments had been well understood and if the local fluxes could have been extrapolated to a regional scale. In some cases, trajectory-transport analyses could be improved by the use of 3-dimensional trajectory techniques and the incorporation of information on the precipitation field. Similarly the value of enrichment factors could be improved if information on the elemental composition of the actual aerosol precursors was available. Some isotopes (e.g. Nd, Sr, B) might provide useful information about atmospheric transport in the future although their use is still in its infancy (Grousset et al. 1988).

Finally, we were not able to quantify the scale of man's perturbation of global biogeochemical cycles of trace elements, particularly for the oligotrophic ocean areas where the effects might be the most important. Recent suggestions that atmospherically derived iron might limit primary productivity in certain regions of the oceans (Martin and Fitzwater 1988) or that slight increases in copper concentrations in oceanic surface waters might suppress primary productivity (Knauer and Martin 1983) raised important questions about how perturbations of the atmosphere were affecting global cycles. Such questions will not be answered until more observational and process studies are conducted.

9.8. REFERENCES

Arimoto, R., R. A. Duce, B. J. Ray, and C. K. Unni. 1985. Atmospheric trace elements at Enewetak Atoll: 2. Transport to the ocean by wet and dry deposition. J. Geophys. Res. 90:2391-2408.

Arnold, M. 1985. Géochimie et transport des aerosols metalliques au-dessus de la Mediterranée occidentale. These d'état, Université de Paris, 226 p.

Barrie, L.A., M. P. Olson, and K. K. Oikawa. 1989. The flux of anthropogenic sulphur into the Arctic from mid-latitudes in 1979-1980. Atmos. Environ. 23, in press.

Bergametti, G. 1987. Apports de matiere par voie atmospherique a la Mediterranée occidentale: Aspects géochimiques et meteorologiques. Ph.D. dissert., Univ. of Paris, 296 p.

Bergametti, G., A. L. Dutot, P. Buat-Ménard, R. Losno, and C. Remoudaki. 1989. Seasonal variability of the elemental composition of atmospheric aerosol particles over the northwestern Mediterranean. Tellus, in press.

Bodhaine, B. A. 1983. Aerosol measurements at four background sites. J. Geophys. Res. 88:10,753-10,768.

Boutron, C. F. 1979. Trace element content of Greenland snows along an east-west transect. Geochim. Cosmochim. Acta 43:1253-1258.

Boutron, C. F., and F. M. Batifol. 1985. Assessing laboratory procedures for the decontamination of Polar snow or ice samples for the analysis of toxic metals and metalloids. Ann. Glaciol. 7:7-11.

Boyle, E.A., S. D. Chapnick, X. X. Bai, and A. Spivack. 1985. Trace metal enrichments in the Mediterranean Sea. Earth Planetary Sci. Ltrs. 74:405-419.

Boyle, E.A., S. D. Chapnick, G. T. Shen, and M. P. Bacon. 1986. Temporal variability of lead in the western North Atlantic. J. Geophys. Res. 91:8573-8593.

Byrd, J. T., and M. O. Andreae. 1982. Tin and methyl tin species in seawater: Concentration and fluxes. Science 218:565-569.

Chester, R., E. J. Sharples, G. S. Sanders, and A. C. Saydam. 1984. Saharan dust incursion over the Tyrrhenian sea. Atmos. Environ. 18:929-935.

Church, T. M., J. M. Tramontano, J. R. Skudlark, T. D. Jickells, J. J. Tokos, and A. H. Knap. 1984. The wet deposition of trace metals to the western Atlantic Ocean at the mid-Atlantic coast and on Bermuda. Atmos. Environ. 18:2657-2664.

Dehairs, F., C. E. Lambert, R. Chesselet, and N. Risler. 1987. Biological production of marine suspended barite and the barium cycle in the Western Mediterranean Sea. Biogeochemistry 4:119-139.

Duce, R. A., G. L. Hoffman, B. J. Ray, I. S. Fletcher, G. T. Wallace, J. L. Fasching, S. E. Piotrowctz, P. R. Walsh, E. H. Hoffman, J. M. Miller, and J. L. Hefter. 1976. Trace metals in the marine atmosphere: Sources and fluxes. In Marine Pollutant Transfer (H. L. Windom and R. A. Duce, eds.) Lexington, MA:Heath, 77-119.

Duce, R. A., C. K. Unnni, B. J. Ray, J. M. Prospero, and J. T. Merrill. 1980. Long-range atmospheric transport of soil dust from Asia to the tropical North Pacific: Temporal variability. Science 209: 1522-1524.

Duce, R. A., R. Arimoto, B. J. Ray, C. K. Unni, and P. J. Harder. 1983. Atmospheric trace elements at Enewetak Atoll: 1. Concentrations, sources and temporal variability. J. Geophys. Res. 88:5321-5342.

Dulac, F. 1986. Dynamique du transport et des retombées d'aerosols metalliques en Mediterranée occidentale. Ph.D. dissert., Univ. of Paris, 241 pp.

Dulac, F., P. Buat-Ménard, D. Martin, A. L. Dutot, G. Bergametti, R. Delmas, and U. Ezat. 1985. Atmospheric pathways of trace metals to the western Mediterranean Sea. In Procs., 5th Int. Conf., Heavy Metals in the Environment (T. Lekkas, ed.) Edinburgh:C. E. P. Consultants, Ltd., 1:72-74.

Dulac, F., P. Buat-Ménard, M. Arnold, U. Ezat, and D.Martin. 1987. Atmospheric input of trace metals to the western Mediterranian Sea: 1. Factors controlling the variability of atmospheric concentrations. J. Geophys. Res. 92:8437-8453.

Dulac, F., P. Buat-Ménard, U. Ezat, S. Melki, and G. Bergametti. 1989. Atmospheric input of trace metals to the western Mediterranean: 2. Uncertainties in modelling dry deposition from cascade impactor data. Tellus , in press.

Elderfield, H., and M. J. Greaves. 1982. The rare earth elements in seawater. Nature 296:214-219.

Elliot, W. P., and R. K. Reed. 1984. A climatological estimate of precipitation for the world ocean. Climate Appl. Meteorol. 23:434-439.

Fitzgerald, W. F. 1986. Cycling of mercury between the atmosphere and oceans. In The Role of Air-sea Exchange in Geochemical Cycling (P. Buat-Ménard, ed.) NATO ASI Series C, Vol. 185, Dordrecht:Reidel, 363-408.

Fitzgerald, W. F., G. A. Gill, and J. P. Kim. 1984. An equatorial Pacific source of atmospheric mercury. Science 224:597-599.

Fogg, T. R., and R. A. Duce. 1985. Boron in the troposphere: Distribution and fluxes. J. Geophys. Res. 90:3781-3796.

Fowler, S.W., P. Buat-Ménard Y. Yokoyama S. Ballestra, E. Holm and H. V. Nguyen. 1987. Rapid removal of Chernobyl fallout from Mediterranean surface waters by biological activity. Nature 329:56-58.

Gagosian, R. B., E. T. Peltzer, and J. T. Merrill. 1987. Long-range transport of terrestrially derived lipids in aerosols from the South Pacific. Nature 325:800-803.

Galasyn, J. F., K. L. Tschudy, and B. J. Huebert. 1987. Seasonal and diurnal variability of nitric acid vapor and ionic aerosol species in the remote free troposphere at Mauna Loa, Hawaii. J. Geophys. Res. 92:3105-3113.

Galloway, J. N., and D. M. Whelpdale. 1987. WATOX-86 overview and western North Atlantic Ocean S and N atmospheric budgets. Global Biogeochem. Cycles 1:261-281.

GESAMP (Joint Group of Experts on the Scientific Aspects of Marine Pollution). 1985. Atmospheric Transport of Contaminants into the Mediterranean Region. WMO Rept./Studies No. 26, 54 pp.

Grousset, F. E., P. E. Biscaye, A. Zindler, J. M. Prospero, and R. Chester. 1988. Nd isotopes as tracers in marine sediments and aerosols: North Atlantic. Earth Planetary Sci. Ltrs. 87:367-378.

Hastie, D. R., H. I. Schiff, D. M. Whelpdale, R. E. Peterson, W. H. Zoller, D. L. Anderson, and T. M. Church. 1988. Description and intercomparison of techniques to measure N and S compounds in the Western Atlantic Ocean Experiment. Atmos. Environ. 22:2393-2399.

Herron, M. M., C. C. Langway, H. Weiss, and J. H. Cragin. 1977. Atmospheric trace metals and sulfate in the Greenland ice sheet. Geochim. Cosmochim. Acta. 41:915-920.

Hydes, D. J. 1985. Dissolved aluminium in the Mediterranean Sea. Terra Cognita 5:189.

Jickells, T. D., A. H. Knap, T. M. Church, J. N. Galloway, and J. M. Miller. 1982. Acid rain on Bermuda. Nature 297:55-57.

Jickells, T. D., T. M. Church, and W. G. Deuser. 1987. A comparison of atmospheric inputs and deep-ocean particle fluxes for the Sargasso Sea. Global Biogeochem. Cycles 1:117-130.

Kim, J. P., and W. F. Fitzgerald. 1986. Sea-air partitioning of mercury in the equatorial Pacific Ocean. Science 231:1131-1133.

Knauer, G. A., and J. H. Martin. 1983. Trace elements and primary production: Problems, effects and solutions. In Trace Metals in Sea Water (C. S. Wong, E. Boyle, K. W. Bruland, J. D. Burton, and E. D. Goldberg, eds.) New York:Plenum, 825-840.

Krell, U., and E. Roeckner. 1988. Model simulation of the atmospheric input of lead and cadmium into the North Sea. Atmos. Environ. 22: 375-381.

Lantzy, R. J., and F. T. Mackenzie. 1979. Atmospheric trace metals: Global cycles and assessment of man's impact. Geochim. Cosmochim. Acta 43:511-525.

Lowenthal, D. H., and K. A. Rahn. 1985. Regional sources of pollution aerosol at Barrow, Alaska, during winter 1979-1980 as deduced from elemental tracers. Atmos. Environ. 19:2011-2024.

Maring, H., D. M. Settle, P. Buat-Ménard, F. Dulac, and C. C. Patterson. 1987. Stable lead isotope tracers of air-mass trajectories in the Mediterranean region. Nature 330:154-156.

Mart, L. 1983. Seasonal variations of Cd, Pb, Cu and Ni levels in snow from the eastern Arctic Ocean. Tellus 35B:131-141.

Martin, J. H., and S. E. Fitzwater. 1988. Iron deficiency limits phytoplankton growth in the north-east Pacific subarctic. Nature 331: 341-343.

Miller, J., D. Martin, and B. Strauss. 1987. A comparison of results from two trajectory models used to produce flow climatologies to the Western Mediterranean. NOAA Tech. Memo. ERL/ARL-151, Silver Springs, MD.

Mosher, B. W., and R. A. Duce. 1983. Vapor phase and particulate selenium in the marine atmosphere. J. Geophys. Res. 88:6761-6768.

Mosher, B. W., and R. A. Duce. 1987. A global atmospheric selenium budget. J. Geophys. Res. 92:13,289-13,298.

Mosher, B. W., R. A. Duce, J. M. Prospero, and D. L. Savoie. 1987. Atmospheric selenium: Geographical distribution and ocean-to-atmosphere flux in the Pacific Ocean. J. Geophys. Res. 92:13,277-13,287.

NILU (Norwegian Institute for Air Research). 1984. Emission Sources in the Soviet Union. NILU Rept. 0.8147, Lillestrøm, Norway, np.

Nriagu, J. O. 1979. Global inventory of natural and anthropogenic emissions of trace metals to the atmosphere. Nature 279:409-411.

Pacyna, J. M. 1984. Estimations of the atmospheric emissions of trace elements from anthropogenic sources in Europe. Atmos. Environ. 18:41-50.

Pacyna, J. M. 1986. Atmospheric trace elements from natural and anthropogenic sources. In Toxic Metals in the Atmosphere (J. O. Nriagu and C. I. Davidson, eds.) New York:Wiley, 33-52.

Parrington, J. R., W. H. Zoller, and N. K. Aras. 1983. Asian dust: Seasonal transport to the Hawaiian Islands. Science 220:195-197.

Raatz, W. E. and G. E. Shaw. 1984. Long-range tropospheric transport of pollution aerosol into the Alaskan Arctic. Climate Appl. Meteorol. 23:1052-1064.

Rahn, K. A. 1981. The Mn/V ratio as a tracer of large scale sources of pollution aerosol for the Arctic. Atmos. Environ. 15:1457-1464.

Rahn, K. A. 1982. On the causes, characteristics and potential environmental effects of aerosol in the Arctic Ocean. Procs., Conf. on the Arctic Ocean: The Hydrographic Environment and the Fate of Pollutants, 163-195.

Rahn, K. A., and D. H. Lowenthal. 1984. Elemental tracers of distant regional gas pollution aerosols. Science 223:132-139.

Rahn, K. A., C. Brosset, B. Ottar, and E. M. Patterson. 1982. Black and white episodes, chemical evolution of Eurasian air masses and long-range transport of carbon to the Arctic. In Particulate Carbon: Atmospheric Life Cycle (G. T. Wolff and R. L. Klimisch, eds.) New York:Plenum, 327-342.

Rasmussen, R. A., M. A. K. Khalil, R. Gunewardena, and S. D. Hoyt. 1982. Atmospheric methyl iodide (CH_3I). J. Geophys. Res. 87:3086-3090.

Seghaier, A. 1984. Abondance et origine de quelques metaux (Al, Fe, Zu, Cu, Cd, Pb) dans l'aerosol marine de la Mediterranée occidentale. Ph.D. dissert., Univ. of Paris, 148 pp.

Settle, D. M., and C. C. Patterson. 1982. Magnitudes and sources of precipitation and dry deposition fluxes of industrial and natural leads to the North Pacific at Enewetak. J. Geophys. Res.87:8857-8869.

Shaw, G. E. 1980. Transport of Asian desert aerosol to the Hawaiian Islands. Appl. Meteor. 19:1254-1259.

Shen, G. T. 1986. Lead and cadmium geochemistry of corals: Reconstruction of historic populations in the upper ocean. Ph.D. dissert., Mass. Inst. Technol., Cambridge, MA.

Slinn, W. G. N. 1983. Air-to-sea transfer of particles. In Air-Sea Exchange of Gases and Particles (P. S. Liss and W. G. N. Slinn, eds.) NATO ASI Series C, Vol. 108, Dordrecht:Reidel, 299-396.

Sturges, W. T., and L. A. Barrie. 1989. Stable lead isotope ratios in Arctic aerosols: Evidence for the origin of Arctic air pollution. Atmos. Environ. 23, in press.

Sturges, W. T., and R. M. Harrison. 1986. Bromine in marine aerosols and the origin, nature and quantity of natural atmospheric bromine. Atmos. Environ. 20:1485-1496.

Uematsu, M., R. A. Duce, J. M. Prospero, L. Chen, J. T. Merrill, and R. L. McDonald. 1983. Transport of mineral aerosol from Asia over the North Pacific Ocean. J. Geophys. Res. 88:5343-5352.

Uematsu, M., R. A. Duce. and J. M. Prospero. 1985. Deposition of atmospheric mineral particles in the North Pacific Ocean. Atmos. Chemistry 3:123-138.

Veron, A. 1988. Dynamique du transport du plomb de pollution dans l'ocean Atlantique Nord Est depuis l'atmosphere jusqu'au sediment. PhD. diss., University of Paris, np.

Veron, A., C. E. Lambert, A. Isley, P. Linet, and F. Grousset. 1987. Evidence of recent lead pollution in deep northeast Atlantic sediments. Nature 326:278-281.

196

Weiss, H., M. M. Herron, and C. C. Langway. 1978. Natural enrichment of elements in snow. Nature 274:352-353.

Wiersma, G. B., and C. I. Davidson. 1986. Trace metals in the atmosphere of remote areas. In Toxic Metals in the Atmosphere (J. O. Nriagu and C. I. Davidson, eds.) New York:Wiley, 201-266.

Whelpdale, D. M., A. Eliassen, J. N. Galloway, H. Dovland, and J. M. Miller. 1988. The transatlantic transport of sulfur. Tellus 40B:1-15.

Wolff, G.T., N. A. Kelly, M. A. Ferman, M. S. Ruthovsky, D. P. Stoup, and P. E. Korsog. 1986. Measurements of sulphur oxides, nitrogen oxides, haze and fine particles at a rural site on the Atlantic Coast. Air Pollut. Control Assoc. 36:585-591.

Zoller, W.H., J. R. Parrington, and J. M. Phelan-Kotra. 1983. Iridium enrichment in airborne particles from Kilauea volcano. Science 222:1118-1121.

10. THE LONG-RANGE TRANSPORT OF MINERAL AEROSOLS: GROUP REPORT

Lothar W. Schutz, Rapporteur
Institut fuer Meteorologie
Johannes Gutenberg-Universitaet
Postfach 39 80
D-6500 Mainz
Federal Republic of Germany

Patrick Buat-Ménard
Centre des Faibles Radioactivités
Laboratoire Mixte CNRS-CEA
Avenue de la Terrasse
91198 Gif sur Yvette, France

Renato A. C. Carvalho
Instituto Nacional de
  Meteorologia e Geofisica
Rua C do Aeroporta de Lisboa
1700 Lisboa, Portugal

Antonio Cruzado
Consell Superior d'Investi-
  gacions Cientifiques (CSIC)
Centro de Estudios Avanzados
  de Blanes
Cami de Santa Barbara
17300 Blanes (Girona), Spain

Joseph M. Prospero, Co-Rapporteur
Rosenstiel School of Marine and
  Atmospheric Science
4600 Rickenbacker Causeway
Miami, FL 33149-1098

Robert Harriss
NASA, Langley Research Center
Mail Stop 483
Hampton, VA 23665

Neils Z. Heidam
Miljøstyrelsen
Luftforureningslaboratoriet
Forsøgsanlaeg Risø
4000 Roskilde, Denmark

Ruprecht Jaenicke
Institut fuer Meteorologie
Johannes Gutenberg-Universitaet
Postfach 39 80
D-6500 Mainz
Federal Republic of Germany

10.1. INTRODUCTION

Mineral dust is ubiquitous in our world. The generation, transport, and deposition of dust are normal geological processes that have helped to shape and create many features on the surface of the earth. Often this normal process is accelerated by man's activities.

At the meeting of our working group in Bermuda, we focused on the cyclic aspect of dust transport. We discussed the entire cycle of sources, transport, and deposition as aspects of a unified process. We agreed that only by investigating the entire process could we begin to understand its component parts. The primary objectives of this effort were to assess the current state of our knowledge and to identify critical research needs.

A. H. Knap (ed.), The Long-Range Atmospheric Transport of Natural and Contaminant Substances, 197–229.

10.2. SOURCES AND EMISSIONS

10.2.1. Source Processes

The arid and semiarid regions of the world are the major sources of atmospheric dust. These sources are depicted in a general way in Figure 10-1. This figure also shows the principal trajectories that transport dust over long distances. Most arid regions are in the low and midlatitudes of the Northern Hemisphere; furthermore, most fall within a contiguous area extending across North Africa through the Middle East and deep into Asia. Many of the main transport paths carry dust over the oceans.

The generation and transport of mineral-dust particles in the atmosphere over arid regions depends on three basic processes: the production of fine-grain soil material by weathering processes; the mobilization of this fine material by transfer of momentum from winds to the surface; the atmospheric mixing processes that carry the material to the free troposphere where the dust can then be dispersed by wind over great distances. We must know more about these source-relevant functions to estimate dust production and to use transport models.

10.2.1.1. Soil-related Processes. Only particles with diameters of less than about 10 μm can remain airborne for days or longer so that they can be transported over a 1000 km or more (Tsoar and Pye 1987). We refer to this size component as the long-range-transport (LRT) soil fraction. To understand the global sources of LRT dust, it is necessary to know at the

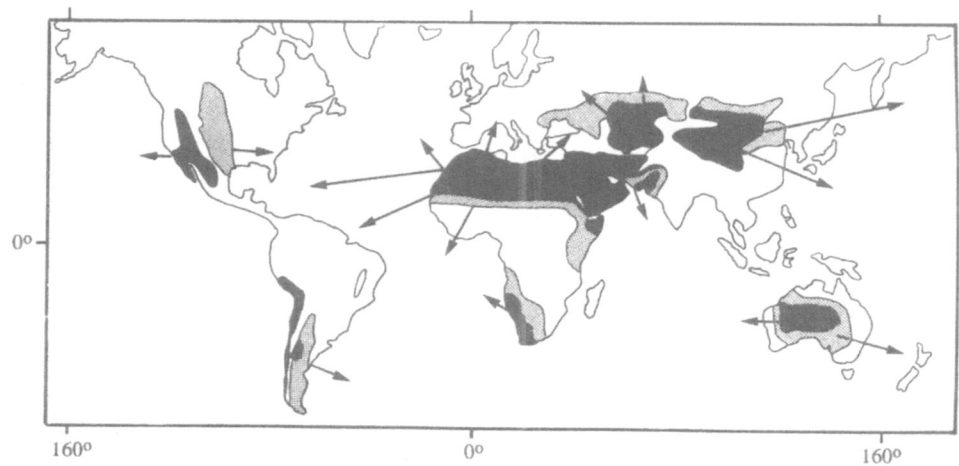

Figure 10-1. The distribution of arid regions and the principal transport directions for mineral dust; the **solid areas** are arid, bordered by semi-arid regions (from Pye 1987 modified after Coude-Gaussen 1984).

very least the fraction of readily deflated LRT particles in the different soil types from the representative source regions. The appropriate size data for various soil types can then be extrapolated to other regions using existing soil maps and the potentiality of a region's being a dust source can be estimated.

For example, sand dunes are a conspicuous feature in many deserts and they can cover a very large fraction of the desert surface, up to 28% of the Sahara (Cooke and Warren 1973). However, dunes are very much depleted of particles in the small-size range and are not usually a major source of LRT material (Fig. 10-2). In contrast, wadis (dry washes) have a small areal coverage (worldwide less than 5% and only 1% in the Sahara) but they are highly enriched in small-particle materials because they are frequently supplied with freshly weathered soil materials (D'Almeida and Schutz 1983). Other sedimentary deposits are also good sources of transport material: for example, dry river and lake beds, weathered and unconsolidated surfaces at the foot of mountains, and playas (Middleton et al. 1986). Unfortunately, there is very little additional information available on the LRT mass fraction for different

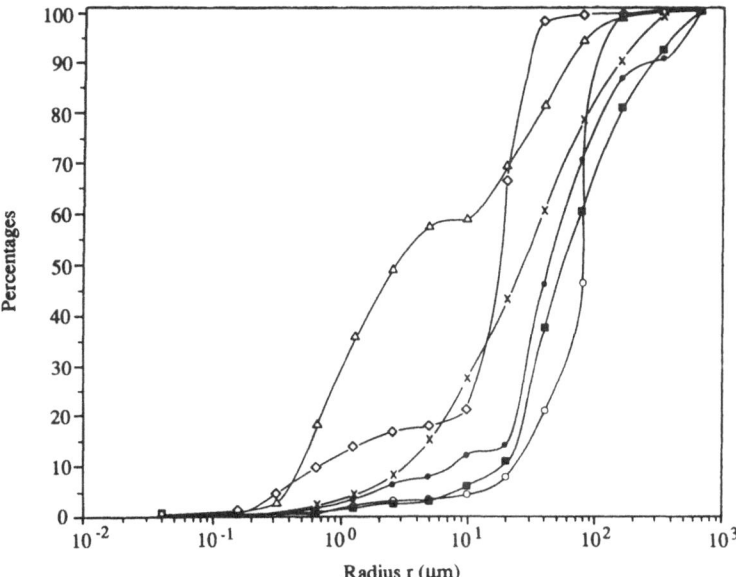

Figure 10-2. Cumulative mass distributions for Saharan soils. High mass fractions in the LRT size range of soils are found in river flood plans (—△—Goundam, Mali, and —◇— Matam, Senegal) and areas of active weathering (—×— Nubian Desert, Sudan). In contrast, highly winnowed desert soils and dune-like materials contain only small amounts of material in the LRT size range (—○— Dar Albeida, Mali; —■—Achegour, Niger; —•— Tamanrasset, Algeria). (D'Almeida and Schutz 1983)

soils. Most measurements of soil particle size are made by soil scientists. Their interests focus on the larger particle sizes; consequently, measurements of the finest particle sizes tend to be crude.
Furthermore, the more detailed studies of soil particle size tend to
center on agricultural soils, not soils in remote arid regions.

10.2.1.2. **Micrometeorology**. Measurements of soil-size distribution
alone only determine the maximum LRT fraction in soils. The various
physical and chemical properties of the soil are more important in determining the actual degree of mobilization of small particles from a terrain. Because these properties cannot be easily predicted from maps of
soil types, the deflation characteristics of different soils must be
directly studied. Most useful for this purpose are the portable wind
tunnels that can be taken to test sites (Gillette et al. 1982, Nickling
1989). With such devices, we have learned much about the production of
dust (Gillette 1981, Gillette et al. 1982). For example, we know that
very large particles, although airborne only briefly, are important in
the deflation of LRT particles. Soil creep, especially the saltation
(bouncing) of particles with diameters larger than about 40 μm, destroys
the soil surface thereby releasing smaller dust particles from the
ground. Although we have learned much from these dust-production studies, the large scatter of the data (Borrmann and Jaenicke 1987) suggests
that further field experiments with wind tunnels are greatly needed.
     For a specific set of soil conditions, deflation will only occur
when the wind velocity exceeds a certain minimum value, the threshold
velocity (Gillette 1981; Gillette et al. 1980, 1982). Although threshold velocities have been measured in only a few places, mostly in the
western United States, they clearly vary greatly from region to region
(Pye 1987). This variability is also reflected in the statistics of
dust-storm occurrence as a function of wind velocity at meteorological
sites in arid regions (Goudie 1981, Helgren and Prospero 1987, Middleton
et al. 1986, Morales 1979). Deflation wind speeds in the Sahara are
typically about 5-15 m/s (Helgren and Prospero 1987). In many regions
deflation can start at relatively low wind speeds, even those that occur
routinely under normal meteorological conditions.
     The character of the soil surface is critically important. Disturbed
surfaces deflate at significantly lower wind speeds than undisturbed
soils. Soils affected by anthropogenic activities (Hyers and Marcus
1981) are generally more readily eroded by wind. To assess man's impact
on mobilization, Gillette (1988) reported fugitive dust-emission factors
that they experimentally determined for different non-natural dust-
emission conditions. This aspect has to be considered when estimating
dust-production rates because human activity in the desert and around the
desert fringes has greatly increased in the past few decades and conti-
nues to increase.
     As previously stated, from the standpoint of LRT, it is important to
know the size distribution of dust mobilized close to the soil surface.
Model simulations of the mobilization of soil material (Westphal et al.
1987) show that the original soil-size distribution is modified during
mobilization both by size-dependent lifting mechanisms and by mixing with
advected aerosol components. Thus, the size distribution of soil

aerosols in source areas must be measured for the initial deflation conditions in the immediate area of the deflation event. Very few data on this subject were available (D'Almeida and Schutz 1983, Gillette et al. 1972, Schutz and Jaenicke 1974).

10.2.2. Climatology and Meteorology of Source Processes

The variability of dust concentrations in the atmosphere is critically dependent on meteorological conditions over a wide range of time and space scales over the source areas.

10.2.2.1. **Climatology**. A primary criterion for deflation is that the soil must be dry (Gillette 1981). The amount of rainfall is important especially as related to soil type as well as the condition of the surface.

Arid regions are conventionally regarded as good sources of dust. However, deflation is not inversely related to rainfall amounts as one might expect; indeed, regions with climatological rainfall amounts of 100-200 mm/year, not hyperarid regions, have the most dust storms (Goudie 1981). Moreover, short-term droughts in agricultural regions can cause dramatic increases in dust mobilization (McCauley et al. 1981). After several years of reduced rainfall, dust-bowl conditions can develop in the vast farming areas of the central United States where climatological rainfall amounts are generally adequate for agricultural purposes. For example, in Dodge City, Kansas, during a drought the number of dust-storm days increased from about 10 a year to over 100 (Middleton et al. 1986). Similarly, as a result of the Virgin Lands scheme in the Soviet Union in the 1950s, the frequency of dust storms increased by a factor of 2.5. In both the central United States and the Soviet Union, the great increase in dust-deflation events was caused by the disturbance of the natural soil surface; once the soils become desiccated, they are easily deflated. Uncultivated soils covered with natural vegetation are much more resistant to wind erosion during dry periods.

10.2.2.2. **Meteorology**. To understand the long-range transport of dust, we need to understand the relationship between dust-deflation events in specific source areas and the controlling meteorological conditions. When soil and micrometeorology conditions are appropriate for dust generation, long-range transport can only occur if the eroding winds are accompanied by meteorological conditions that will carry the dust to high altitudes and if the long-range wind field has a strong horizontal component.

Many researchers have investigated the relationship of dust storms to meteorology. Most are retrospective case studies that trace spectacular LRT-dust events back in time and space by trajectory techniques and that infer meteorological conditions in the source area from conventional synoptic-chart analysis. On the basis of these studies we have reached some general conclusions about the meteorological settings of dust storms in different source areas (D'Almeida 1986, Iwasaka et al. 1983, Kalu 1979, Merrill et al. 1985, Morales 1979; for a general review see Chapter 5 in Pye 1987) and developed a general understanding of the synoptic

meteorology of dust events globally. These studies clearly show that
conditions vary greatly from one region to another as depicted in
Figure 10-3.

Such retrospective studies, however, have shortcomings, one of which
is that major storms often occur in regions about which there are few
data. These studies also fail to document the conditions for less spectacular dust events or to explain why the skies over many arid regions
are dust-laden almost all the time. This very important ''background''
dust could constitute a substantial if not major fraction of the global
dust flux. Unfortunately, we have a very poor understanding of the meteorological conditions that produce this background dust because of the
paucity of any detailed systematic studies of dust-storm generation in
source areas. The studies of dust events have focused on a very limited
number of localities and include few measured parameters. Such studies
do not provide the necessary information to extrapolate the data to
broader time and space scales.

Although focused studies of deflation and transport are required to
understand LRT, we could learn much from compilations of meteorological
data from existing data bases. For example, detailed information and
statistics on a wide range of meteorological parameters (e.g., air

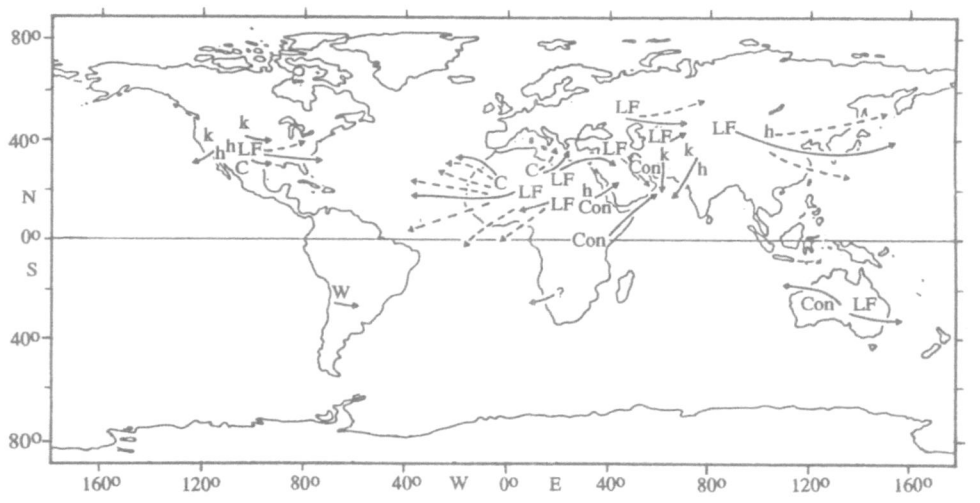

Figure 10-3. The principal synoptic meteorological conditions associ-
     ated with dust events and the main trajectories for dust transport.
     The **solid arrows** indicate primary trajectories, the **broken arrows**,
     secondary or seasonal trajectories. **LF** = low-pressure fronts; **C**,
     surface cyclones; **Con**, convergence (monsoonal low and anticy-
     clones); **k**, katabatic winds (foehn); **h**, haboobs; **W**, upper level
     westerlies. (Middleton et al. 1986)

temperature and humidity, rainfall, wind, horizontal visibility, thunder-
storms, wind-velocity classes, etc.) in conjunction with statistics on
various dust phenomena including haze and dust storms would be extremely
helpful. Past studies (e.g., Jaenicke 1979) have gained some understanding of dust-storm events and haze phenomena from conventional meteorological data. However, individual researchers cannot easily obtain such
data except for specific targeted studies.

Meteorological data that can provide information about deflation and
vertical transport in the boundary layer are especially important (Tsoar
and Pye 1987). Data are needed on the frequency of occurrence of winds
in the range of and greater than the deflation threshold (5-15 m/s) and
on the thermal structure of the lower troposphere. Statistics on the
stability of the different layers of the lower troposphere need to be
established, particularly those describing the occurrence, intensity, and
altitude of inversion layers. Wind-field statistics for the lower troposphere over arid and desert regions and their surroundings are needed
for use in simple mineral-aerosol LRT models. For more elaborate studies, 3-dimensional wind fields (for instance, such as those generated
every six hours by the European Centre for Medium Range Weather Fore-
casts) meet most requirements of the transport studies of specific events
or for seasonal time scales.

10.2.3. Large-scale Monitoring of Desert Source Regions and Source
Fluxes

Even if we understood the source processes, we would not be able to estimate large-area source fluxes quantitatively. Such estimates require a
systematic long-range monitoring program. The only satisfactory measurement technique for this purpose is sun photometry.

Sun photometers are simple hand-held instruments that measure the
spectral attenuation of direct solar radiation by the atmosphere (i.e.,
the aerosol optical depth, or turbidity). In the visible spectrum, measurements are most sensitive to attenuation by particles in the size
range of some tenths of a micrometer to several micrometers. Thus,
photometers do not fully cover the size range required for mineral aerosols. However, turbidity readings can be converted to equivalent aerosol
mass concentrations by correlation measurements with high-volume aerosol
samplers (D'Almeida 1986); ideally, correlations are best done in conjunction with aircraft measurements of vertical distribution of aerosols.
Once calibrated, photometers can provide information about the dust load
in the entire atmospheric column. Several investigators have successfully monitored soil aerosols using sun photometers (D'Almeida 1986,
Jaenicke and Schutz 1978, Prospero and Carlson 1972, Prospero et al.
1979, Volz 1970).

Photometer measurements should be supplemented with standard meteorological surface observations including radiosondes. Especially important are temperature soundings since they provide information about the
vertical extent of the dust layer.

At present we have data from monitoring programs in only a few
places in North Africa. Similar programs are badly needed in other arid

regions. Photometers are used in several networks around the world, including BAPMoN. However, few of these stations are in arid regions.

Sun photometer data will become increasingly important over the next few years as remote-sensing systems become operational. Such data will be required to provide ground truth for the satellite measurements.

10.3. TRANSPORT AND TRANSFORMATION

It is difficult to characterize long-range transport based on experimental field data alone. Most field studies consisted of ground-based network measurements of limited parameters. The density of network stations was so low that one could only characterize the very broadest features of the temporal and spatial variability of transport. Furthermore, vertical distribution data were hopelessly inadequate.

This problem might be alleviated by combining aerosol data with supporting meteorological data. Transport trajectories, although not quantitative, provide information on possible source locations. However, for a more quantitative picture of LRT, we need elaborate models that incorporate the entire process from deflation to deposition. Remote-sensing products could help interpret such data; eventually satellites will provide quantitative measurements of large-scale fluxes.

The general subject of the meteorological aspects of long-range atmospheric transport is thoroughly reviewed by Merrill (1986) and is further discussed by Whelpdale and Moody in Chapter 1 (p. 3). Our working group at the Bermuda Biological Station workshop dealt only with those aspects immediately concerned with mineral aerosols.

10.3.1. Trajectories

Trajectories are mostly used to trace the movement of mesoscale-pollution episodes and, in this context, they have been somewhat successful. For case studies conducted near sources or for transport times of less than a week, isobaric backward and forward trajectories based on weather maps can easily be used for areas with a reasonable coverage of meteorological observations. They are valuable tools that allow one generally to distinguish different air masses and to identify in a general way different source regions. Using Saharan aerosols sampled during ship cruises in 1969 and 1973 on the tropical northern Atlantic Ocean, this simple trajectory analysis (Jaenicke et al. 1971, Schutz 1977, respectively) allowed for the differentiation between air masses originating from separate desert regions. Over the western Mediterranean Sea, Bergametti (1987) used 3-dimensional air-mass trajectories and satellite imagery to identify three major source regions in northern Africa. There was a seasonal transport pattern that was related to the seasonal atmospheric circulation in the western Mediterranean. Wolff et al. (1987) report large variations in trace-element concentrations on Bermuda; these were related to trajectory classes that were associated with distinctly different North American and North African sources.

Long-range-transport studies generally concern events that cover one to two weeks. This places severe demands on trajectory analyses. It is generally acknowledged that, for such long transport times, isentropic trajectory techniques provide the most consistent results. These techniques have been successfully used for tracing the movement of dust clouds from various sources to the northern Pacific (Merrill et al. 1985) and South Pacific (Gagosian et al. 1987) where transport time from the source is usually close to two weeks. Although the accuracy of using trajectories to pinpoint source areas is difficult to verify, the general character of the trajectory is usually consistent with the composition of the concurrently collected aerosol sample. Also, seasonal changes in trajectory paths are reflected in a corresponding change in aerosol concentration and composition. For example, the spring maximum in dust concentration over the central and tropical North Pacific coincides with midlevel trajectories emanating from the midlatitudes in Asia; the summer minimum coincides with boundary-layer trajectories originating in the eastern north Pacific, primarily in the low latitudes (Merrill et al. 1985).

Although trajectories are useful in identifying possible source areas, they do not provide quantitative data on the LRT process. In particular they could not provide information on source processes nor do they incorporate the effects of atmospheric diffusion and removal.

10.3.2. Transport Models

Because of the lack of source data, we must rely on models developed to study LRT. Even simple models, based on mean values (Schutz 1980), provided useful information about the transport process. Schutz et al. (1981) presented a simple model that yields the flux of dust across the west coast of North Africa, the progressive deposition rates across the northern Atlantic, and the remaining flux in the overlying atmosphere (see Fig. 10-4). Their results agree reasonably well with estimates of transport and deposition based on actual field measurements. Note in particular the very high deposition rates close to the source in Figure 10-4.

More complicated LRT models are based on atmospheric physical principals. These models try to account for the temporal and areal variabilities of dust outbreaks and the vertical structure of the atmosphere. With such models, one can partially compensate for the lack of detailed data and at least some observed transport features can be explained as related to meteorological parameters and synoptic patterns.

The most advanced dust transport model was developed and tested for the Saharan case (Westphal et al. 1987). This model combines a limited-area, dynamical model with a predictive aerosol model and simulates the spatial and temporal distributions of aerosol particles with diameters from 0.2 μm to over 100 μm. Because it is initialized with mean wind and temperature profiles, it realistically simulates the main features over the desert within the planetary-boundary layer. The model aerosol-size distributions and mass distributions agree reasonably well with actual measurements, although mass concentrations within the submicron and large-sized-particle range are overestimated and those for the ultra-giant particles are underestimated. Despite these shortcomings, models such as this provide a useful depiction of mobilization and transport

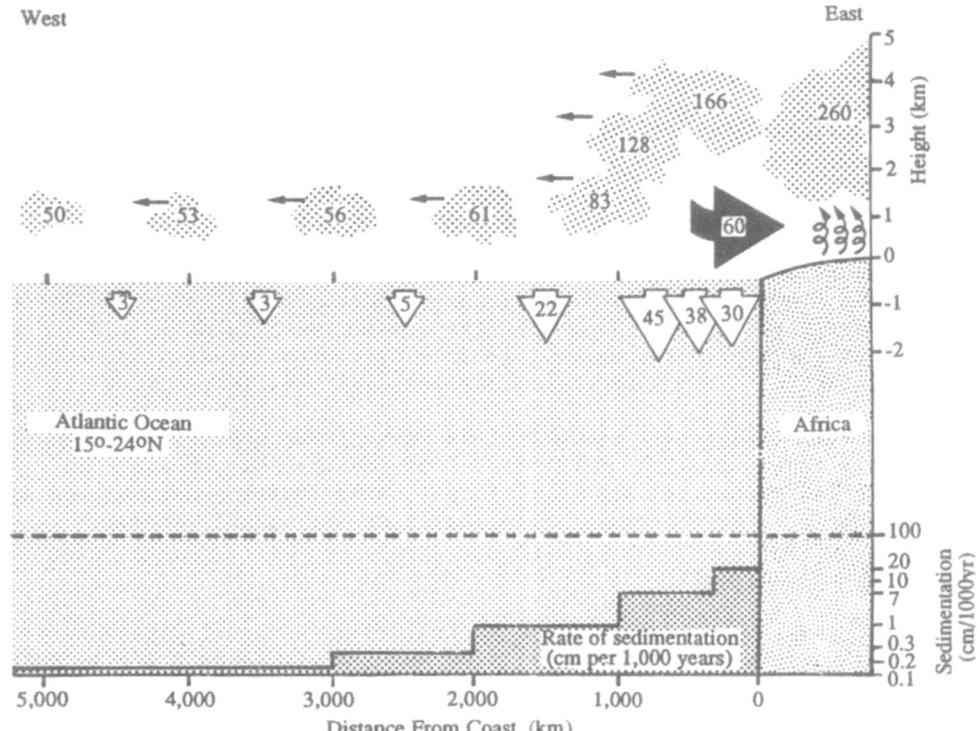

Figure 10-4. The transport and deposition of Saharan dust over the tro-
    pical North Atlantic. The airborne flux (millions of tonnes a year)
    is shown in the **stippled areas**; the deposition flux (millions of
    tonnes a year), in the **downward-pointing arrows**; the aerial return
    (millions of tonnes a year) to Africa, by the **solid arrow**. The
    computed aeolian sediment accumulation rates are presented in centi-
    meters per thousand years. (Middleton et al. 1986, based on model
    calculations from Schutz et al. 1981)

processes. More sophisticated models are being developed, including a 3-
dimensional version of the Westphal model (Westphal et al. 1989). These
models are useful because they are adaptable to many desert environments
in addition to the Sahara.

On a larger scale global models (Joussaume and Sadourny 1989) simulate the mean intercontinental dispersion patterns of dust. Such models
are useful in assessing global-scale effects, such as those associated
with climate change.

As we stated earlier, because of the importance of particles with
diameters below about 10 μm, LRT models must be designed to deal effectively with this size class. However, models must also be able to

simulate the transport of ultra-giant particles with diameters of 50 μm
to 100 μm and greater. Such large particles have been detected not only
over the desert during sandstorms but also thousands of kilometers from
their sources and in ocean-water columns and deep-sea sediments (Bergametti 1987, Carder et al. 1986, Leinen 1989). The presence of such particles is unexpected because of their large settling velocities. We
suspect that such large particles could only be transported over great
distances through the periodic mixing action of large updrafts near
clouds and frontal systems; consequently, accurate models should include
parameterizations of cloud physical and convective features so as to
simulate these lifting processes.

Our working group concluded that model development was relatively
advanced, and there seemed to be no urgent need for new models. Rather,
we felt that the major need was for measurements of source-relevant parameters essential to validate existing models—a necessity that must be
met before the present models can be fully developed and extended.

10.3.3. Modification During Transport

Mineral-aerosol particles are relatively inert and do not undergo drastic
chemical changes, such as those experienced by species in the nitrogen
and sulfur cycle. However, during transport mineral aerosol undergoes
dramatic physical changes during wet and dry removal.

The formation of loess deposits is an example of the effect of size-
dependent removal processes (for a general discussion of loess, see Pye
1987). The parent material is generally believed to be soils washed out
at the edges of glaciers. Winds carry away material from the glacial
outwash. Larger particles (10's of microns in diameter) are redeposited
within several hundreds of kilometers of the source (Pye 1984). The
resulting loess deposit is a fine-textured soil that is devoid of fine
particles and coarse particles.

Schutz (1980) has shown the evolution of aerosol-size distribution
during a dust storm and the subsequent transport over oceans. The mean
radius of the aerosol mass distribution rapidly shifts towards smaller
sizes as transport progresses; this process is schematically depicted in
Figure 10-5. These size effects are primarily caused by gravitational
settling and are well understood. However, little is known about changes
raised by cloud and rain in the size distribution. Especially important
is the formation of agglomerates of many small particles and various
salts (Andreae et al. 1986, Jaenicke et al. 1971). These agglomerates
could influence the vertical flux of the dust and the optical properties
of the aerosol.

With liquid water present, chemical transformations can take place
on dust particles. As an example, Glaccum and Prospero (1980) and Schutz
and Sebert (1987) observed large needlelike gypsum crystals in Saharan
air samples collected over the northern Atlantic Ocean. These needles,
up to 100 μm long, could be the product of wetted calcium carbonate compounds, such as $CaCO_3$ transformed through the uptake of SO_2 from the air.
Other chemical transformation processes are possible, including the
adsorption of gases onto dry and wetted particles. The adsorption of
gaseous organic compounds could produce surface coatings that could

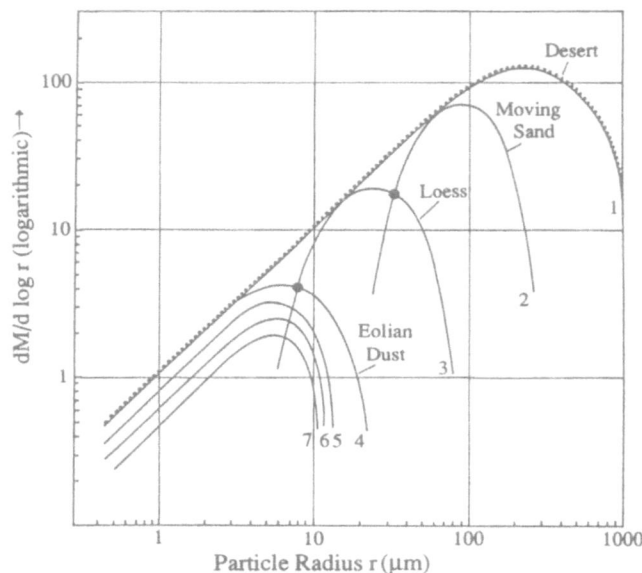

Figure 10-5. Sand fractionation processes by wind. The original sand
    distribution is fractionated into the major fractions **2** moving
    sand, **3** loess, and **4** aeolian dust as a function of distance from the
    source. Curves **5, 6,** and **7** depict the change in concentration due
    to both wet and dry removal from the atmosphere. The sum of curves
    2, 3, and **4** should be equal to the original curve **1**, the desert.
    (Junge 1979)

affect the uptake and release of water during cycles of changing humidity. However, in some cases, these chemical conversions could be an
artifact generated after collection on the filter (Glaccum and Prospero
1980).

  Finally, the cloud cycling process can greatly alter the solubility
characteristics of many species in mineral dust. During the condensa-
tion/evaporation cycles, the mineral particle is exposed to many pH conditions and is coated with an aqueous solution that can range from
extremely dilute to brinelike. This exposure can leach many relatively
inert species out of the mineral matrix and render them soluble (see
section 10.5.3.2. below).

10.3.4. Source-Specific Tracers

In studies of long-range transport, aerosol samples are collected at
sites far from sources. Although attempts are made to identify sources
by tracing air-mass movements, source identification is often ambiguous
because the back trajectory of the air mass usually passes over many
regions, any one of which could have been the source. It is, therefore,

quite helpful to have markers or tracers in the aerosol that can uniquely
identify a specific region. Unfortunately, such source ''fingerprints''
are very difficult to find.

The elemental composition of an aerosol is one possible tracer.
Lowenthal and Rahn (1985) have identified many elemental tracers of
anthropogenic sources. This same technique has been attempted for soil
aerosols. Schutz and Rahn (1982) showed that the composition of LRT dust
particles is quite similar to the composition of those at the source.
Unfortunately, their investigation also shows that the composition of
particles in the LRT-size range is close to that of the average crustal
material (Fig. 10-6). Thus, our working group agreed that mineral aerosols did not have obvious source-tracing characteristics.

Although the composition of individual samples may not provide a
clear source signal, the analyses of larger sample sets can yield important evidence. For example, Heidam (1984, 1985) studied the enrichment
factors of Arctic aerosols using statistical multivariate methods.
Because Mn yielded an enrichment factor ranging from 0.5-5 over a year, a
purely crustal origin is implied. However, there is a systematic seasonal variation in the enrichment. Although this variation is only a

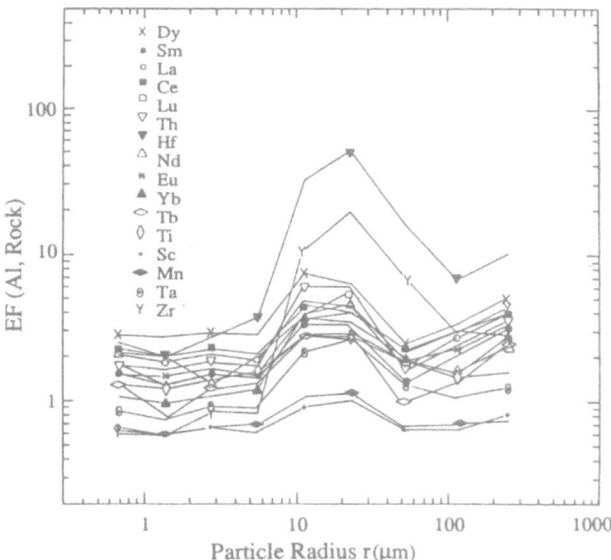

Figure 10-6. Enrichment factors (relative to average crustal material)
for various soils as a function of soil particle size. Enrichment
factors are calculated relative to the Al concentration in average
crustal rock. The largest enrichments were obtained for particles
larger than about 20 μm in diameter. Particles in this size range
could not normally be transported great distances. Particles in
the LRT size range (several microns in diameter) have compositions
close to that of average crustal rocks. (Schutz and Rahn 1982)

small part of the total Mn concentration signal, it shows that the non-crustal Mn probably originates at a distant ferro-industrial source. Heidam (1984) drew the same conclusion about Fe, which had an even lower enrichment but the same seasonal pattern.

These results suggest that source-specific identifiers might be found for LRT mineral aerosols if multivariate methods were to be applied to time series for a suitable set of elemental ratios. Bergametti (1987) successfully used this technique to study Si/Al and Fe/Al ratios in aerosols collected at Corsica during dust events.

The isotopic ratios of selected elements might also be used to characterize source regions. Grousset et al. (1988), in studies of sediments and aerosols in and over the northern Atlantic, applied isotopic ratios of Nd to characterize the origin of the aeolian component. Over the southern Atlantic, Biscaye et al. (1974), by using 87Sr/86Sr ratios, were able to distinguish dusts from the Sahara from those from the Kalahari Desert of southwestern Africa.

Traditional mineralogical methods may also be useful. Bucher and Lucas (1984) and Schutz and Sebert (1987) found that calcite and palygorskite are tracers of mineral dust from the northern Sahara. Buat-Ménard et al. (1983) found that the amounts of illite and kaolinite from mineral aerosols collected in Enewetak in the Pacific help to distinguish sources in Asia from those in North America.

Sources can also be identified by the micromorphological investigation of particles using the scanning electron microscope (see, for example, Coude-Gaussen 1989). This method is applicable to single mineral particles with diameters larger than about 5-10 mm. However, this procedure is not useful with smaller particles because the morphology can only be defined in terms of very broad structural features; and energy-dispersive, x-ray-spectrometer studies can only provide qualitative information on the major elemental composition.

Another approach is to take advantage of the unique molecular signatures of various classes of organic compounds, such as lipids, from marine and terrestrial plants and animals. Plant materials are often mobilized with soil particles and carried great distances. These compounds can be excellent tracers (Gagosian 1986, Simoneit 1977).

Pollen may also be potential markers because each species has a very different form and structure and consequently could be used to characterize a source. Under the appropriate conditions, pollen can be injected into the atmosphere along with the aerosol particles. Thus, the pollen might reveal the connection between source areas, transport systems, and receptor sites. Any climatic gradient between subregions at a source would be reflected in the vegetation characteristics and thus in the pollen produced in that area. In studies of marine sediments off northwest Africa, Hooghiemstra et al. (1986) identified seven groups of pollen that could be used to characterize five distinct latitudinal vegetation zones in northwest Africa.

Finally, elemental carbon produced during biomass burning is also a potential tracer of soil dust emitted from semi-arid regions, such as the Sahel. For example, Andreae et al. (1984) found detectable concentrations of elemental carbon from biomass burning in Africa over the tropical Atlantic. However, this tracer is not always useful because biomass burning does not necessarily take place in the season that the most dust is generated.

10.3.5. Remote Sensing of Large-Scale Atmospheric Transport Processes

Ultimately, remote sensing will play a critical role in the quantitative study of LRT. However, most current literature uses satellite-derived images in qualitative investigations of the magnitude, frequency, and movements of large dust storms, intense pollution episodes, volcanic dust clouds, and haze layers produced by fires (see, e.g., Chung and Le 1984, Fishman et al. 1987, Flannigan and Vonder Haar 1986, Middleton et al. 1986,). Figure 10-7 illustrates the movement of a dust cloud across the tropical northern Atlantic from North Africa towards the Caribbean using low resolution, daily images from a geostationary satellite. The dust concentration in the region of Barbados during this particular event was about 80 µg/m3 on June 14-16, 1984, compared to about 5 µg m/3 on June 18-19 after the dust had passed (unpublished data available from Dr. J. M. Prospero, RSMAS, University of Miami, FL 33149-1098).

Despite the usefulness of satellites, at present such data cannot be used to estimate the total amount of dust transported from a source area or the amount deposited to an ocean. Truly quantitative techniques are yet to be developed. In this section we have briefly assessed recent and anticipated developments in remote-sensing technologies that offer particular promise for the quantification of long-range transport. A program in which remote sensing would be properly integrated with intensive in situ measurements would answer many scientific issues not only for mineral dust but also for other aerosol species.

Lasers are very useful in the remote sensing of the atmosphere (Killinger and Menyuk 1987). The simplest laser technology is surface-based lidar, which can provide high-frequency time-series measurements of vertical distributions of aerosols over a permanent monitoring station (Iwasaka et al. 1983). This technique is most effective when it is used as a part of a comprehensive experiment involving in situ measurements of aerosol and trace gas species and detailed meteorological analyses of the study area. Such measurements provide a detailed picture of transport processes that will not be rivaled by any technology on the horizon. Unfortunately, most lidar studies are, in effect, isolated experiments with very little backup except for conventional meteorological data.

Airborne lidar is used in intensive field campaigns to continuously characterize the spatial distributions of atmospheric aerosol associated with such phenomena as dust storms (Talbot et al. 1986), regional haze from biomass burning (Andreae and Andreae 1988, Andreae et al. 1988), and air-pollution episodes moving from a continent to an ocean (Harriss et al. 1984). At the present time, such data are useful as a survey tool to map distributions so that detailed aircraft studies can be planned in real time.

High-frequency satellite imagery was also available from geostationary and polar-orbiting satellites with multispectral passive-imaging systems (Carlson 1979). These data are used to infer various characteristics about aerosol concentrations and optical effects (Fraser et al. 1984, Griggs 1983) and are used to produce global maps of aerosol optical

SAHARAN DUST TRANSPORT

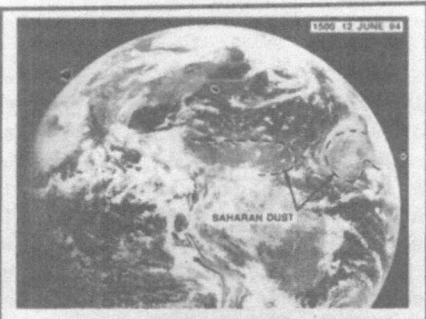

Figure 10-7

The movement of a dust cloud across the tropical North Atlantic from North Africa towards the Caribbean using low resolution, daily images from a geostationary satellite (courtesy of the National Oceanic and Atmospheric Administration, Washington, D. C.)

thickness over the oceans. Maps of weekly mean values are now being routinely distributed to the atmospheric chemistry community. These maps clearly illustrate the impact of the major arid regions and also of major anthropogenic sources.

Remote-sensing techniques, such as high-altitude airborne multispectral imaging, high-resolution satellite imaging (e.g., Landsat and SPOT), space-shuttle photography, and synthetic-aperture radar (SAR), are employed for specialized investigations of soil erosion, regional pollution, and biomass burning and other land disturbances as well as to identify regional geomorphological features (National Academy of Sciences 1986, 1988; National Oceanographic and Atmospheric Administration 1987).

The use of radar and microwave remote sensing to investigate the long-range transport of natural and contaminant materials was, to our knowledge, largely unexplored. Both satellite and aircraft data sets are available.

The future holds much promise for the further development of an increasingly powerful array of remote-sensing tools devoted to earth-science studies. Advanced lidar technology for use on high-altitude aircraft, the space shuttle, and the US ESA/Japan Earth Observing System (EOS) will be of particular interest for the studies recommended by this workshop. With data from these systems, it should be possible to develop relatively quantitative global maps of aerosol distributions essentially on a daily basis and to follow aerosol cloud movements over very long distances. Thus, these lidars would provide us with two critically needed data components for our studies.

10.4. DEPOSITION

Aeolian deposits are prominent in many parts of the world both on the continents and in the oceans (for reviews, see Prospero 1981, Pye 1987) and play a major role in geological processes, especially in the formation of soils and deep-sea sediments. To understand the complete cycle of mineral dust, we must have accurate and extensive deposition measurements. Such data are vitally important to the computing of deposition fluxes. With an appropriate measurement strategy, it should be possible to make good estimates of deposition rates, especially over the oceans. This is in contrast to the assessment of sources, where we can never hope to have anything better than very rough estimates of upward fluxes.

Nonetheless there have been few systematic studies of the actual deposition rates of mineral-aerosol particles, especially over remote areas and most particularly over oceans. Recent deposition studies in several ocean areas (Bergametti 1987, Prospero et al. 1987, Uematsu et al. 1985) affirm the importance of mineral-aerosol fluxes to ocean chemistry and to sediment-forming processes. These studies also show that dry- and wet-deposition rates vary considerably in time and space and that a major fraction of the total annual deposition occurs in only a few days of the year. Consequently, accurate deposition-flux data can only be obtained from long-term, continuous-time-series measurements.

Normally, precipitation (wet process) removes about 80-90% of the aerosol mass in the atmosphere. Wet deposition is comparatively easy to measure using automatic collectors that are only open during precipitation. The situation with mineral dust is different because of the high concentration of large particles. Because of the high gravitational settling rate of large particles, dry deposition accounts for a significant fraction of the total deposited flux. Data gathered over the oceans (Bergametti 1987, Prospero et al. 1987, Uematsu et al. 1985) suggest that about 20% to 40% of the mineral-aerosol flux is attributable to dry deposition. The dry-deposition rates for dust are much higher than those for other submicron aerosol components (such as radionuclides, nitrate, and sulfate) especially at locations close to the dust sources.

Unfortunately, dry deposition to the soil or to water surfaces cannot be directly measured accurately (Giorgi 1986, McMahon and Denison 1979, Sehmel 1980, Slinn 1983). Active aerosol-measurement techniques (such as electo-optical spectrometers) are hampered by slow instrument-response times and poor counting sensitivity. Passive techniques have commonly used, for example, the measurement of deposition to a surrogate surface, such as a plastic plate (Arimoto et al. 1985, Jonas and Heinemann 1985, Lane and Stukel 1978). Such techniques are criticized because they do not mimic the characteristics of natural surfaces (for example, in their aerodynamics, roughness, or adhesion properties; Hicks et al. 1980). Nonetheless, the values obtained with such devices could be accurate within a factor of 2 or 3 (Arimoto et al. 1985). Although such accuracy leaves much to be desired, at the very least these collectors do provide a good picture of temporal variations.

Another approach is to compute the flux using dry-deposition models (Arimoto and Duce 1986, Giorgi 1986). This requires accurate measurements of the particle-size distribution together with other parameters, such as relative humidity and wind speed (Slinn 1983). Because of the large temporal variability of the aerosol, aerosol properties must be measured on time scales of a day or less.

The coupling of precipitation and aerosol studies can also provide valuable information on removal mechanisms. Such integrated data would be useful in validating LRT models, which eventually must include a parameterization of the aerosol-removal process. Removal processes could not be realistically incorporated into LRT models because deposition rates are dependent on many factors, including particle concentration, composition, and size, for which the necessary data were not readily available. Therefore, we still had to rely on experimental field measurements to determine fluxes to the ground and sea surface (Buat-Ménard 1986a, 1986b).

10.5. THE SIGNIFICANCE AND IMPACT OF THE LONG-RANGE TRANSPORT OF DUST

Mineral particles play a role in many physico-chemical processes of the atmosphere, hydrosphere, and geosphere, some of which are only poorly understood. As a consequence, mineral dust can impact such diverse systems as weather and climate and can affect the well-being of living organisms (for a general discussion, see Pye 1987). Some of these effects, which were a major concern of the participants of our NATO workshop, are discussed in the following sections.

10.5.1. The Impact of Mineral Aerosol on Clouds and Rain

Mineral particles play an important role in forming rain. Usually, the formation of ice crystals is the first and most critical step in the initiation of precipitation. Ice nucleation depends on several aerosol properties, including the size and mineralogy of dust particles. Research conducted over the past decade has made it clear that deserts are a major source of mineral dust for huge areas of the world. It is paradoxical that mineral dust from arid regions may play an important role in triggering the rain formation process on a global scale. However, studies of the large-scale impact of mineral dust on precipitation processes have not been done.

Mineral dust is often derived from sources that contain basic compounds (Schutz and Sebert 1987). The incorporation of dust in rainwater tends to shift the pH towards alkalinity. As a consequence, rainwater collected in Europe during Saharan dust episodes exhibits a pH of 6.2 as compared to the usual 4.6 (Schutz and Kramer 1987). Indeed, in some areas of the world, such as India (Khemani et al. 1987), there is no acid-rain problem--the rain is basic because of alkaline dust.

10.5.2. Radiative Effects of Mineral Aerosol

Aerosols interact with solar radiation. Most obvious are the optical effects of volcanic eruptions, such as El Chichon, that produce colorful sunsets. Beautiful sunset effects are also produced by aerosols generated by man's activities.

Are aerosol radiative effects severe enough to influence weather and climate? Aerosol could affect climate by changing the albedo of the earth and by altering the thermal structure of the atmosphere. Increasing the albedo causes radiative energy to be scattered back into space; consequently, there is a net energy loss to the earth and subsequent cooling. Conversely absorption retains radiative energy and converts it into heat; dust layers can cause a redistribution of heating in the atmosphere, producing a net cooling at the surface and a net heating at higher altitudes. This redistribution of heating could strongly effect a variety of processes, including cloud formation and distribution.

An extreme example of the climatic effects of aerosols is that postulated for ''nuclear winter.'' Absorbing particles (soot) at high altitudes could cause cooling and produce an inversion of the temperature structure of the atmosphere. This inversion produces the harsh climatic effects that are hypothesized. It is noteworthy that the aerosol concentrations (and aerosol optical depths) employed in some of the less severe nuclear-winter scenarios (National Academy of Sciences 1985) are comparable to those observed in major dust storms in West Africa.

At present, the link between aerosols and climate is poorly understood. Many studies have tried to correlate volcanic activity and global climate; however, results are ambiguous because of the large natural variability of the earth's surface temperature. On the other hand, using a long time series of atmospheric turbidity measurements, Helmes and Jaenicke (1988) showed that there is a statistically significant

correlation between the aerosol optical depth and the average temperature
of the Northern Hemisphere over the last 100 years.

It is well established that dust storms cause localized thermal
effects. Carlson and Prospero (1972) and Carlson and Caverly (1977)
estimate that the mineral-aerosol content of the trade-wind region off
the Sahara desert could cause a temperature reduction of a few degrees in
the region's surface air. Over the desert during sandstorms, temperature
reductions of more than 5° C are common. Measurable reductions in surface temperature are also observed over land at considerable distances
from the dust storm areas, such as in Nigeria (Brinkman and McGregor
1983).

There is speculation that aerosols can effect the radiation balance
sufficiently to have an impact on the meteorology and climate in arid
regions. Charney et al. (1975) suggest that high-mineral-aerosol concentrations could cause sufficient heating aloft to suppress convection
and thereby reduce cloud processes and rainfall. Under such a scenario
deserts might become self-propagating. On the other hand, desert aerosol
might be self-limiting: in aerosol free air, the sun's energy heats the
desert surface causing convection and the uplifting of surface material.
With increased amounts of mineral aerosol, solar energy would be scattered back into space, the heating of the desert surface would be reduced,
and the generation and lifting of dust would be impeded.

Aside from the large-scale aspects of dust and climate, there is
interest in the radiative effects of high concentrations of dust aerosols
on the dynamics of the atmosphere and the local meteorology (Carlson and
Caverly 1977, Karyampudi and Carlson 1989, Westphal et al. 1987, 1989).
Their work suggests that the impact of high-dust concentrations is indeed
significant.

To date, most radiation models are based on the radiative properties
of bare mineral particles. Such models also incorporate the effects of
clouds, especially with regard to albedo. However, there is speculation
that the presence of mineral aerosol might modify the optical and radiative properties of clouds. Once a mineral particle is incorporated into
a cloud droplet, the droplet becomes much more absorbing. Thus, the
radiative properties of clouds in dusty regions might be markedly different from conventional clouds. However, there are no radiative models
that account for clouds with mineral particles within the droplets.

Although the ultimate radiative effects of dust were uncertain,
there is sufficient evidence to suggest that the effects of mineral dust
must be considered when modeling the radiation balance. At the very
least, these effects should be estimated for the lower latitudes, which
are the major areas of intense dust production and transport. These
regions are also the location of the major energy exchanges of the
earth's climate system.

10.5.3. Ocean Chemistry and Sediment Formation

Aeolian mineral particles are a major source of the particulate and dissolved materials found in many areas of the open ocean (see, for example,
Chester 1986, Prospero 1981, Pye 1987). Particles deposited at the sea
surface fall to the bottom, and, while settling, undergo important

changes in their morphology and chemical composition. These changes affect the chemical composition of the surrounding water and the organisms and biogenic materials that are also settling into the sediments. The exchange of soluble compounds and elements between such particles and the water can control the concentrations of trace elements in open ocean waters.

10.5.3.1. Formation of Ocean Sediments. Aeolian inputs can be a significant source of nonbiogenic material in ocean sediments. This is not only true for waters off major arid regions (Africa, Asia, Australia) but also for central oceanic areas far from any continents. Various approaches have been used to identify and quantify the aeolian contribution (Prospero 1981). Biscaye (1965) has studied fine-fraction (clay) mineralogy and Rea and Janecek (1982) and Rea et al. (1985) have measured quartz particles. Stable isotope measurements do allow us to differentiate between various continental source regions (Biscaye et al. 1974, Grousset et al. 1988). Model-derived estimates and direct measurements of atmospheric deposition rates indicate that, in the northern Pacific Ocean and the tropical northern Atlantic, the post-glacial accumulation rates of deep-sea sediments can be explained primarily by aeolian inputs (Prospero et al. 1987, Schutz et al. 1981, Uematsu et al. 1985).

A significant fraction of the deposited mineral particles is grazed by zooplankton (Chester 1982). This has great importance for both organisms and particles. As a particle passes through the digestive tract of a larger zooplankton organism, it undergoes strong acid attack that can result in increased dissolution rates inside the organism or, once excreted, in the ocean if the mineral particle is exposed once again to the water column. However, in the gut the insoluble portion of the particle is usually incorporated in a fecal pellet (Chester 1982), which is much larger than the original particle and consequently has a much higher settling velocity. The organic covering on a fecal pellet provides a degree of protection against further dissolution, at least until the pellet itself is disaggregated or until it too is ingested by another organism. During its passage to the ocean floor, a particle may pass through many such ingestion/excretion cycles and, as a consequence, could be almost completely dissolved. In the sediment, the particle could undergo further processing by burrowing organisms and bacteria; this could bring about the further dissolution of the more resistant materials, albeit on a much longer time scale.

Large fecal pellets have high Stokes settling velocities (30-100 m/day). If a significant fraction of the mineral particles are incorporated into fecal pellets, then the geographic pattern of atmospheric deposition should not be greatly affected by the subsequent transport by currents and the concentration distribution of minerals in sediments should reflect that of the aerosol at the time of deposition. Under such conditions, aeolian mineral components in sediments could provide information about paleoclimatic conditions, such as wind patterns. However, before aeolian component data can be used for such purposes, we need more quantitative information on the settling velocity of mineral-aerosol particles in the oceanic water column and on the proportion of aerosol materials that reach the ocean bottom. Such studies would require the use of sediment traps and large-volume *in situ* filtration devices, among others. It is, therefore, important that future ocean-flux studies include systematic measurements of the flux of aeolian material in various ocean basins. For maximum effectiveness, the water-column studies should be done concurrently with atmospheric measurements of aerosol properties. Because of the expected temporal variability of both the atmospheric deposition and biogenic fluxes in the water column, long time-series measurements are also needed.

10.5.3.2. Release of Soluble Compounds and Elements to the Water Column.
The few studies of the dissolution properties of aerosol particles in aqueous solutions suggest that particle dissolution of aeolian materials could be important. The dissolution of particles in seawater is a rate-controlled process that depends greatly on the size and structure of the settling particle and its history while settling to the ocean floor, especially the role of organisms. Of all the elements in mineral particles, those that constitute the structural matrix (Al, Si, and Fe among others) are the most resistant to dissolution. Even so, a significant fraction of these elements can dissolve in seawater in a matter of hours because of their previous weathering history in the soil and in the atmosphere and also because of the strong electrochemical forces acting on the particle surface when it is exposed to water.

Studies show that up to 10% of the total aluminum and iron contents of aerosol particles dissolves when in contact with seawater (Duce 1986, Maring and Duce 1987, Prospero et al. 1987). Similar amounts of silica are presumably leached out of the seawater as well. This process could control the concentration distributions of these elements in the surface waters of the northern Atlantic and northern Pacific (Landing and Bruland 1987, Measures et al. 1984, Orians and Bruland 1987).

Other elements more weakly linked to the aluminosilicate matrix (Mn, As, P, heavy metals) may also be leached to the surrounding water. However, the low solubility of many such elements in oxidizing environments prevents them from remaining in solution and they may be largely retained by the settling particles.

The impact of mineral dissolution on the marine environment varies from element to element (Moore et al. 1984). Perhaps the most important impact is on the growth rates of marine organisms. Such development is basically controlled by the flux of nutrients and of selected elements. These fluxes are normally dependent on inputs from deeper waters. Whenever vertical fluxes are restricted, only sparse populations of primary producers (photosynthetic plankton) can be supported. In oligotrophic areas, such as the Sargasso Sea and the central northern Pacific gyre, new photosynthetic productivity depends on the eddy diffusion of nutrients through the nutricline and is consequently limited to a thin layer at the bottom of the euphotic zone.

In some oligotrophic ocean waters, the atmospheric inputs of nutrients and other biologically important elements can be of the same magnitude as those from deep waters (Duce 1986). Even in regions that are not normally regarded as oligotrophic, the atmospheric input of essential trace metals, such as iron, may be essential to phytoplankton growth (Martin and Fitzwater 1988). Thus, LRT can make a substantial contribution to the overall primary productivity of large regions of the world ocean.

Settling mineral particles are known to be negatively charged and may bind both metallic ions and hydrophobic organic molecules on their surface. Because of the large specific surface of the aerosol particles, they have a large potential for scavenging and, hence, they must be considered as a sink mechanism when dealing with the biogeochemical cycling of some pollutants.

We have a poor understanding of many aspects of the interaction of deposited aerosol particles and the surrounding seawater and organisms. In particular, we have little data on the effect of particle size and morphology on settling velocities, the rates of dissolution of various elements, and the ability of organisms to feed on such particles. Also, there is little data on the importance of aerosol inputs in promoting biological productivity; at that time, the implied impact was based on many assumptions about atmospheric and oceanic processes. Finally, we know little about the effects of biological ''processing'' in regulating the fraction of aerosol particles that reach deep-ocean sediments.

10.5.4. Paleoclimatic Interpretations

The concentration and physical characteristics of the aeolian components in the sediments could provide information about climate on the continents and the strength of the transporting winds. The basic hypothesis is that the concentration of these components would increase with increasing aridity in the source area and with the increasing speed of the transporting winds. For example, in their studies of northern Pacific core sections covering the past 70 million years, Rea et al. (1985) found major changes in particle size, quartz concentration, and mass-accumulation rate. On this basis, they concluded that there have been major changes in climate and the consequent vigor of the global wind circulation systems. Some of these conclusions are contrary to the long-held beliefs of geologists and paleoclimatologists whose conclusions have been based on more conventional data.

Deep-sea sediments undoubtedly contain a strong and varied aeolian signal that must be linked in some way with climatic processes on the continents. Nonetheless, at this time it is difficult to interpret the aeolian signal in the sediments. Before this can be done, the entire process of deflation and long-range transport must be better understood. A major problem in interpreting the oceanic record is that we can use only very simple models to understand the link between particle size and the speed of the transporting winds. Our knowledge is quite good at the micrometeorological level from studies of deflation. However, we have no information on the impact of larger scale atmospheric mixing processes on the modulation of particle size. The most dramatic evidence of this lack of knowledge is that we are unable to arrive at a convincing explanation for the supergiant particles that have been found as far as many thousands of kilometers from possible sources (Carder et al. 1986).

Another difficult question is the relationship between soil deflation and aridity. As we stated earlier, the correlation between dust-storm frequency and aridity is at a maximum in regions where the annual rainfall is between 100 mm and 200 mm, not in hyperarid regions (Goudie 1983). This observation was consistent with the growing evidence that water is essential for the production of fine-sedimentary particles and for creating surficial environments where these weathering products can be readily deflated. In contrast, many hyperarid regions are essentially blown out; in such regions, the principal features of the terrain are rocky plains and dune fields comprised primarily of coarse materials not readily deflated or transported great distances by winds. Therefore, the most abundant producers of soil dust may be those areas that are undergoing a transition from moist to arid (Prospero and Nees 1986).

Another aspect that we considered was the relationship between dust storms and weather. Studies of dust storms in Africa, North America, and Asia show that there is a characteristic set of synoptic conditions unique to each region. Thus, if the climate were to change, the weather could also be expected to change and, with it, the sources of the dust, the dust-generation mechanisms, and the transport paths. Therefore, even if there was a simple relationship between dust size, wind velocity, and transport distance, changes in sediment character could be produced simply by changes in the source areas and the transport paths.

Finally, we know that large quantities of dust can be produced by climatic conditions other than aridity. Most significant is the evidence (Pye 1984) that the immense deposits of loessal soils in North America, Europe, and Asia were formed through the action of winds in conjunction with glaciations. Because of the dearth of fine particles in these deposits, we can surmise that huge quantities of fine-soil material must have been carried away by winds, much of it to the oceans.

10.6. CONCLUSIONS AND RECOMMENDATIONS

10.6.1. Source Terms

Our most urgent need is for information regarding source identification and source processes. The pressing questions are: Where does the dust come from? What is the character of the source soils? Under what conditions is dust generated? What are the controlling meteorological conditions for deflation?

We place a high priority on the study of soil characteristics and the performance of deflation experiments in different soil types, particularly wind-tunnel studies in the field. Our primary objectives are to:

1. Investigate the effect of soil composition and physical properties, especially disturbance effects by man,

2. measure the size distribution of deflation aerosols with respect to micrometeorological and soil conditions, and

3. study mineral and chemical segregation during deflation to obtain data relevant to identifying source tracers.

We also stress the importance of making deflation measurements in widely varying environments around the world since most available data apply only to limited areas of the Sahara and the western United States. This is partially due to a limited research capability; to our knowledge, only two people are currently active in deflation studies. There is essentially no data from sites in the Southern Hemisphere.

There are very few data on the vertical distribution of mineral aerosols in the atmosphere from any location on earth. Data over source areas is essential if we are to bridge the gap between studies of surface-related deflation process and the larger scale meteorology necessary to determine whether a dust event is restricted to a small area or whether it is undergoing LRT. Thus, vertical profiles of mineral dust and related parameters measured from aircraft are of paramount importance.

Because we realized that it would not be possible to carry out field programs in all areas of interest, we suggest that all possible information about mineral dust deflation be extracted from existing data records. For example, solar-intensity data might be used to infer turbidity. Meteorological observations of haze, dust, and visibility could be useful if their relationship to larger scale dust patterns could be established. Also, it might be possible to extract mineral-aerosol concentration data from existing measurements of trace elements, such as Fe, Mn, or Al. Many monitoring programs measure these species for other purposes; nonetheless, at most sites the dominant source of these elements will most likely be soil material.

Because of the increasing interest in LRT, an atlas of meteorological data specifically aimed at dust-deflation studies might be feasible. Specialized atlases are often produced for scientists in specific study areas, e.g., the monthly El Nino bulletin published by NOAA (US Department of Commerce). Many national and international agencies compile meteorological data atlases, however, they are assembled without regard to the needs of those studying dust deflation.

Although the major objective of our work is to understand the factors controlling dust deflation from the standpoint of LRT, these studies could have other applications as well. For example, eventually it should be possible to predict the location, duration, and intensity of dust storms. This predictive capability will benefit local populations since the storms greatly reduce visibility and impact on everyday activities. Therefore, we suggest that field programs make one of their goals the improvement of this predictive capability.

10.6.2. Transport Terms

We must have aerosol and precipitation data from those areas impacted by LRT materials. Such data is necessary for quantifying the magnitude of the transport and of the deposition. Because of the sporadic nature of dust events, aerosol concentration and deposition rates are highly variable. Thus, quantitative estimates of fluxes can only be obtained by long-term monitoring. Such data are also necessary to validate the many transport models that are under development as are measurements of aerosol size, concentration, and composition.

Sun photometer networks should be established to monitor the aerosol optical depth on a global scale. These measurements would provide important data on aerosol transport of all kinds. Photometers should be placed at aerosol monitoring stations in an effort to relate surface aerosol concentrations to vertical column loadings. In addition to providing directly useful data for LRT studies, these measurements would serve as ground truth for satellite systems that will become active in the 1990s.

In the end, the quantification of LRT will depend on the development of remote-sensing techniques, especially those aboard satellites. Over the next decade, several systems—both passive multispectral ones and active ones based on lidar—will be capable of providing such data. Although at the time of our meeting at the Bermuda Biological Station, the primary purpose of many of these systems was not directly concerned with aerosol measurements of interest to us, these systems could be enhanced so that the product would be more suitable to our needs. Closer contact between the aerosol community and the satellite-system designers should be encouraged. At the very least, efforts should be made to assure that the data processing will be carried out in such away that the data of greatest interest to the LRT community are readily accessible.

10.6.3. Transformation

The physiochemical properties of mineral aerosols should be investigated. In the past the role of mineral particles in the cloud-nucleation processes has been extensively studied. However, to our knowledge, no studies have dealt with the changes brought about in the physical and chemical properties of aerosols after prolonged exposure to the atmosphere and especially during cloud cycling. Of special interest to us would be the surface sorptive properties of the aerosol and the solubility of the aerosol components.

10.6.4. Deposition

For paleoclimatic interpretations, it is important to understand the link between the properties of mineral particles deposited at a distant site and the processes occurring in the source regions. Especially important is the study of mechanisms for the transport and deposition of large, giant, and ultragiant particles in the atmosphere and the relationship between the transporting wind speed and the particle-size spectrum. With such information, we will be better able to interpret the climatic record in deep-sea sediments and in ice cores.

10.6.5. General

It was the consensus of our group that it is time to begin planning an intensive LRT field experiment in West Africa and over the tropical northern Atlantic. The experiment should include 1) ground stations in source areas, 2) aircraft over the sources and the tropical North Atlantic, 3) a network of islands and ships (which would also study water column processes), 4) remote sensing, and 5) modelers. This program could be carried out in 5-10 years and would provide valuable data to the entire community studying long-range transport.

10.7. REFERENCES

Andreae, M. O., and T. W. Andreae. 1988. The cycle of biogenic sulfur compounds over the Amazon basin: 1. Dry season. J. Geophys. Res. 93: 1487-1497.

Andreae, M. O., T. W. Andreae, R. J. Ferek, and H. Raemdonck. 1984. Long-range transport of soot carbon in the marine atmosphere. Sci. Total Environ. 36:73-80.

Andreae, M. O., R. J. Charlson, F. Bruynzeels, H. Storms, and R. Vergrieken. 1986. Internal mixture of sea salt, silicates, and excess sulfate in marine aerosols. Science 232:1620-1622.

Andreae, M. O., E. V. Browell, M. Garstang, G. L. Gregory, R. C. Harriss, G. F. Hill, D. J. Jacob, M. C. Pereira, G. W. Sachse, A. W. Setzer, P. L. Silva Dias, R. W. Talbot, A. L. Torres, and S. C. Wofsy. 1988. Biomass-burning and associated haze layers over Amazonia. J. Geophys. Res. 93:1509-1527.

Arimoto, R., and R. A. Duce. 1986. Dry deposition models and the air/sea exchange of trace elements. J. Geophys. Res. 91:2787-2792.

Arimoto, R., R. A. Duce, B. J. Ray, and C. K. Unni. 1985. Atmospheric trace elements at Enewetak Atoll: 2. Transport to the ocean by wet and dry deposition. J. Geophys. Res. 90:2391-2408.

Bergametti, G. 1987. Apports de matiere par voie atmospherique a la Mediterranee Occidentale: Aspects geochimiques et meteorologiques. Ph.D. dissert., Univ. of Paris, 296 pp.

Biscaye, P. E. 1965. Mineralogy and sedimentation of recent deep-sea clay in the Atlantic Ocean and adjacent seas and oceans. Geol. Soc. Am. Bull. 76:803-832.

Biscaye, P. E., R. Chesselet, and J. M. Prospero. 1974. Rb-Sr, 87Sr/86Sr isotope system as an index of provenance of continental dusts in the open Atlantic Ocean. Rech. Atmos. 8:819-829.

Borrmann, S., and R. Jaenicke. 1987. Wind tunnel experiments on the resuspension of sub-micrometer particles from a sand surface. Atmos. Environ. 21:1891-1898.

Brinkman, A. W., and J. McGregor. 1983. Solar radiation in dense Saharan aerosol in northern Nigeria. Quart. Roy. Meteorol. Soc. 109:831-847.

Buat-Ménard, P. 1986a. Air to sea transfer of anthropogenic trace metals. In The Role of Air-Sea Exchange in Geochemical Cycling (P. Buat-Ménard, ed.), NATO ASI Series C, Vol. 185, Dordrecht:Reidel, 497-529.

Buat-Ménard, P. 1986b. The ocean as a sink for atmospheric particles. In The Role of Air-Sea Exchange in Geochemical Cycling (P. Buat-Ménard, ed.), NATO ASI Series C, Vol. 185, Dordrecht:Reidel, 165-183.

Buat-Ménard, P., V. Ezat, and A. Gaudichet. 1983. Size distribution and mineralogy of aluminosilicate dust particles in tropical Pacific air and rain. In Precipitation Scavenging, Dry Deposition and

Resuspension, Vol. 2 (H. R. Prupracher, R. G. Semonin, and W. G. N. Slinn, eds.) New York:Elsevier, 1259-1269.

Bucher, A., and C. Lucas. 1984. Sedimentation eolienne intercontinentales, poussieres sahariennes et geologie. Bull. Centre Rech. Explor. Product. Elf Aquitaine 8:151-165.

Carder, K. L., R. G. Steward, P. R. Betzer, D. L. Johnson, and J. M. Prospero. 1986. Dynamics and composition of particles from an aeolian input event to the Sargasso Sea. J. Geophys. Res. 91:1055-1066.

Carlson, T. N. 1979. Atmospheric turbidity in Saharan dust outbreaks as determined by analysis of satellite brightness data. Mon. Weather Rev. 107:322-355.

Carlson, T. N., and R. S. Caverly. 1977. Radiative characteristics of Saharan dust at solar wavelengths. J. Geophys. Res. 82:3141-3152.

Carlson, T. N., and J. M. Prospero. 1972. The large-scale movement of Saharan air outbreaks over the northern equatorial Atlantic. Appl. Meteorol. 11:283-297.

Charney, J., P. H. Stone, and W. J. Quirk. 1975. Drought in the Sahara: A biogeophysical feedback mechanism. Science 187:434-435.

Chester, R. 1982. Particulate aluminum fluxes in the eastern Atlantic. Marine Chemistry 11:1-16.

Chester, R. 1986. The marine mineral aerosol. In The Role of Air-Sea Exchange in Geochemical Cycling (P. Buat-Ménard, ed.), NATO ASI Series C, Vol. 185, Dordrecht:Reidel, 443-476.

Chung, Y.-S., and H. V. Le. 1984. Detection of forest-fire smoke plumes by satellite imagery. Atmos. Environ. 18:2143-2151.

Cooke, R. U., and A. Warren. 1973. Geomorphology in Deserts London: Batsford, 375 pp.

Coude-Gaussen, G. 1984. Le cycle des poussieres eoliennes desertiques actualles et la sedimentation des loess peridesertiques quaternaires. Bull. Centre Rech. Explor. Product. Elf Aquitaine 8:167-182.

Coude-Gaussen, G. 1989. Discussion of the sedimentary logical, characterization of proximal or distal, atmospheric or deposited, Saharan dusts. In Paleoclimatology and Paleometeorology, Modern and Past Patterns of Global Atmospheric Transport NATO ASI·Series C, Dordrecht:Kluwer.

D'Almeida, G. A. 1986. A model for Saharan dust transport. Climate Appl. Meteorol. 25:903-916.

D'Almeida, G. A., and L. Schutz. 1983. Number, mass and volume distribution of mineral aerosol and soils of the Sahara. Climate Appl. Meteorol. 22:233-243.

Duce, R. A. 1986. The impact of atmospheric nitrogen, phosphorous and iron species on marine biological productivity. In The Role of Air-Sea Exchange in Geochemical Cycling (P. Buat-Ménard, ed.), NATO ASI Series C, Vol. 185, Dordrecht:Reidel, 497-529.

Fishman, J., F. M. Vukovitch, D. R. Cahoon, and M. C. Shipman. 1987. The characterization of an air pollution episode using satellite total ozone measurements. Climatol. Appl. Meteorol. 26:1638-1654.

Flannigan, M. D., and T. H. Vonder Haar. 1986. Forest fire monitoring using NOAA satellite AVHRR. Can. Forest Res. 16:975-982.

Fraser, R. S., Y. J. Kaufman, and R. L. Mahoney. 1984. Satellite measurements of aerosol mass and transport. Atmos. Environ. 18:2577-2584.

Gagosian, R. B. 1986. The air-sea exchange of particulate organic matter: The sources and long-range transport of lipids in aerosols. In The Role of Air-Sea Exchange in Geochemical Cycling (P. Buat-Ménard, ed.), NATO ASI Series C, Vol. 185, Dordrecht:Reidel, 409-442.

Gagosian, R. B., E. T. Peltzer, and J. T. Merrill. 1987. Long-range transport of terrestrially derived lipids in aerosols from the South Pacific. Nature 325:800-803.

Gillette, D. A. 1981. Production of dust that may be carried great distances. In Desert Dust: Origin, Characteristics, and Effects on Man (T. L. Pewe, ed.) Spec. Paper 186, Boulder:Geol. Soc. Am., 11-26.

Gillette, D. A. 1988. Threshold friction velocities for dust production for agricultural soils. J. Geophys. Res. 93:12,645-12,662.

Gillette, D. A., I. M. Blifford, Jr., and C. R. Fenster. 1972. Measurements of aerosol size distributions and vertical fluxes of aerosols on land subject to wind erosion. Appl. Meteorol. 11:977-987.

Gillette, D. A., J. Adams, A. Endo, D. Smith, and R. Kihl. 1980. Threshold velocities for input of soil particles into the air by desert soils. J. Geophys. Res. 85:5621-5630.

Gillette, D. A., J. Adams, D. Muhs, and R. Kihl. 1982. Threshold friction velocities and rupture moduli for crushed desert soils for input of soil particles into the air. J. Geophys. Res. 87:9003-9015.

Giorgi, F. 1986. A particle dry-deposition parameterization scheme for use in tracer transport models. J. Geophys. Res. 91:9794-9806.

Glaccum, R. A., and J. M. Prospero. 1980. Saharan aerosols over the tropical North Atlantic--mineralogy. Marine Geol. 37:295-321.

Goudie, A. S. 1981. Aeolian processes, landforms, and the spatial distribution of aridic soils. Paper presented, Int. Symp. Aridic Soils, Jerusalem.

Goudie, A. S. 1983. Dust storms in space and time. Prog. Phys. Geog. 7:502-530.

Griggs, M. 1983. Satellite measurements of tropospheric aerosols over the tropical North Atlantic. Adv. Space Res. 2:109-118.

Grousset, F. E., P. E. Biscaye, A. Zindler, J. M. Prospero, and R. Chester. 1988. Nd isotopes as tracers in marine sediments and aerosols: North Atlantic. Earth Planetary Sci. Ltrs. 87:367-378.

Harriss, R. C., E. V. Browell, D. I. Sebacher, G. L. Gregory, R. R. Hinton, S. M. Beck, D. S. McDougal, and S. T. Shipley. 1984. Atmospheric transport of pollutants from North America to the North Atlantic Ocean. Nature 308:722-724.

Heidam, N. Z. 1984. The components of the Arctic aerosol. Atmos. Environ. 18:329-343.

Heidam, N. Z. 1985. Crustal enrichments in the Arctic aerosol. Atmos. Environ. 19:2083-2097.

Helgren, D. M., and J. M. Prospero. 1987. Wind velocities associated with dust deflation events in the western Sahara. Climate Appl. Meteorol. 26:1147-1151.

Helmes, L., and R. Jaenicke. 1988. Long-term series of atmospheric turbidity, estimated from records of sunshine duration and cloud cover. In Aerosols and Climate (P. V. Hobbs and M. P. McCormick, eds.) Hampton, VA:Deepak Publishing (Environ. Sci. Technol.).

Hicks, B. B., M. L. Wesely, and J. L. Durham. 1980. Critique of Methods to Measure Dry Deposition. EPA-600/9-80-050, Res. Triangle Park, NC:Environ. Sci. Res. Lab., n.p.

Hooghiemstra, H., O. C. A. Chiori, and H.-J. Beug. 1986. Pollen and spore distribution in recent marine sediments: A record of NW African seasonal wind patterns and vegetation belts. Meteor. Forschungs Ergebnisse C40:87-135.

Hyers, A. D., and M. G. Marcus. 1981. Land use and desert dust hazards in central Arizona. In Desert Dust: Origin, Characteristics, and Effect on Man (T. L. Pewe, ed.) Special Paper 186, Boulder:Geol. Soc. Am., n.p.

Iwasaka, Y., H. Minoura, and K. Nagaya. 1983. The transport and spacial scale of Asian dust-storm clouds: A case study of the dust-storm event of April 1979. Tellus 35B:189-196.

Jaenicke, R. 1979. Monitoring and critical review of the estimated source strength of mineral dust from the Sahara. In Saharan Dust: Mobilization, Transport, Deposition (C. Morales, ed.) New York:Wiley, 233-242.

Jaenicke, R., and L. Schutz. 1978. Comprehensive study of physical and chemical properties of the surface aerosols in the Cape Verde Islands region. J. Geophys. Res. 83:3585-3599.

Jaenicke, R., C. Junge, and H. J. Kanter. 1971. Messungen der Aerosol-grossenverteilung uber dem Atlantic. Meteor. Forsch. Ber. B7:1-54.

Jonas, R., and K. Heinemann. 1985. Studies on the dry deposition of aerosol particles on vegetation and plane surfaces. Aerosol Sci. 16:463-471.

Joussaume, S., and R. Sadourny. 1989. Desert dust and climate: Investigation using an atmospheric general circulation model. In Paleoclimatology and Paleometeorology: Modern and Past Patterns of Global Atmospheric Transport, NATO ASI Series, Dordrecht:Kluwer.

Junge, C. 1979. The importance of mineral dust as an atmospheric constituent. In Sahara Dust: Mobilization, Transport, Deposition (C. Morales, ed.) New York:Wiley, 49-60.

Kalu, A. E. 1979. The African dust plume: Its characteristics and propagation across West Africa in winter. In Saharan Dust: Mobilization, Transport, and Deposition. (C. Morales, ed.) New York:Wiley, 95-118.

Karyampudi, V. M., and T. N. Carlson. 1988. Analysis and numerical simulations of the Saharan air layer and its effect on easterly wave disturbances. Atmos. Sci. 45:3102-3136.

Khemani, L. T., G. A. Momin, M. S. Naik, P. S. Prakasa Rao, P. D. Safai, and A. S. R. Murty. 1987. Influence of alkaline particulates on pH of cloud and rain water in India. Atmos. Environ. 21:1137-1145.

Killinger, D. K., and N. Menyuk. 1987. Laser remote sensing of the atmosphere. Science 235:37-45.

Landing, W. M., and K. W. Bruland. 1987. The contrasting biogeochemistry of iron and manganese in the Pacific Ocean. Geochim. Cosmochim. Acta 51:29-43.

Lane, D. D., and J. J. Stukel. 1978. Aerosol deposition on a flat plate. Aerosol Sci. 9:191-197.

Leinen, M. 1989. The late quaternary record of atmospheric transport to the northwest Pacific from Asia. In Paleoclimatology and

Paleometeorology: Modern and Past Patterns of Global Atmospheric Transport, NATO ASI Series, Dordrecht:Kluwer.

Lowenthal, D. H., and K. A. Rahn. 1985. Regional sources of pollution aerosol at Barrow, Alaska, during winter 1979-1980 as deduced from elemental tracers. *Atmos. Environ.* 19:2011-2024.

Maring, H. B., and R. A. Duce. 1987. The impact of atmospheric aerosols on trace metal chemistry in open ocean surface seawater: 1. Aluminum. *Earth Planetary Sci. Ltrs.* 84:381-392.

Martin, J. H., and S. E. Fitzwater. 1988. Iron deficiency limits phytoplankton growth in the north-east Pacific subarctic. *Nature* 331: 341-343.

McCauley, J. F., C. S. Breed, M. J. Grolier, and D. J. MacKinnon. 1981. The US dust storm of February 1977. In *Desert Dust: Origin, Characteristics, and Effects on Man* (T. L. Pewe, ed.) Special Paper 186, Boulder:Geol. Soc. Am., 123-148.

McMahon, T. A., and P. J. Denison. 1979. Empirical atmospheric deposition parameters--a survey. *Atmos. Environ.* 13:571-585.

Measures, C. I., B. Grant, M. Khadem, D. S. Lee, and J. M. Edmond. 1984. Distribution of Be, Al, Se and Bi in the surface waters of the western North Atlantic and Caribbean. *Earth Planetary Sci. Ltrs.* 71: 1-12.

Merrill, J. T. 1986. Atmospheric pathways to the oceans. In *The Role of Air-Sea Exchange in Geochemical Cycling* (P. Buat-Ménard, ed.), NATO ASI Series C, Vol. 185, Dordrecht:Reidel, 35-63.

Merrill, J. T., R. Bleck, and L. Avila. 1985. Modeling atmospheric transport to the Marshall Islands. *J. Geophys. Res.* 90:12,927-12,936.

Middleton, N. J., A. S. Goudie, and G. L. Wells. 1986. The frequency and source areas of dust storms. In *Aeolian Geomorphology* (W. G. Nickling, ed.) New York:Allen and Unwin, 237-259.

Moore, R. M., J. E. Milley, and A. Chatt. 1984. The potential for biological mobilization of trace elements from aeolian dust in the ocean and its importance in the case of iron. *Oceanol. Acta* 7:221-228.

Morales, C. 1979. The use of meteorological observations for studies of the mobilization, transport and deposition of Saharan soil dust. In *Saharan Dust: Mobilization, Transport, Deposition.* (C. Morales, ed.) New York:Wiley, 119-131.

National Academy of Sciences (National Research Council). 1985. *The Effects on the Atmosphere of a Major Nuclear Exchange.* Washington: National Academy Press, 193 pp.

National Academy of Sciences. 1986. *Remote Sensing of the Biosphere.* Washington:National Academy Press, 135 pp.

National Academy of Sciences. 1988. *Space Science in the Twenty-First Century.* Washington:National Academy Press, 84 pp.

National Oceanic and Atmospheric Administration. 1987. *Space-based Remote Sensing of the Earth.* Washington:US Govern. Printing Office, 123 pp.

Nickling, W. 1989. Particulate emissions from desert soils. In *Paleoclimatology and Paleometeorology: Modern and Past Patterns of Global Atmospheric Transport*, NATO ASI Series, Dordrecht:Kluwer.

Orians, K. J., and K. W. Bruland. 1986. The biochemistry of aluminium in the Pacific Ocean. *Earth Planetary Sci. Ltrs.* 78:397-410.

Prospero, J. M. 1981. Aeolian transport to the world ocean. In The Sea: Vol. 7. The Oceanic Lithosphere (C. Emiliani, ed.) New York:Wiley, 801-874.

Prospero, J. M., and T. N. Carlson. 1972. Vertical and areal distribution of Saharan dust over the western equatorial North Atlantic Ocean. J. Geophys. Res. 77:5255-5265.

Prospero, J. M., and R. T. Nees. 1986. Impact of the North African drought and El Nino on mineral dust in the Barbados trade winds. Nature 320:735-738.

Prospero, J. M., D. L. Savoie, T. N. Carlson, and R. T. Nees. 1979. Monitoring Saharan aerosol transport by means of atmospheric turbidity measurements. In Saharan Dust: Mobilization, Transport, Deposition (C. Morales, ed.) New York:Wiley, 171-186.

Prospero, J. M., R. T. Nees, and M. Uematsu. 1987. Deposition rate of particulate and dissolved aluminum derived from Saharan dust in precipitation at Miami, Florida. J. Geophys. Res. 92:14,723-14,731.

Pye, K. 1984. Loess. Prog. Phys. Geography 8:176-217.

Pye, K. 1987. Aeolian Dust and Dust Deposits New York:Academic Press, 334 pp.

Rea, D. K., and T. R. Janecek. 1982. Late Cenozoic changes in atmospheric circulation deduced from North Pacific eolian sediments. Marine Geol. 49:149-167.

Rea, D. K., M. Leinen, and T. R. Janecek. 1985. Geologic approach to the long-term history of atmospheric circulation. Science 227:721-725.

Schutz, L. 1977. Die Saharastaub-Komponente Uber dem Subtropischen Nord-Atlantik. Ph.D. dissert., Univ. of Mainz, FRG, 153 pp.

Schutz, L. 1980. Long-range transport of desert dust with special emphasis on the Sahara. In Aerosols: Anthropogenic and Natural, Sources and Transport (T. J. Kneip and P. J. Lioy, eds.) New York:Ann. NY Academy Sci. 338:515-532.

Schutz, L., and R. Jaenicke. 1974. Particle number and mass distribution above 10^{-4} cm radius in sand and aerosol of the Sahara desert. Appl. Meteorol. 13:863-870.

Schutz, L., and M. Kramer. 1987. Rainwater composition over a rural area with special emphasis on the size distribution of insoluble particulate matter. Atmos. Chemistry 5:173-184.

Schutz, L., and K. A. Rahn. 1982. Trace-element concentrations in erodible soils. Atmos. Environ. 16:171-176.

Schutz, L., and M. Sebert. 1987. Mineral aerosols and source identification. Aerosol Sci. 18:1-10.

Schutz, L., R. Jaenicke, and H. Pietrer. 1981. Saharan dust transport over the North Atlantic Ocean. In Desert Dust: Origin, Characteristics, and Effects on Man (T. L. Pewe, ed.) Special Paper 186, Boulder:Geol. Soc. Am., 87-100.

Sehmel, G. A. 1980. Particle and gas dry deposition: A review. Atmos. Environ. 14:983-1011.

Simoneit, B. R. T. 1977. Organic matter in eolian dust over the Atlantic Ocean. Marine Chemistry 5:443-467.

Slinn, W. G. N. 1983. Air-to-sea transfer of particles. In Air-Sea Exchanges of Gases and Particles (P. S. Liss and W. G. N. Slinn, eds.), NATO ASI Series C, Vol. 108, Dordrecht:Reidel, 299-405.

Talbot, R. W., R. C. Harriss, E. V. Browell, G. L. Gregory, D. I. Sebacher, and S. M. Beck. 1986. Distribution and geochemistry of aerosols in the tropical North Atlantic troposhere: Relationship to Saharan dust. J. Geophys. Res. 91:5173-5182.

Tsoar, H., and K. Pye. 1987. Dust transport and the question of desert loess formation. Sedimentology 34:139-153.

Uematsu, M., R. A. Duce, and J. M. Prospero. 1985. Deposition of atmospheric mineral particles in the North Pacific Ocean. Atmos. Chemistry 3:123-138.

Volz, F. E. 1970. Spectral skylight and solar radiation measurements in the Caribbean: Maritime aerosols and Saharan dust. Atmos. Sci. 27: 1041-1047.

Westphal, D. L., O. B. Toon, and T. N. Carlson. 1987. A two-dimensional numerical investigation of the dynamics and microphysics of Saharan dust storms. J. Geophys. Res. 92:3027-3049.

Westphal, D. L., O. B. Toon, and T. N. Carlson. 1989. A case study of mobilization and transport of Saharan dust. Atmos. Sci., in press.

Wolff, G. T., T. M. Church, J. N. Galloway, and A. H. Knap. 1987. An examination of SO_x, NO_x and trace metal washout ratios over the western Atlantic Ocean. Atmos. Environ. 21:2623-2628.

11. THE LONG-RANGE TRANSPORT OF SULFUR AND NITROGEN COMPOUNDS

Hiram Levy, II, Rapporteur
NOAA/Geophysical Fluid
  Dynamics Laboratory
Princeton University
Princeton, NJ   08540

James N. Galloway
Department of Environmental
  Sciences
University of Virginia
Charlottesville, VA   22903

Anton Eliassen
The Norwegian Meteorological
  Institute
P. O. Box 43, Blindern N-0313
Oslo 3, Norway

Bernard E. A. Fisher
Central Electricity Research
  Laboratories
Kelvin Avenue
Letherhead
Surrey KT22 7SE, United Kingdom

Krystyna Gorzelska
Department of Environmental
  Sciences
University of Virginia
Charlottesville, VA   22903

Donald R. Hastie
Chemistry Department
York University
4700 Keele Street
North York, Ontario
Canada M3J 1P3

Jennie L. Moody
Department of Environmental
  Sciences
University of Virginia
Charlottesville, VA   22893

Alexey G. Ryaboshapko
Institute of Applied Geophysics
Glebovskaya Street 20B
Moscow 107258, USSR

Dennis Savoie
Rosenstiel School of Marine and
  Atmospheric Science
University of Miami
4600 Rickenbacker Causeway
Miami, FL   33149-1098

Douglas M. Whelpdale
Atmospheric Environment Service
4905 Dufferin Street
Downsview, Ontario
Canada M3H 5T4

11.1.   INTRODUCTION

In our working group, we considered the long-range transport (extending beyond 1-2 days or 1,000 km and reaching a global scale) of reactive nitrogen and sulfur species from either anthropogenic or natural sources (NO, NO_2, SO_2, H_2S, DMS, NH_3; see Chapter 4, p. 87) and their conversion

A. H. Knap (ed.), The Long-Range Atmospheric Transport of Natural and Contaminant Substances, 231–257.
© 1990 Kluwer Academic Publishers.

products ($SO_4^=$, NO_3^-, HNO_3, PAN, and a broad group of organic nitrate compounds).

Although the long-range transport of sulfur and nitrogen species serves as a tracer of atmospheric motion, it also influences both the chemical reactivity of the troposphere and the overall biogeochemical cycles of nitrogen and sulfur. The concentrations of reactive nitrogen compounds, particularly NO and NO_2, determine the level of ozone production and indirectly influence OH, two species that control the chemical reactivity of the troposphere (National Academy of Sciences 1984, World Meteorological Organization 1985).

The principal role of long-range transport in the biogeochemistry of S and N is to transport both species from major source regions to oceans and sparsely populated continents. For much of the globe with limited sources of S or N, this long-range transport supplies the biogeochemical cycle. Annually ~ 80 Tg of sulfur are deposited over land areas and ~ 260 Tg are either deposited or recycled over the world's oceans. Budget calculations, which assume a well-mixed atmosphere, indicate that approximately half the anthropogenic sulfur is transported from continents to oceans (Brimblecombe et al. 1988).

Logan (1983) estimated that 40% of the nitrogen emitted over eastern North America is advected over the ocean and Levy and Moxim (1987) calculated the same fate for 25% of the total North American emissions. Recent numerical simulations for the global emission of anthropogenic nitrogen (Levy and Moxim 1989) find that 25% of the emissions are transported out of the source regions, mainly over the oceans.

11.2. SOURCES

Generally we expected anthropogenic sources of sulfur and nitrogen in areas of either high population or high per capita fossil-fuel consumption to have the greatest impact on long-range transport. Natural sources, except possibly volcanoes, are generally quite diffuse and play a lesser role. In remote regions of the Southern Hemisphere with no local or regional sources, biogeochemical cycles may be dominated by long-range transport from more diffuse sources (see Section 11.7, p. 236).

The major natural sources of sulfur are volcanic emissions (~ 20 Tg S/yr), biogenic emissions of reduced sulfur compounds (~ 65 Tg S/yr), and sea-salt emissions (> 140 Tg S/yr) (Ryaboshapko 1983). Sea-salt sulfur is primarily confined to large particles within the marine-boundary layer and is relatively unimportant to the global budget. Brimblecombe et al. (1988) have stated that the global biogeochemical sulfur cycle has been modified by industrial emissions which they have estimated to be ~ 90 Tg S/yr, of which ~ 80 Tg S/yr is sulfur dioxide, ~ 10 Tg S/yr is primary sulfate, and a few Tg S/yr are reduced sulfur compounds. Ryaboshapko (1983) reported that human activities have also doubled aeolian emissions to as much as 20 Tg S/yr.

Natural emissions of nitrogen are ~ 8 Tg N/yr from lightning and ~ 8 Tg N/yr from biological action in soils with smaller contributions from the biological processes in oceans, ammonia oxidation, and input from the stratosphere (Logan 1983). Anthropogenic nitrogen emissions have also had a major impact on the global biogeochemical nitrogen cycle. The estimated total global NO_x input is approximately 50 Tg N/yr of which ~ 21 Tg N/yr is from fossil-fuel combustion and ~ 12 Tg N/yr is from agricultural biomass burning (Logan 1983).

11.3. FAVORABLE TRANSPORT CONDITIONS

The seasonally averaged circulation of the atmosphere identifies potential pathways for long-range transport (see Chapter 1, p. 3). However, fluctuations about mean-flow patterns can occur on many different spatial and temporal scales and may enhance or inhibit long-range transport.

For example, over several months, the El Nino southern oscillation causes winds in the equatorial region to reverse direction. Over days, synoptic-scale events, such as a blocking high pressure in the midlatitudes, can alter the path of extratropical cyclones. On a still smaller scale—over a period of hours and an area of only a few kilometers—intense vertical transport can vent material out of the boundary layer into the free troposphere.

These meteorological events by their episodic nature do not appear explicitly in the average atmospheric circulation. However, they strongly influence individual instances of long-range transport; and, depending on their strength and frequency, they may influence the long-term average of long-range transport. These variations in meteorology as well as the spatial and temporal variability of sources, the deposition processes, and the chemical transformations all determine the most favorable conditions for long-range transport.

11.4. VERTICAL TRANSPORT

Within the boundary layer, the sulfur and nitrogen species are efficiently removed by dry deposition to the surface and by scavenging by sea-salt aerosol. Boundary-layer decay times of less than a day have been reported for nitrogen oxides and nitric acid (Hastie et al. 1988) and similar decay times have been reported for sulfur dioxide in both source regions and near coasts (Hastie et al. 1988, Korolev and Ryaboshapko 1980, Nguyen et al. 1975, Prahm et al. 1976). Therefore, the long-range transport of sulfur and nitrogen is greatly increased when the species are lifted above the atmospheric boundary layer into the more rapidly flowing free troposphere. This can be achieved through (1) cumulus convection, (2) synoptic-scale vertical motion (from a front or cyclone), (3) dry convection, or (4) the breakdown of a well-mixed atmospheric boundary layer by cooling from below. The first two mechanisms are associated with condensation and may result in the removal of soluble species by wet deposition. The latter two are essentially dry processes that take place over land during the day when the surface is heated, in the evening when the surface cools, or when air is advected from a warmer to a colder surface. The first three mechanisms feed rapidly into the middle troposphere causing the fastest transport. Once in the middle troposphere, the S and N compounds are free from surface loss and less subject to wet deposition. The air at these altitudes is not efficiently circulated in precipitation-forming systems. Therefore, it is available for the long-range transport of the compounds until carried back to the boundary layer by systematic downward motion either behind a cold front or associated with a high-pressure system or by subsidence associated with deep convection.

11.5. CHEMICAL TRANSFORMATIONS

Only species with long chemical lifetimes can be transported far. Because of the rapid photochemistry normally found in an emission region, we included both the immediate area of emission and the first few hundred kilometers of the boundary layer in our definition of an effective source region for long-range transport. HNO_3, particulate nitrate, organic nitrates (including peroxyacyl nitrates if the temperature is low), and sulfate are the species available for long-range transport. For the relevant chemistry, see Findlayson-Pitts and Pitts (1986) and NAS (1984).

Most anthropogenic nitrogen is emitted as nitric oxide, which can be oxidized to other reactive odd nitrogen compounds, such as NO_2, N_2O_5 or HONO. None of these has a sufficiently long lifetime to survive long-range transport, except in the absence of sunlight. However, another end product, nitric acid, has thermal stability, a small photodissociation cross section in the troposphere, and slow reactions with atmospheric species and is available for long-range transport above the boundary layer.

Peroxyacetyl nitrate (PAN) is the best known of the peroxyacyl nitrates that can be generated where there are high concentrations of hydrocarbons with more than two carbons. PAN has a low photodissociation rate and reacts very slowly with HO and O_3. Its major chemical-loss mechanism is thermal decomposition; i.e., its lifetime is temperature dependent (1.7 hr at 20° C, 50 hr at 0° C, 14 days at -10° C, and 105 days at -20° C). Thus, it is stable at the low temperatures found at high altitudes and latitudes.

Anthropogenic sulfur is emitted largely as SO_2 and then oxidized to sulfate. Gas-phase oxidation occurs through reaction with HO radicals to form SO_3 and subsequent hydrolysis to give H_2SO_4. Aqueous-phase oxidation is equally or more important. Aqueous S(IV) can be oxidized to sulfate by O_2, O_3, or H_2O_2. The relative importance of different mechanisms depends on the particular atmospheric condition.

The major biogenic sulfur species are dimethyl sulfide (DMS), H_2S, methyl mercaptan, and OCS. (OCS was not considered because of its long lifetime and uniform concentration.) DMS oxidation by OH in unpolluted marine air yields primarily SO_2, which is then oxidized to $SO_4^=$ and MSA. Reactions with NO_3 or IO may contribute to DMS oxidation in regions of high nitrogen-oxide concentration or high productivity; dimethyl sulfoxide or dimethyl sulfone may also be produced. The oxidation of the other species results in SO_2 but CS_2 oxidation may produce OCS as well.

11.6. DEPOSITION PROCESSES

Wet and dry deposition, by removing sulfur and nitrogen from the atmosphere, control the long-range transport of sulfur and nitrogen species
and supply biogeochemical cycles in regions where there are no significant sources.

The wet deposition of sulfur and nitrogen depends on the effective
solubility and concentration of the species as well as on the nature of
the wet deposition. Dry deposition, which combines meteorological, physical, and chemical processes, depends on the near-surface, vertical
atmospheric motion or stability, on the underlying surface (e.g., roughness, moistness, vegetation) and on the sulfur and nitrogen compounds
themselves (i.e., their phase, chemical reactivity, solubility) (see
Tables 11-1 and 11-2).

The following are examples in which the absence of a deposition
process strongly influences long-range transport.

1. In the Arctic winter, ineffective scavenging and the ''smooth''
    cold surface inhibit removal; long-range transport is effec-
    tive into and within the Arctic.

Table 11-1. Effectiveness of removal processes by region and season.

Region	Season	Removal Process Wet	Dry
Arctic/Antarctic	Summer	Moderately effective	Moderately effective
	Winter	Ineffective	Ineffective
Temperate zones			
Continent	Summer	Effective	Effective
	Winter	Effective	Moderately effective
Ocean	Summer	Moderately effective	Moderately effective
	Winter	Effective	Moderately effective
Tropics/Subtropics			
Continent	Summer	Effective	Effective
	Winter	Effective	Effective
Ocean	Summer	Effective	Effective
	Winter	Effective	Effective
Tropical forests		Effective	Effective
Mountainous		Effective	Effective
Arid (deserts)		Ineffective	Ineffective
Coastal zone			
Offshore flow	Cold/warm*	Effective	Effective
	Warm/cold	Moderately effective	Moderately effective
Onshore flow	Cold/warm	Effective	Effective
	Warm/cold	Moderately effective	Moderately effective

*Air temperature/underlying surface temperature.

Table 11-2. Effectiveness of removal processes for sulfur and nitrogen species.

Chemical Species	Removal Process	
	(Wet)	(Dry)
Ammonia, HCl, HNO$_3$	Effective	Effective
Fine aerosols with SO$_4$, metals, ammonium, fine dust	Effective	Ineffective
NO	Ineffective	Ineffective
NO$_2$, ozone	Ineffective	Moderately effective
SO$_2$: Near field	Moderately effective	Effective
Far field	Effective	Effective
SO$_4$ particles	Effective	Ineffective

2. Reactive species, such as ammonia, HCl, or HNO$_3$ gas, can only be transported very long distances in substantial quantities if both dry- and wet-removal pathways are inhibited (see Section 11.7).

3. Sulfur injection into the stratosphere by volcanoes with no wet or dry removal results in the long-range transport of sulfur until the sulfur enters the lower troposphere.

Even when the removal processes are effective, some small fraction of the emission may be transported very long distances (see Section 11.7.6). If local sources are small enough, long-range transport from major sources (sulfur or nitrogen oxides from fuel combustion) may still be detectable and important to local budgets. In other circumstances, long-range transport may be difficult to detect because of large nearby sources. For example, S and N transported from North America to the European coast may be obscured by recirculating S and N compounds of European origin (Levy and Moxim 1987, also see Section 11.7.3.). A distinction should be made between those cases where long-range transport has a dominant role in global budgets and biogeochemical cycles and those cases, such as Chernobyl, where it only provides a tracer of meteorological motions.

11.7. CASE STUDIES

11.7.1. Introduction

The existence of large sources and favorable transport patterns and the capability of being transformed into stable species suggest that the long-range transport of S and N eastward from North America, westward from Africa, and eastward from Asia is probable. Sulfur and nitrogen concentrations found thousands of kilometers away from any possible major

source support this prediction (see Chapter 4, p. 87). However, except
for the transport of sulfur (<10%, Whelpdale et al. 1988) and nitrogen
(<5%, Levy and Moxim 1987) from eastern North America to Europe, data
were generally too limited to estimate fluxes.

At times the discovery of long-range transport can be accidental.
Consider, for example, the first observations of high concentrations of
mineral aerosol at the SEAREX sampling site on Enewetak in the spring of
1979 (Duce et al. 1980) that led to the Asian dust network over the Pacific Ocean. This network data revealed not only more extensive transport
of mineral aerosols than was previously believed but also substantial
transport of $SO_4^=$ and NO_3^- (Prospero et al. 1985).

The case studies mentioned in this chapter and those discussed in
Chapter 4 (p. 87) have documented the existence and, in the case of the
North Atlantic Ocean, the magnitude of the long-range transport of S and
N. The following seven case studies, which involve mostly unpublished
data and calculations from several regions of the world, further support
the long-range transport of S and N.

11.7.2. Sulfur and Nitrogen Transport From Asia to the North Pacific

Measurements of inorganic sulfate and nitrate (particulate NO_3^- + vapor
phase HNO_3) concentrations in the boundary layer over the North Pacific
strongly indicate substantial transport of the two constituents or their
precursors from Asia (Prospero et al. 1985). The monthly mean concentrations of nitrate and non-sea-salt (nss) sulfate in the marine boundary
layer at Midway (Fig. 11-1) in the North Pacific (based on nearly seven
years of measurements as part of the SEAREX [Sea-Air Exchange] program)
of nss-sulfate concentrations in the free troposphere at the Mauna Loa
Observatory on Hawaii (Parrington and Zoller 1984), and of Asian dust in
the North Pacific (Uematsu et al. 1983) all show a maximum from February
through May. The seasonal cycles are similar because they are driven by
the same meteorology: strong vertical lifting at the source and favorable
winds aloft. However, since neither the nitrate nor the nss-sulfate
concentrations are significantly correlated with the soil component, both
are believed to be primarily derived from anthropogenic sources in Asia
rather than from soil material (unpublished data available from
Dr. Savoie, RSMAS, University of Miami, 4600 Rickenbacker Causeway,
Miami, FL 33149-1098).

Nitrate concentrations from three remote tropical and subtropical
South Pacific SEAREX stations showed little seasonal variability for
1983-1987 (Fig. 11-2) with average values of 0.117 $\mu g/m^3$ (±0.011) at
Funafuti; 0.116 $\mu g/m^3$ (±0.008) at American Samoa; and 0.117 $\mu g/m^3$
(±0.010) at Rarotonga. Savoie et al. (1987) reported a comparable, albeit somewhat higher, concentration of 0.16 $\mu g/m^3$ for Southern Hemisphere
air over the Indian Ocean. In comparison, even the lowest monthly means
at Midway and Oahu, 0.18 $\mu g/m^3$ and 0.25 $\mu g/m^3$, respectively, are higher
and the mean concentrations of 0.289 $\mu g/m^3$ at Midway and 0.359 $\mu g/m^3$ at
Oahu are factors of 2.5-3 greater than ''background.'' The South Pacific
values seem to represent a natural marine ''background'' for nitrate
derived from the stratosphere, lightning, and possibly unknown oceanic
sources.

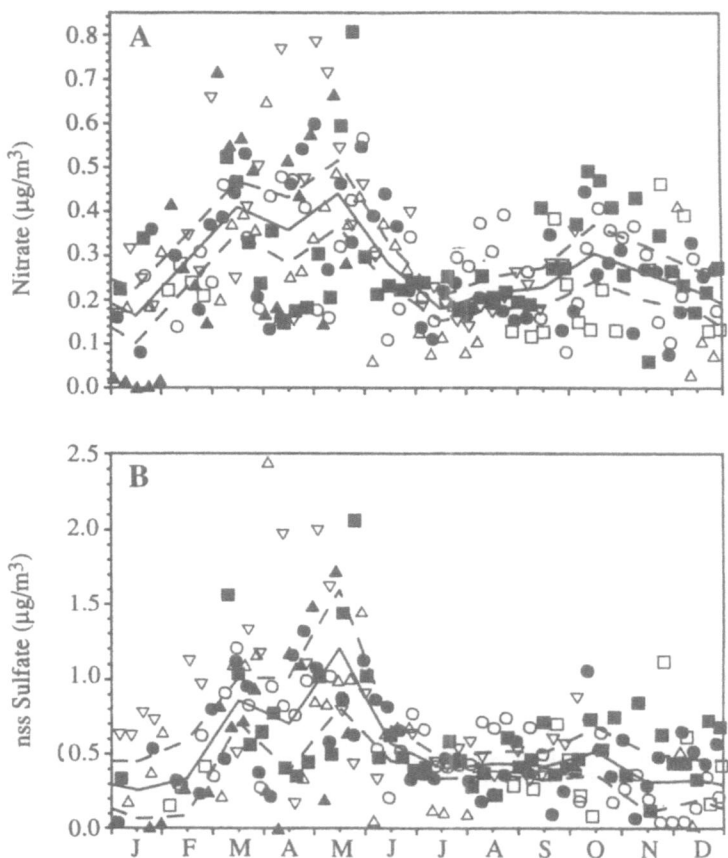

Figure 11-1. Composited seasonal cycles of nitrate and non-sea-salt
    sulfate in the atmospheric boundary layer at Midway Island for ○
    1981, ● 1982, △ 1983, ▲ 1984, □ 1985, ■ 1986, and ▽ 1987. The **solid
    lines** connect the monthly mean concentrations; the **dashed lines**
    indicate the upper and lower 95% confidence bounds of the monthly
    means.

    Fluctuations in the primary productivity of and the resulting emis-
sion of dimethyl sulfide (DMS) cause large seasonal and geographic variabilities in sulfate concentrations in air over the ocean. Consequently,
it is harder to determine the average marine-derived nss-sulfate than to
determine the background nitrate. However, Saltzman et al. (1985) found
that the ratio of methanesulfonate (MSA) to nss-sulfate, the two major
end products from the atmospheric oxidation of DMS, is quite constant
(MSA/nss sulfate = 0.07) in the tropical and subtropical marine regions
unaffected by the transport of sulfate from continents. Using this

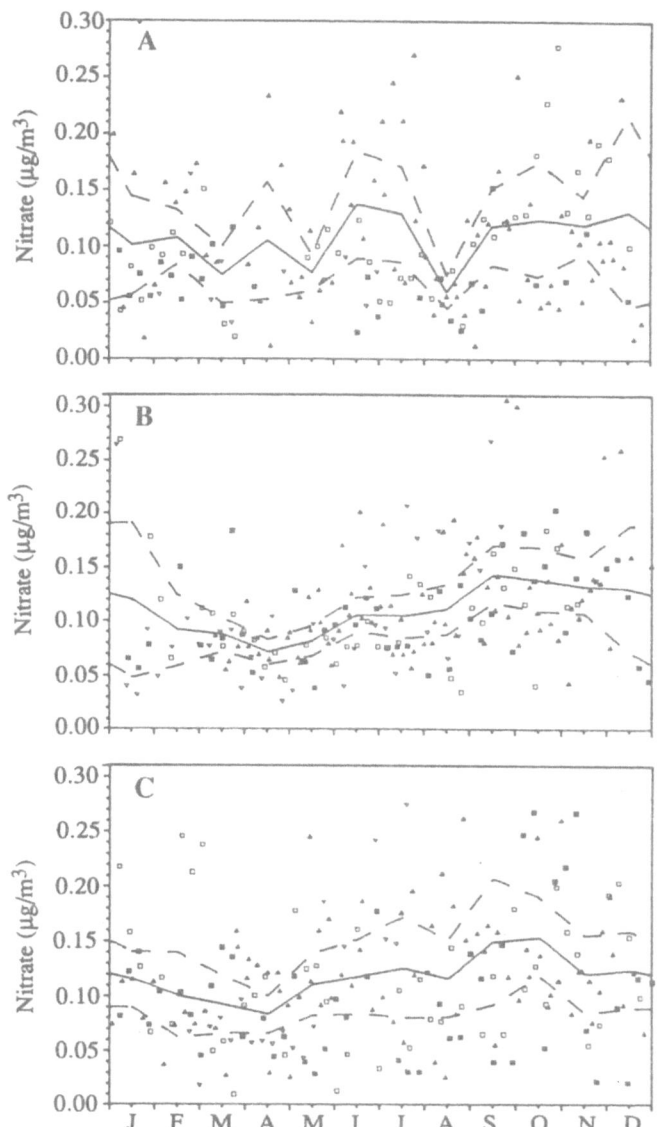

Figure 11-2. Composited seasonal cycles of nitrate at **A** Funafuti, **B** American
Samoa, and **C** Rarotonga based on data from 1983 ($\triangle$), 1984 ($\blacktriangle$), 1985
($\square$), 1986 ($\blacksquare$), and most of 1987 ($\triangledown$). Each symbol represents a
week-long sample. The **solid lines** connect the mean concentrations
for all the data collected during a given month, regardless of the
year. The **dashed lines** indicate the upper and lower 95% confidence
bounds for those monthly means.

ratio, the annual average MSA concentration at Midway of 0.029 $\mu g/m^3$
(Saltzman et al. 1986) gives an average biogenic nss-sulfate concentration of 0.41 $\mu g/m^3$ or 75% of the annual average total of 0.551 $\mu g/m^3$.
Given that the MSA concentrations at American Samoa, New Caledonia, and
Norfolk Island are comparable to those at Midway (Saltzman et al. 1986),
we assumed the same for Oahu and calculated that the biogenic source
there accounted for about 83% of the 0.493 $\mu g/m^3$ nss-sulfate.

Subtracting the estimated biogenic contribution from the mean nss-
sulfate concentration and the marine ''background'' from the mean nitrate
concentrations, we were left with an estimate of the contribution from
long-range transport of about 0.17 μg NO_3^-/m^3 and 0.14 μg $SO_4^=/m^3$ at
Midway, and 0.25 μg NO_3^-/m^3 and 0.08 μg $SO_4^=/m^3$ at Oahu.

Despite the potential uncertainties (see Section 11.8.), deposition
fluxes based on these values provided a rough estimate of the S and N
transported to the midlatitudes of the North Pacific. Using a total
(wet-plus-dry) deposition velocity of 1 cm/sec, we calculated deposition
fluxes of 54-79 kg $NO_3^-/km^2/yr$ and 25-44 kg $SO_4^=/km^2/yr$. By applying
this to the entire North Pacific north of 10°N and west of 120°W
(area = 54.3 x 10^6 km^2), we estimated total fluxes of 2.9-4.3 Tg/yr for
nitrate and 1.4-2.4 Tg/yr for sulfate.

The above fluxes represented 15%-20% of the total nitrogen and 3-5%
of the total sulfur emitted in Japan and China. Although the major
impact of this anthropogenic material is on the chemistry of the marine
troposphere, it is also possible that the nitrate represents a significant nutrient input to some regions of the North Pacific.

11.7.3. The Long-Range Transport of Sulfur to Ireland

The role played by North American emissions in the transport and deposition of sulfur to Bermuda during a westerly flow is well documented. In
1987 Galloway and Whelpdale estimated that about 25% to 40% of the sulfur
advected to the western North Atlantic Ocean atmosphere is transported
west of 60°W. Whelpdale et al. (1988), using both observations and a
mean flow model, estimated that a low percentage of the sulfur advected
east from North America reaches Europe.

To determine if there was evidence of the long-range transport of
sulfur from North America to Europe, we compared two years of sulfur-
deposition data from Adrigole on the southwest coast of Ireland to data
from Bermuda (where it is known that there is a North American influence)
and from Amsterdam Island, assumed to be representative of the remote
marine environment (Galloway and Gaudry 1984).

The volume-weighted average concentration of nss-$SO_4^=$ at Ireland is
similar to that of Bermuda and larger than that at Amsterdam Island and
other remote marine areas (Table 11-3.). The similarities in concentration suggested similar continental impacts at Bermuda and Adrigole.
However, to document the impact of the long-range transport of continental material to Adrigole, the influence of European sources must first be
eliminated. This task is accomplished by using an isobaric air-mass
trajectory model (Harris 1982). With this model a data set containing
only storms associated with direct westerly flow can be created. The
volume-weighted average nss-$SO_4^=$ concentration of this data set is

Table 11-3. Non-sea-salt $SO_4^=$ in precipitation from Ireland, Bermuda, and Amsterdam Island.

	Volume-Weighted Average Concentration (μeq/l)	Deposition (g S m²/yr)	Source
Ireland	13.9	0.50	
Bermuda	13.1	0.23	Galloway et al. 1989
Amsterdam Island	4.9	0.1	Galloway & Gaudry 1984
Remote Marine Areas	2-6.3	0.066-0.13	Galloway 1985

10.0 μeq/l, again larger than in remote marine areas. The use of the volume-weighted average concentration to compare data sets did not allow a comparison of variability among the data sets. Therefore, the remaining comparisons were of distributions of nss-$SO_4^=$ deposition. Specifically, the variability of deposition at Ireland, Bermuda, and Amsterdam Island was compared using plots that rank and normalize the events in increasing order of their per-event deposition. We used deposition because of the effect that differences in precipitation amount between the two sites have on the concentration (Galloway et al. 1989).

Figure 11-3 shows the normalized rankings for the entire Bermuda data set (BDA:ALL), and the data sets that have only events associated with westerly flow (BDA: WEST FLOW) and southeasterly flow (BDA: SE FLOW). We compared these three to the entire data set from Amsterdam Island (AMI). The impact of continental sources on Bermuda was clearly evident. The deposition was higher for air from the west although the normalized ranking of the southeasterly flow was still greater than that of Amsterdam Island, possibly as a result of the long-range transport from Africa and Europe (Galloway et al. 1989).

The next figure compares the nss-$SO_4^=$ deposition at Ireland with that at Bermuda and Amsterdam Island (Fig. 11-4). One explanation for the remarkable similarity in the normalized rankings at Ireland and Bermuda, both of which are greater than Amsterdam Island, was that they are both impacted to the same degree by continental emissions. In the case of Ireland, these emissions may be from Europe. To eliminate the direct influence of Europe, the normalized ranking was also calculated for only those storms associated with westerly flow.

The impact of European emissions on Adrigole was evident in the comparison of the normalized rankings for events with westerly flow (IRE: WEST FLOW, Fig. 11-5.) versus all events (IRE:ALL). Again, note that the westerly flow events had a much higher normalized ranking than that from Amsterdam Island. The Bermuda and Ireland data sets both suggested that precipitation, when flow was from the marine sector (''clean'') and not directly from the upwind continent, still might be influenced by anthropogenic emissions.

We compared the ''clean'' sectors for both Bermuda and Ireland to Amsterdam Island (Fig. 11-6). Although the normalized ranking for

Figure 11-3

Precipitation events at Amsterdam Island and Bermuda ranked by increasing deposition. ——●—— (AMI) shows the normalized rankings for the total data set, Amsterdam Island (Galloway and Gaudry 1984); ——○—— (BDA: ALL), the total data set from Harbor Radio Tower, Bermuda (Galloway et al. 1989); ——■—— (BDA:SE FLOW), the Bermuda data set, events associated with southeasterly airflow (Galloway et al. 1989); and ——□—— (BDA: WEST FLOW), the Bermuda data set, events associated with westerly airflow (Galloway et al. 1989). Upper values of normalized ranks for BDA:ALL and BDA:WEST FLOW were cut off to enhance differences in the lower portion of the rank scale. The actual maximum value for both data sets was 210 neq/ cm²/event.

Figure 11-4

Precipitation events at Amsterdam Island, Bermuda, and Ireland ranked by increasing deposition. ——●—— (AMI) shows the normalized rankings for the total data set, Amsterdam Island (Galloway and Gaudry 1984); ——○—— (BDA:ALL), the total data set, Harbor Radio Tower, Bermuda (Galloway et al. 1989); and ——■—— (IRE:ALL), the total data set, Adrigole, Ireland. The upper value of the normalized rank for BDA:ALL was cut off to enhance the differences in the lower portion of the rank scale. The actual maximum value for the data set was 210 neq/cm²/event.

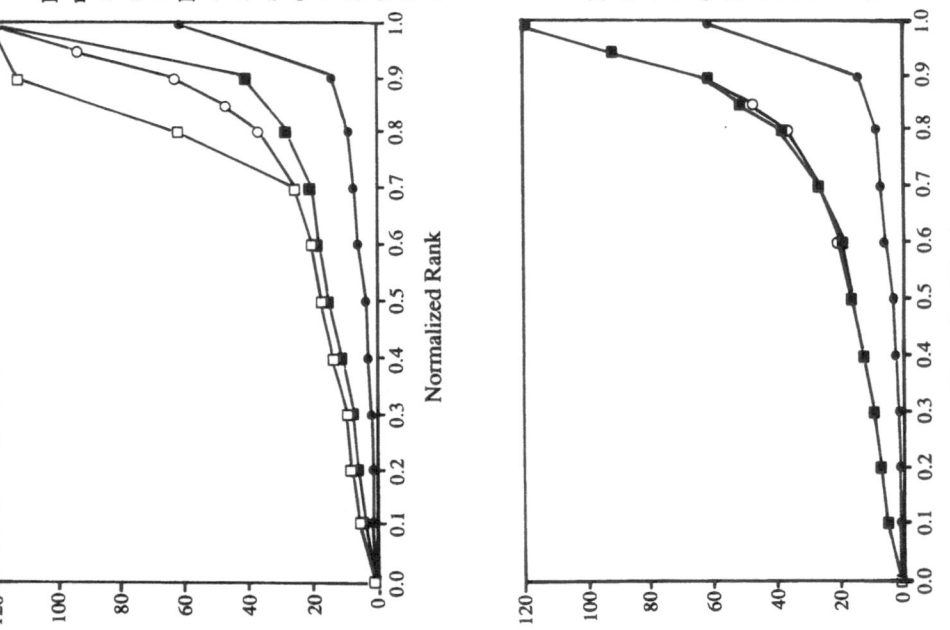

Figure 11-5

Precipitation events at Amsterdam Island and Ireland ranked by increasing deposition. —●— (AMI) represents the normalized rankings of the total data set, Amsterdam Island (Galloway and Gaudry 1984); —■—(IRE:WEST FLOW), the Irish data set, events associated with westerly airflow; and —○— (IRE:ALL), the total data set, Adrigole, Ireland.

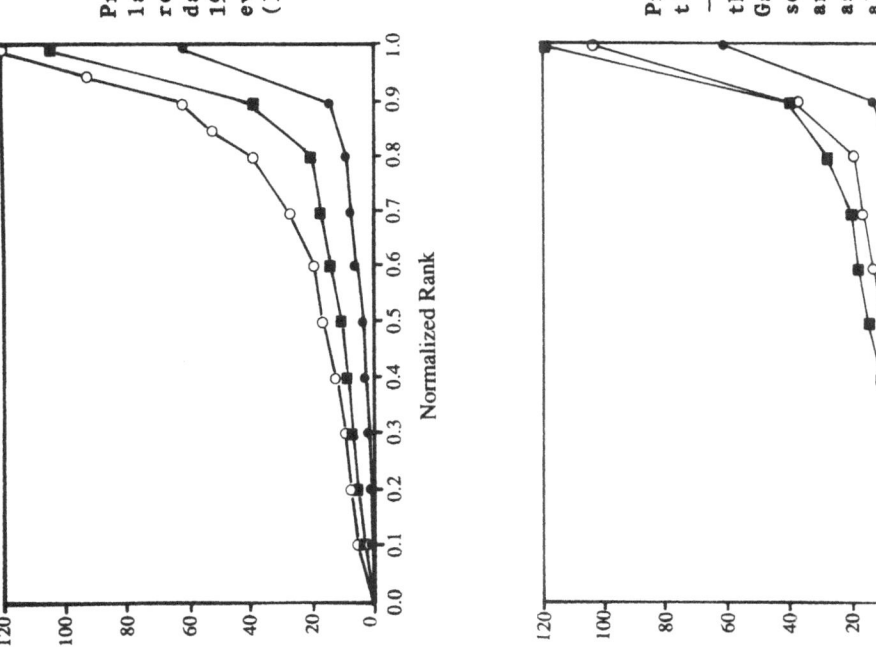

Figure 11-6

Precipitation events ranked by increasing deposition at Amsterdam Island, Ireland, and Bermuda. —●— (AMI) represents the normalized rankings of the total data set, Amsterdam Island (Galloway and Gaudry 1984); —○—(IRE:WEST FLOW), the Irish data set, events associated with westerly airflow; and —■— (BDA:SE FLOW), the Bermuda data set, events associated with southeasterly airflow (Galloway et al. 1989).

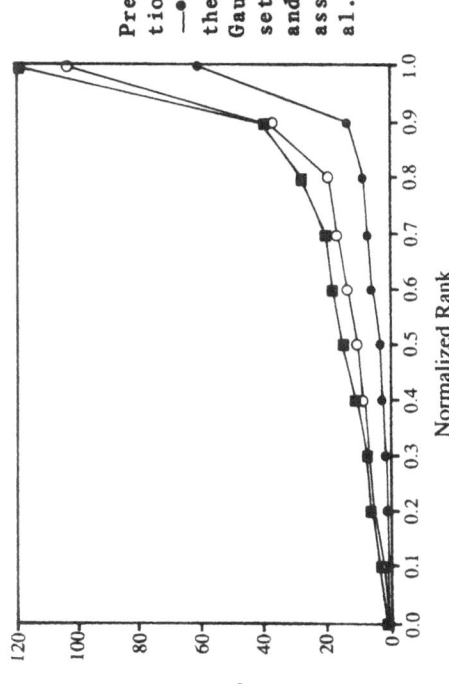

deposition associated with westerly (clean) flow at Adrigole was slightly smaller than that for the southeasterly (clean) flow at Bermuda, both were still significantly greater than the distribution at Amsterdam Island. This suggested a continental influence at Adrigole, Ireland, and Bermuda even when the flow was from the open ocean. The entire atmosphere of the North Atlantic Ocean might be influenced by continental emissions from not only North American but also from Europe and Africa. This explanation was supported by the conclusions of Galloway et al. (1989), Levy and Moxim (1987, 1989), and Savoie (1984).

In summary, our analysis of two years of data on nss-$SO_4^=$ in precipitation from Adrigole, Ireland, suggested that long-range transport of continental sulfur occurs over the Atlantic Ocean. Whelpdale et al. (1988) estimated that the increase in the concentration of nss-$SO_4^=$ in precipitation on the west coast of Europe from North American emissions is ~ 5 μeq/l, roughly the difference between the volume-weighted concentration of nss-$SO_4^=$ at Adrigole during periods of westerly flow (10 μeq/l) and that observed at Amsterdam Island (4.1 μeq/l). However, at the time of our NATO ARW, it was not possible to state whether the originating continent was North America, Africa, or perhaps Europe.

11.7.4. Impact of Fuel-Combustion Emission on the Nitrogen Chemistry of the Southern Hemisphere

The largest single source of nitrogen in the troposphere appears to be the combustion of fossil fuel (Logan 1983). Levy and Moxim (1989) used a general-circulation transport model to simulate the global spread and deposition of this nitrogen. The reactive nitrogen species are transported collectively as a single tracer and the chemical transformation of oxides to soluble acid and nitrate end products is only present implicitly in the group coefficients for dry and wet removal. Levy and Moxim (1987) have already used this model to simulate the transport and deposition of North American nitrogen emissions.

We compared the model's nitrate deposition and surface concentration of total reactive nitrogen with observations from the Northern Hemisphere (Tables 11-4 and 11-5). There was reasonable agreement with the values for the simulated depositions at locations both close to and far from major sources and the observed values in Table 11-4 as there was with the model's surface concentrations (Table 11-5) over the North Pacific. We subtracted the background nitrate of 0.12 μg/m^3 (see Section 11.7.2.) from the observed surface concentrations in Table 11-5.

Although the surface measurements from the Pacific exclude insoluble compounds, past observations of PAN, NO, NO_2, (Davis et al. 1987, McFarland et al. 1979, Ridley et al. 1987, Singh et al. 1986) are generally less than 20 ppt in that region. The model's transport of combustion emissions in the Northern Hemisphere accounted for at least 50% to 66% of the observed nitrogen in the North Pacific. Part of the model's shortfall might be its lack of explicit transport of PAN and other insoluble organic nitrates, but deficiencies in the model's transport and other sources were also likely. Recent simulations of deposition to the North Pacific give 1.0-1.2 tg N/yr (unpublished data available from Dr. Levy,

Table 11-4. Model-versus-observed values for nitrogen in wet deposition
at remote locations.

Location	Precipitation		Wet Deposition	
	Observed	Model	Observed	Model
	(cm/yr)		(mMole N/m^2)	
Poker Flats, Alaska	29	30-50*	0.5	0.1-0.2*
Bermuda	140	189	6.3	5.4
Mauna Loa, Hawaii	47	97	0.5	0.3
Adrigole, Ireland	200	90	11.2	5.3
Bay de Expoir, Newfoundland	149	93	9.1	10.0
Nova Scotia	138	116	16.0	15.0

Sources: Observed values for Alaska, Bermuda, Hawaii, and Ireland are
from unpublished data available from Dr. Galloway, Department of Environmental Sciences, University of Virginia, Charlottesville; data for
Newfoundland and Nova Scotia are from Vet et al. 1986.

*Because of a steep gradient in model precipitation, neighboring boxes
were included.

Table 11-5. Model-versus-observed concentration values of surface
R(NO_y)/yr from the Pacific (pptv).

Location	Observed		Model
	Total	Total minus Background of 38 pptv	
Enewetak	54	16	20
Fanning	58	20	11
Funafutti	40	2	3
Midway	105	67	43
Nauru	58	20	4
New Caldonia	76	38	10-50*
Norfolk	65	27	28
Oahu, Hawaii	130	92	(50-60)* 33**
Samoa	43	5	3
Shemya	94	56	74
Rarotonga	42	4	4

Sources: Observed values from unpublished data available from D. Savoie,
RSMAS, University of Miami, 4600 Rickenbacker Causeway, Miami, FL
33149-1098.

*Because of the presence of a local source, data were taken from upwind
boxes.

**The local Hawaiian source was eliminated in this simulation.

NOAA/Geophysical Fluid Dynamics Lab, Princeton University, Princeton, NJ
08540).

We found that the model transported almost no combustion nitrogen
from source regions in the Northern Hemisphere to the Southern Hemisphere
and emissions already present could have accounted for less than 10% of
the background nitrogen discussed in Section 11.7.2. We concluded that
fossil-fuel combustion had little impact on the nitrogen budget in most
of the Southern Hemisphere. Even a factor of two increase in the model's
long-range transport or the inclusion of PAN was not expected to alter
these conclusions significantly. However, the transport of nitrogen compounds down from the stratosphere into the troposphere may be quite
important in the Southern Hemisphere (Levy et al. 1980).

11.7.5. Arctic Haze

Over the last ten years many investigations (Barrie 1986; Heintzenberg
and Larssen 1983; Heintzenberg et al. 1981; Hov et al. 1984; Isaksen et
al. 1985; Pacyna and Ottar 1985; Rahn et al. 1977, 1980) have clearly
shown that substantial amounts of air pollution are being transported
into the Arctic region several thousand kilometers from major anthropogenic source areas. Concentrations generally peak from January to April,
with a minimum from June to August. This behavior is a result of seasonal variations in both transport and deposition. The flow pattern
associated with the wintertime Siberian high-pressure cell produces a
persistent transport from the major anthropogenic source areas in Europe
towards the Arctic. Furthermore, residence times of sulfur and nitrogen
in these areas increase in winter and early spring. Dry deposition is
inhibited by a dry, cold snow cover with a large surface resistance and
by a smooth surface and stable boundary layer giving a large aerodynamic
resistance. Wet deposition is small because there is little
precipitation.

Although European emissions are persistently transported into the
Arctic at this time of the year, most of the North American emissions are
carried eastwards over the Atlantic Ocean and are subjected to frequent
precipitation. Therefore, much of the sulfur and nitrogen is wet deposited before it reaches Europe or the Arctic; however, some is transported into the free troposphere by synoptic-scale vertical motion and
may reach the Arctic. Modeling studies (Barrie et al. 1989, Iversen
1987, Levy and Moxim 1989) have confirmed that most contaminants found in
the Arctic haze in the atmospheric boundary layer are from Europe.

11.7.6. The Atmospheric Chemical Climate Over the North Atlantic

The chemical climate of the atmosphere over the North Atlantic is comprised of a long-term mean and of fluctuations about the mean. It is
influenced by natural sources from the ocean, from continents, and from
the atmosphere itself; by anthropogenic sources from North America and
Europe; and by the region's meteorological climatology.

A natural radioactive tracer like ^{222}radon, which is emitted from
sources on land and advected to the oceanic atmosphere, is excellent for
tracking continental influence. Some continental air masses have radon

concentrations higher than 20 pCu/m^3 (Bonsang et al. 1980); radon concentrations lower than 4 pCu/m^3 represent less continental influence and
concentrations of 1-2 pCu/m^3 or less are only found in clean air masses
that have not contacted land for 10 days (unpublished data available from
Dr. A. G. Ryaboshapko, Institute of Applied Geophysics, Moscow).

Seasonal variations of ^{222}radon measured on board research vessels
in the central part of the North Atlantic (53°N, 35°W) showed a maximum
in the fall and winter although four years of nss-sulfate and nitrate
measurements did not (Fig. 11-7, unpublished data available from Dr.
Ryaboshapko, Institute of Applied Geophysics, Moscow). Over the North
Atlantic (Fig. 11-8), nss-sulfate, nitrate, and ^{222}radon all have minimum concentrations between 20°W and 30°W longitude. European emissions
influence the eastern Atlantic out to 20°W longitude and the influence
of North America extends out to 40°W longitude. The value for the longitude belt 60°-70°W is based on very few measurements and is not representative. The very high variability of the nss-sulfate data is caused
by both variations in nss-sulfate itself and variations in the chloride
concentrations used for sea-salt corrections.

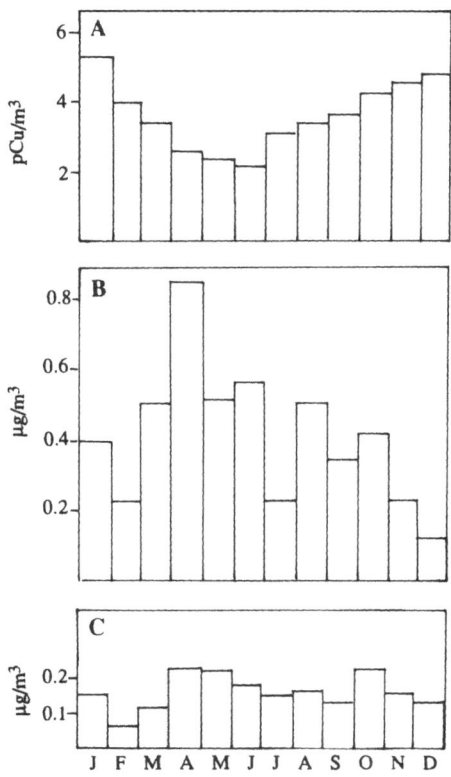

Figure 11-7. Seasonal variations of **A** ^{222}radon, **B** particulate nss-
sulfate, and **C** particulate nitrate over the central part of the
North Atlantic (53°N,35°W).

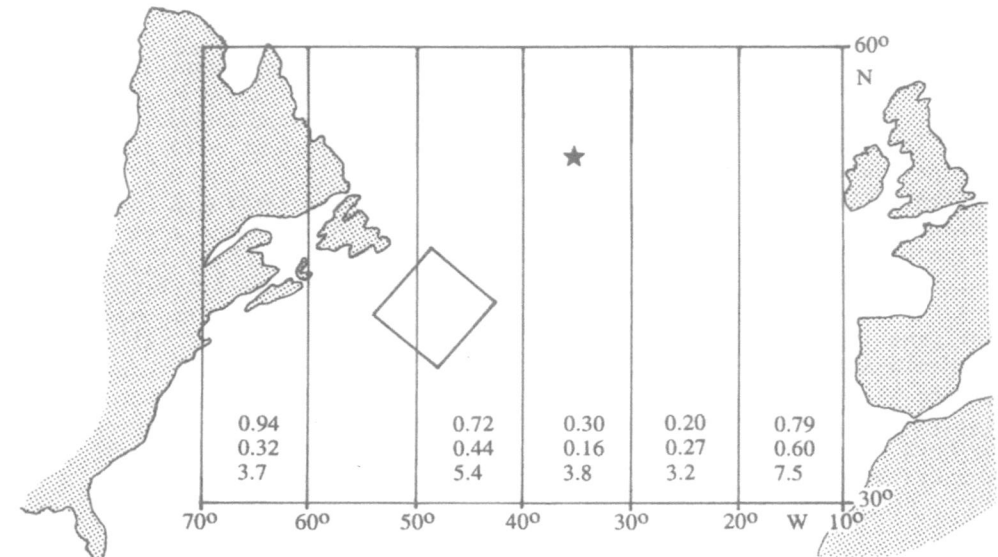

Figure 11-8. Longitudinal distribution of particulate nss–sulfate
($\mu g/m^3$), particulate nitrate (pg/m^3), and ^{222}radon (pCu/m^3) over
the North Atlantic. The ''Charley'' point is noted by an asterisk.
The data obtained between 40°W and 60°W are representative of the
inset square.

Ryaboshapko et al. (1987) calculated the average surface air concentrations of sulfur dioxide for 10×10–degree squares over the North
Atlantic from shipboard measurements (Fig. 11-9). Most data were
obtained in the region of point ''Charley'' (Fig. 11-8), a North Atlantic
oceanic station at 53°N, 35°W, where the average sulfur-dioxide concentration is about 0.1 μg SO_2/m^3. Similar values ranging from 0.03 μg
SO_2/m^3 to 0.15 μg SO_2/m^3 are found between 20°-60°W. These results
agree reasonably well with the background concentrations determined in
other regions of the world's oceans (Nguyen et al. 1983). Any continental influence on sulfur dioxide in the boundary layer over the North
Atlantic Ocean seems to be restricted within an area no farther than 1000
km from land.
Ryaboshapko et al. (1987) reported similar shipboard measurements for
nitrogen dioxide (Fig. 11-10). Concentrations range from 0.25 μg NO_2/m^3
to 0.5 μg NO_2/m^3 for the central part of the ocean and show the influence
of anthropogenic and natural continental sources up to 1000-1500 km from
the coast.
The greatest impact of continental emissions is expected to be in
coastal regions. The Western Atlantic Ocean Experiment (WATOX) supports
these conclusions for SO_2, NO_2, and related species (the gases PAN, NO,
HNO_3, and NO_3 and SO_4 in aerosols and precipitation) (Galloway et al.
1989, see Chapter 4, p. 87). The WATOX precipitation data for $SO_4^=$ and

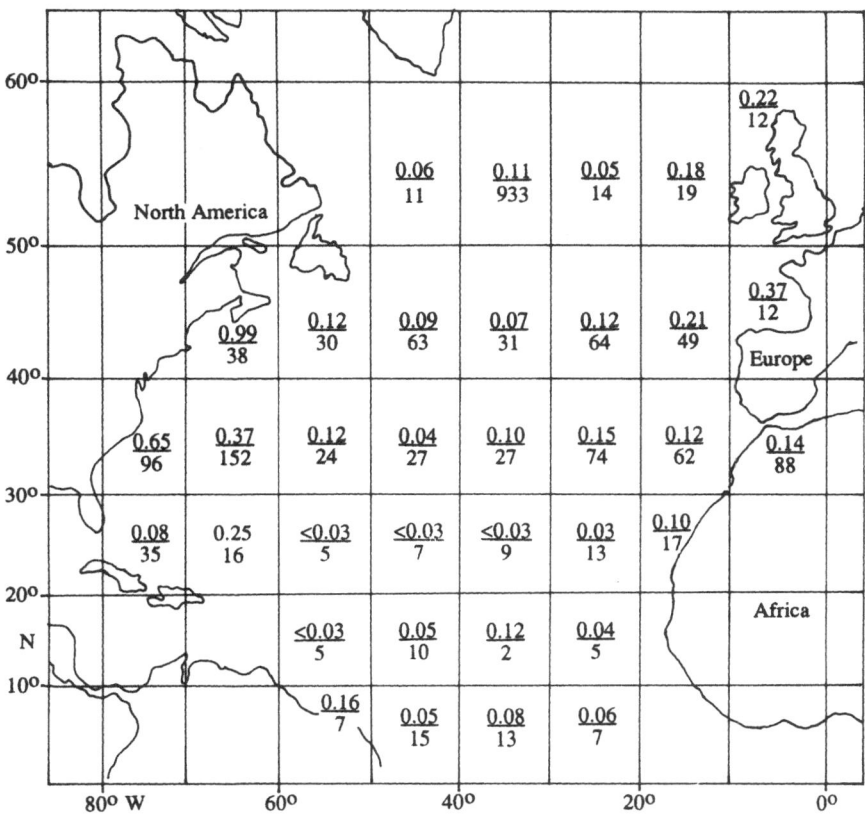

Figure 11-9. Atmospheric concentrations of SO_2 over the North Atlantic.
The numerators are the concentration values ($\mu g/m^3$) and the
denominators, the numbers of measurements.

NO_3^- from Adrigole, Ireland, show that European emissions have an impact
on S and N in the atmosphere of the eastern North Atlantic Ocean (Section
11.7.3.). The data also support the conclusions of Levy and Moxim (1987)
and Whelpdale et al. (1988) that a small percentage of the North American
emissions of SO_2 and NO_x, respectively, can be observed in the atmosphere
of the eastern Atlantic Ocean.

11.7.7. The Long-Range Transport of PAN Over the North Atlantic Ocean

Recently, Galloway and Whelpdale (1987) and Levy and Moxim (1987) estimated that 1.0-1.5 Tg N/yr are advected from North America over the North
Atlantic Ocean. This is a significant fraction (~ 25%) of the total NO_x
emissions in eastern North America. The latest studies indicate that
most advected nitrogen is deposited close to the North American continent
(Hastie et al. 1988, Luke and Dickerson 1987, Misanchuk et al. 1987).

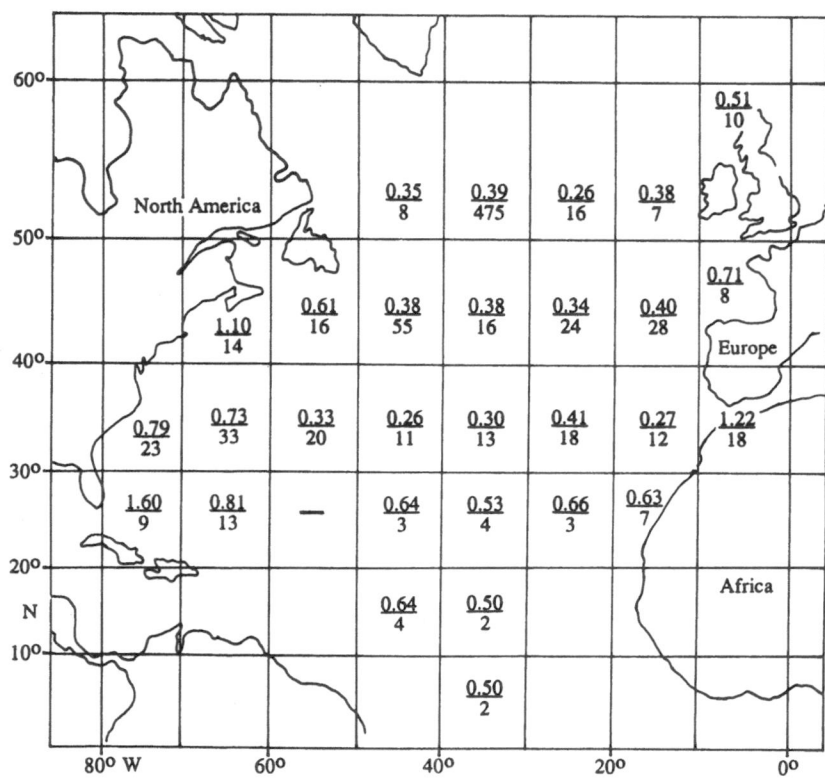

Figure 11-10. Atmospheric concentrations of NO₂ over the North Atlantic. The **numerators** are the concentration values ($\mu g/m^3$) and the **denominators**, the numbers of measurements.

Probably less than 10% of the nitrogen is transported east of 60°W (Galloway et al. 1989) and no more than 5% may reach Europe (Levy and Moxim 1987). However, since these estimates do not take into account the transport of nitrogen in forms other than NO_2, NO_3 or HNO_3, the values of the species being transported may be underestimated.

The potentially important species not considered is peroxyacetylnitrate (PAN) (Crutzen 1979, Nielsen et al. 1981). The formation and decomposition cycle of PAN involves NO_2 and provides a mechanism for storage and release of reactive nitrogen.

$$CH_3 - \overset{\overset{\textstyle O}{\|}}{C} - O_2 + NO_2 \quad \underset{k_2}{\overset{k_1}{\longleftarrow\!\!\longrightarrow}} \quad CH_3 - \overset{\overset{\textstyle O}{\|}}{C} - O_2NO_2,$$

Peroxyacetyl radical $\longleftarrow\!\!\longrightarrow$ PAN.

The temperature dependence of the dissociation constant (k_2 = [-13,543/T]S) makes PAN very stable at low temperatures (see Section 11.5). Thus, the atmospheric lifetime of PAN in cooler regions should be long enough to facilitate long-range transport. The subsequent entrance into a warmer environment should cause the thermal decomposition of PAN and release of NO_2.

The prevailing atmospheric transport paths over the western North Atlantic Ocean are cool enough to stabilize PAN, and no effective sink mechanism is active in this region. Oceans and precipitation are minor sinks because of the low solubility of PAN (Singh et al. 1986); losses over snow and ice surfaces are low (Bottenheim et al. 1986, Brice et al. 1984); and degradation through reactions with OH or Cl radicals should be unimportant.

Because the experimental data pertaining to PAN levels in the western North Atlantic region were scarce when we met in Bermuda, we had to consider the proposed long-range transport of PAN as likely but unproven. Bottenheim and Gallant (1987) reported highly variable levels of PAN ranging between 20-510 ppt in the boundary layer and 25-230 ppt in the free troposphere 250 km off the east coast of North America. The observed PAN:(NO_y) ratios are typical of continental background air (Bottenheim and Gallant 1987) and suggest long-range transport. There was further evidence from measurements made during the STRATOZ III trans-Atlantic flights that have consistently high (60-180 ppt) levels of PAN in the northern latitudes (53-60°N) (Rudolph et al. 1987). The high concentrations of PAN (~200 ppt) observed at Alert (82°N/60°W) in air masses that originated from the North Atlantic region (Bottenheim et al. 1986) lend support to the concept of long-range transport even though the source region may be Europe, not North America.

The chemistry, meteorological conditions, and indirect data strongly suggest that PAN is transported far from its source. However, based on information available at the time of our workshop, we were not able to quantify the transport or its effect on the distribution of nitrogen oxides in remote regions, particularly the Southern Hemisphere.

11.8. UNCERTAINTIES, GAPS, AND RECOMMENDATIONS

Having summarized the basic processes that underlie long-range transport and discussed seven case studies specific to sulfur and nitrogen, we then examined the sources of uncertainty in these and future studies, identified the major gaps in our knowledge, and suggested possible future research directions.

Many of our uncertainties involved attempts to identify and quantify specific sources either from an analysis of specific transport events or from an analysis of an ensemble of data from specific receptor locations. Both cases are discussed in the following sections.

11.8.1. Natural Variations

The major difficulty in identifying the impact of distant sources with a long-term sampling program is the variability or potential variability of the natural source strengths. The following examples illustrate our problem.

At three remote stations in the tropical and subtropical South Pacific, the annual mean nitrate concentrations are all 0.11 $\mu g/m^3$ ± 0.01 (see Section 11.7.2.). This created some major questions: Was this value actually representative of a natural marine ''background'' in the region of these stations? If so, What were the major sources for this nitrate: the stratosphere, lightning, or some unknown oceanic process? The answer to the latter question was critically important since the sources and their spatial and temporal variabilities would determine our ability to model their inputs. Did they vary substantially from one geographical region to another; for example, from the equatorial regions to the poles or from productive to oligotrophic regions of the ocean? Did the background in the Northern Hemisphere differ substantially from that in the Southern Hemisphere? We only knew the ''apparent'' background levels at three relatively closely spaced stations.

For sulfate another series of questions arose. We knew that a major natural marine source of sulfate is the oxidation of dimethyl sulfide (DMS) emitted from the ocean. By a different pathway, this oxidation results in another stable end product, methanesulfonate (MSA). Results of studies in the tropics (see Section 11.7.2.) indicate that the $MSA:nss-SO_4$ ratios from this source are fairly constant, on the order of 0.07, at low latitudes. The constancy of this ratio should have allowed us to assess the biogenic portion of the total nss-sulfate concentration by concurrently measuring MSA. Unfortunately, it was not clear that the ratio was indeed constant. Concurrent data from high-latitude sites and ice cores suggested that the ratio might be considerably higher, on the order of 1 or more, in the colder regions. If we were to make use of MSA as a tracer of biogenic sulfate, the variables that control the ratio had to be known so that we could obtain precise estimates of its value at a given time and location.

The biogenic source strength could also vary with temperature and insolation. Although their variation is relatively small in the tropics, it becomes very large in the high latitude and polar regions especially where many areas may be ice covered for significant periods. There were few data on the year-to-year variations although they could be substantial in some regions. For example, the productivity in the equatorial Pacific might change drastically during times of El Nino--southern oscillation events.

Other natural oceanic or atmospheric sources might also be contributing significantly to the normal levels of nss-sulfate in the marine troposphere. If so, questions similar to those posed for the ''background'' nitrate needed to be answered.

11.8.2. Interpretation Uncertainties for Individual Events

A powerful tool for identifying specific source regions and transport paths is trajectory analysis, either directly using detailed meteorological data or indirectly using surrogate inert tracers. If the emission density of the measured species, or its precursor, is very low in the measurement area, it may be possible to track material back to its source region (Levy 1987). However, any procedure for calculating the trajectory of air parcels contains inherent errors that grow exponentially with travel time. Two trajectories starting close together begin to diverge after several days travel, since the velocity of their endpoints is different and the difference generally increases as they separate. This is more pronounced in regions where the pressure field has high curvature, such as in cyclones, or where the velocity field is highly variable, such as in high-pressure systems. It would, therefore, be preferable to consider some form of ''ensemble'' averaging of the data, such as sector analysis, to get a measured value that would not be critically dependent on the exact path of a single trajectory.

Ensemble averaging would also reduce the uncertainties resulting from the lack of knowledge of removal processes between the suspected major source region and the receptor site. For a suspected source region, the average fraction surviving to reach the receptor is the product of the fraction of the emission that is not removed within the source region and the fraction that survives removal between source and receptor. In principle, the former can be roughly estimated using regional transport models. The latter depends on the most efficient removal process and can be roughly estimated given a general understanding of atmospheric deposition processes.

With some knowledge of atmospheric deposition processes, the best that we could hope for would be that the estimate was consistent (within a factor of 3 either way) with the measured values. Generally, measured values are likely to represent overestimates of the transport component because of

1. The positive bias in the measurement (no negative values),

2. weak (natural?) sources between the source and receptor neglected in the calculation, and

3. Contributions from nearby sources that errors in the trajectory analysis could not exclude.

An alternative would be to consider measurements of inert, yet source-specific, tracers combined with the species of interest. High or low concentrations of materials from marine or terrestrial sources, or of elements associated with particular industrial processes, could be used to check that back trajectories were accurately identifying the sources of material originating very far from a sampling point.

11.8.3. Gaps and Recommendations

The gaps in our knowledge of sulfur and nitrogen transport and the research needed to provide information were either specific to sulfur or nitrogen or dealt with long-range transport in general.

For both sulfur and nitrogen there was the question of their chemical climatology in regions with no data—for example, the Indian Ocean. As a first step, we recommend shipboard measurements be taken in those regions.

Specific to sulfur was the general question of biogenic emissions and the chemical reactions leading to their oxidation to sulfate and other stable end products. Biogenic emissions are potential sources for long-range transport. More importantly, they may interfere with attempts to quantify the long-range transport of anthropogenically emitted sulfur. We recommend a field-measurement program to quantify the location, strength, and variability of biological sources and laboratory kinetics research to establish the reaction path for DMS oxidation.

An important question regarding nitrogen was the nature of the background nitrogen and its source or sources. To identify these sources, both nitrate and its precursors should be measured in remote regions in both the boundary layer and the free troposphere. A key to understanding background nitrogen and nitrogen transport in general is the chemical reaction path that converts the emitted oxides to the deposition products. Of special interest are PAN and other organic nitrates. We recommend field measurements of nitrate and its precursors, particularly PAN. Measurements of PAN in the winter-free troposphere, specifically at high latitudes, are important. The field measurements needed to unravel the reaction path should be made in emission regions as well as coastal and remote locations. These measurements should then be used to develop chemical models of the NO_x conversion.

The lack of quantitative emission fluxes is a general problem for all natural sources and for many of the anthropogenic sources around the globe. With reactive gases, there is the added difficulty of determining an effective source strength for long-range transport. This involves not only the actual emission strength but also the emitted species' chemical reactivity and the meteorology of the region. The effective source is regional in scale and a complex chemical transport problem on its own. We recommend measurement programs in important source regions, careful world-wide inventories of anthropogenic sources, and the development of regional chemical/transport models to address the question of effective source strengths. Further, we encourage the continued development of sophisticated global atmospheric chemistry models with the potential to treat the fundamental processes of long-range transport realistically.

11.9. REFERENCES

Barrie, L. A. 1986. Arctic air pollution: An overview of current knowledge. Atmos. Environ. 20:643-663.

Barrie, L. A., M. P. Olson, and K. K. Oikawa. 1989. The flux of anthropic sulphur into the Arctic from mid-latitudes in 1979/80. Atmos. Environ. 23, in press.

Bonsang, B., B. C. Nguyen, A. Gaudry, and G. Lambert. 1980. Sulfate enrichment in marine aerosols owing to biogenic gaseous sulfur compounds. J. Geophys. Res. 85:7410-7416.

Bottenheim, J. W., and A. J. Gallant. 1987. The occurrence of peroxyacetyl nitrate over the Atlantic Ocean east of North America during WATOX-86. Global Biogeochem. Cycles 1:369-380.

Bottenheim, J. W., A. G. Gallant, and K. A. Brice. 1986. Measurements of NO_y species and O_3 at 82° N latitude. Geophys. Res. Ltrs. 13:113-116.

Brice, K. A., S. A. Penkett, D. H. F. Atkins, F. J. Sandalls, D. J.
    Bamber, A. F. Tuck, and G. Vaughan. 1984. Atmospheric measurements
    of peroxyacetylnitrate (PAN) in rural, southeast England: Seasonal
    variations, winter photochemistry and long-range transport. Atmos.
    Environ. 18:2691-2702.
Brimblecombe, P., C. Hammer, H. Rodhe, and A. Ryaboshapko. 1988. Changes
    of the sulfur cycle in recent milenia. In Evolution of the Global
    Biogeochemical Sulphur Cycle (P. Brimblecombe and A. Lein, eds.)
    New York:Wiley.
Crutzen, P. J. 1979. The role of NO and NO_2 in the chemistry of the
    troposphere and stratosphere. Ann. Rev. Earth Planetary Sci. 7:443-
    472.
Davis, D. D., J. D. Bradshaw, M. O. Rodgers, S. T. Sandholm, and S.
    KeSheng. 1987. Free tropospheric and boundary layer measurements of
    NO over the Central and Eastern North Pacific Ocean. J. Geophys. Res.
    92:2049-2070.
Duce, R. A., C. K. Unni, B. J. Ray, J. M. Prospero, and J. T. Merrill.
    1980. Long-range atmospheric transport of soil dust from Asia to
    the tropical North Pacific: Temporal variability. Science
    209:1522-1524.
Findlayson-Pitts, B. J., and J. N. Pitts, Jr. 1986. Atmospheric Chemis-
    try. New York:Wiley, 1098 pp.
Galloway, J. N., and A. Gaudry. 1984. The composition of precipitation
    on Amsterdam Island, Indian Ocean. Atmos. Environ. 18:2649-2656.
Galloway, J. N., and D. M. Whelpdale. 1987. WATOX-86 overview and west-
    ern North Atlantic Ocean S and N atmospheric budgets. Global Bio-
    geochem. Cycles 1:261-281.
Galloway, J. N., R. S. Artz, W. C. Keene, T. M. Church, and A. H. Knap.
    1989. Processes controlling the concentration of $SO_4^=$, NH_4^+, H^+,
    $HCOOH^-$, and CH_3COO^- in Bermuda precipitation. Tellus, in press.
Harris, J. M. 1982. The GMCC Atmospheric Trajectory Program. NOAA Tech.
    Memo ERL/ARL 116, Air Resources Lab., Rockville, MD, 30 pp.
Hastie, D. R., H. L. Schiff, D. M. Whelpdale, R. E. Peterson, W. H.
    Zoller, and D. L. Anderson. 1988. Nitrogen and sulfur over the
    western Atlantic Ocean. Atmos. Environ 22:2381-2391.
Heintzenberg, J., and S. Larssen. 1983. SO_2 and SO_4 in the Arctic:
    Interpretation of observations at three Norwegian and Arctic/sub-
    Arctic stations. Tellus 35B:255-265.
Heintzenberg, J., H.-C. Hannsson, and H. Lannefors. 1981. The chemical
    composition of Arctic haze at Ny-Alesund, Spitsbergen. Tellus 33:
    162-171.
Hov, O., S. A. Penkett, I. S. A. Isaksen, and A. Semb. 1984. Organic
    gases in the Norwegian Arctic. Geophys. Res. Ltrs. 11:425-428.
Isaksen, I. S. A., O. Hov, S. A. Penkett, and A. Semb. 1985. Model anal-
    ysis of the measured concentration of organic gases in the Nor-
    wegian Arctic. Atmos. Chemistry 3:3-27.
Iversen, T. 1987. Simulation of the atmospheric transport of sulphur
    dioxide and particulate sulphate to the Arctic. Norwegian Institute
    for Air Res. Rept. No. NILU OR 83/86, Oslo, Norway, n.p.
Korolev, S. M., and A. G. Ryaboshapko. 1980. Investigation of parameters
    of dilution and removal of sulphur dioxide and lead during transport

above the ocean. In *Procs., Long-Range Transport of Pollutants and its Relation to General Circulation Including Stratospheric/Tropospheric Exchange Processes,* WMO Pub. 538, Geneva:WMO, 117–124.

Levy, H., II. 1987. Tracers of atmospheric transport. *Nature* 325:761–767.

Levy, H., II, and W. J. Moxim. 1987. Fate of US and Canadian combustion nitrogen emissions. *Nature* 328:414–416.

Levy, H., II, and W. J. Moxim. 1989. Simulated global distribution and deposition of reactive nitrogen emitted by fossil-fuel combustion. *Tellus* 41B:256–271.

Levy, H., II., J. D. Mahlman, and W. J. Moxim. 1980. A stratospheric source of reactive nitrogen in the unpolluted troposphere. *Geophys. Res. Ltrs* 7:441–444.

Logan, J. A. 1983. Nitrogen oxides in the troposphere: Global and regional budgets. *J. Geophys. Res.* 88:10,785–10,807.

Luke, W. T., and R. R. Dickerson. 1987. The flux of reactive nitrogen compounds from eastern North America to the western Atlantic Ocean. *Global Biogeochem. Cycles* 1:329–344.

McFarland, M., D. Kley, J. W. Drummond, A. L. Schmeltekopf, and R. H. Winkler. 1979. Nitric oxide measurements in the equatorial Pacific region. *Geophys. Res. Ltrs.* 6:605–608.

Misanchuk, B. A., D. R. Hastie, and H. I. Schiff. 1987. The distribution of nitrogen oxides off the east coast of North America. *Global Biogeochemical Cycles* 1:345–355.

National Academy of Science (National Research Council). 1984. *Global Tropospheric Chemistry: A Plan for Action.* Washington:National Academy Press, 194 pp.

Nguyen, B. C., B. Bonsang, G. Lambert, and Z. Z. Pasquier. 1975. Residence times of sulphur dioxide in the marine atmosphere. *Pure Appl. Geophys.* 113:489–500.

Nguyen, B. C., B. Bonsang, and A. Gaudry. 1983. The role of the ocean in the global tropospheric sulfur cycle. *J. Geophys. Res.* 88:10,903–10,914.

Nielsen, J., U. Samuelsson, P. Grennfelt, and E. L. Thomsen. 1981. Peroxyacetyl nitrate in long-range transported polluted air. *Nature* 293:553–555.

Pacyna, J. M., and B. Ottar. 1985. Transport and chemical composition of the summer aerosol in the Norwegian Arctic. *Atmos. Environ.* 19:2109–2120.

Parrington, J. R., and W. H. Zoller. 1984. Diurnal and longer term changes in the composition of atmospheric particles at Mauna Loa, Hawaii. *J. Geophys. Res.* 89:2522–2534.

Prahm, L. P, U. Torp, and R. M. Stern. 1976. Deposition and transformation rates of sulphur oxides during atmospheric transport over the Atlantic. *Tellus* 28:355–372.

Prospero, J. M., D. L. Savoie, R. T. Nees, R. A. Duce, and J. Merrill. 1985. Particulate sulfate and nitrate in the boundary layer over the North Pacific Ocean. *J. Geophys. Res.* 90:10,586–10,596.

Rahn, K. A., R. D. Borys, and G. E. Shaw. 1977. The Asian source of Arctic haze bands. *Nature* 268:713–715.

Rahn, K. A., E. Joranger, A. Semb, and T. J. Conway. 1980. High winter concentrations of SO_2 in the Norwegian Arctic and transport from Eurasia. Nature 287:824-825.

Ridley, B. A., M. A. Carroll, and G. L. Gregory. 1987. Measurements of nitric oxide in the boundary layer and the free troposphere over the Pacific Ocean. J. Geophys. Res. 92:2025-2047.

Rudolph, J., B. Vierkorn-Rudolph, and F. X. Meixner. 1987. Large-scale distribution of peroxyacetylnitrate: Results from the STRATOZ III flights. J. Geophys. Res. 92:6653-6661.

Ryaboshapko, A. 1983. The atmospheric sulphur cycle. In The Global Biogeochemical Sulphur Cycle (M. V. Ivanov and J. R. Freney, eds.) New York:Wiley, 203-296.

Ryaboshapko, A. G., V. I. Lepeshkin, E. D. Podgurskaya, and V. I. Medinets. 1987. Air pollution monitoring over the North Atlantic. In Procs., Intern. Symp., Integrated Global Monitoring of State of the Biosphere (WMO Tech. Document No. 151), (U. A. Izrael, F. Y. Rovinsky, A. V. Tsyban, S. M. Semenov, and V. A. Abakumov, eds.) Geneva:WMO, 261-282.

Saltzman, E. S., D. L. Savoie, J. M. Prospero, and R. G. Zika. 1985. Atmospheric methanesulfonic acid and non-sea-salt sulfate at Fanning and American Samoa. Geophy. Res. Ltrs. 12:437-440.

Saltzman, E. S., D. L. Savoie, J. M. Prospero, and R. G. Zika. 1986. Methanesulfonic acid and non-sea-salt sulfate in Pacific air: Regional and seasonal variations. Atmos. Chemistry 4:227-240.

Savoie, D. L. 1984. Nitrate and non-sea-salt sulfate aerosols over major region of the world ocean: Concentrations, sources and fluxes. Ph.D. dissert., Univ. of Miami, FL, 432 pp.

Savoie, D. L., J. M. Prospero, and R. T. Nees. 1987. Nitrate, non-sea-salt sulfate, and mineral aerosol over the northwestern Indian Ocean. J. Geophys. Res. 92:933-942.

Singh, H. B., L. J. Salas, and W. Viezee. 1986. Global distribution of peroxyacetyl nitrate. Nature 321:588-591.

Uematsu, M., R. A. Duce, J. M. Prospero, L. Chen, J. T. Merrill, and R. L. McDonald. 1983. Transport of mineral aerosol from Asia over the North Pacific Ocean. J. Geophys. Res. 88:5343-5352.

Vet, R. J., W. B. Sukoff, M. E. Still, and R. Gilbert. 1986. CAPMON Precipitation Chemistry Data Summary 1983-1984. Atmospheric Environment Service, Downsview, Ontario, n.p.

Whelpdale, D. M., A. Eliassen, J. N. Galloway, H. Dovland, and J. M. Miller. 1988. The trans-Atlantic transport of sulfur. Tellus 40B:1-15.

World Meteorological Organization. 1985. Atmospheric Ozone 1985: An Assessment of Our Understanding of the Processes Controlling its Present Distribution and Change, Vol. 1, Global Ozone Res. Monitoring Proj. Rept. 16, Geneva:World Meteorological Organization, n.p.

12. THE LONG-RANGE TRANSPORT OF ORGANIC COMPOUNDS

Terry Bidleman, Rapporteur
Department of Chemistry and
  Marine Science Program
University of South Carolina
Columbia, SC 29208

Elliot L. Atlas
Department of Oceanography
College of Geosciences
Texas A & M University
College Station, TX 77843-3146

Roger Atkinson
Statewide Air Pollution
  Research Center
University of California
Riverside, CA 92521

Bernard Bonsang
Centre Des Faibles
  Radioactivites
Laboratoire Mixte CNRS-CEA
91198 Gif-sur-Yvette, Cedex,
France

Kathryn Burns
Bermuda Biological Station
  for Research, Inc.
17 Biological Station Lane
Ferry Reach GEO1, Bermuda

William C. Keene
Department of Environmental
  Sciences
University of Virginia
Charlottesville, VA 22903

Anthony H. Knap
Bermuda Biological Station
  for Research, Inc.
17 Biological Station Lane
Ferry Reach GEO1, Bermuda

John Miller
ARL/NOAA, Room 927
8060 13th Street
Silver Springs, MD 20910

Jochen Rudolph
Institut fuer Atmospharische
  Chemie
Kernforschungsanlage Juelich
  GmbH
D-5170 Juelich, FRG

Shinsuke Tanabe
Department of Environmental
  Conservation
Ehime University
3-5-7, Tarumi
Matsuyama 790, Japan

12.1. INTRODUCTION

When we met at the Bermuda Biological Station for Research, our working
group addressed the question of the long-range transport of organic compounds first in terms of the individual compound classes. The main
groups of compounds we discussed at the NATO workshop had (1) major
sources in continental areas, (2) a long enough residence time in the
atmosphere to enable species to be transported hundreds to thousands of

259

A. H. Knap (ed.), The Long-Range Atmospheric Transport of Natural and Contaminant Substances, 259–301.

kilometers, and (3) a potential to impact remote areas. The classes of compounds we chose to examine were:

- Nonmethane hydrocarbons (C_2–C_{33})(NMHC)

- Low–molecular–weight halocarbons (excluding freons and other long–lived halocarbons, such as CCl_4)

- C_1 and C_2 carboxylic acids

- Polynuclear aromatic hydrocarbon (PAH)

- Chlorinated pesticides and polychlorinated biphenyls (PCBs)

- High–molecular–weight oxygenated compounds

- Polymeric and ''soot'' carbon

We intended to examine the state of knowledge in the scientific community concerning the source emissions, transport, transformations, and deposition processes involving these classes of compounds. In general, we expected the transport of organic compounds to follow the atmospheric paths described in Chapter 1 (p. 3). However, because of the diversity of the organic substances released into the atmosphere, such factors as the vertical–exchange potential, interaction with clouds, recirculation, and other physical processes play varying roles in the transport of these substances. First we discussed the significance of the compound classes stated above and then we evaluated the separate processes involved in the long–range transport of these substances. Finally, we gave our recommendations for future research.

12.2. SIGNIFICANCE AND RATIONALE

Light– and medium–molecular–weight nonmethane hydrocarbons (NMHC) have been found in the remote troposphere in volume–mixing ratios of ppb (10^9) or sub–ppb. They have large sources in the Northern Hemisphere and intermediate atmospheric lifetimes. These compounds are sensitive indicators of anthropogenic influence and of OH–radical reactivity in the troposphere.

Because of the rapid reaction of many NMHC with the OH radical, they can have a considerable impact on photochemical cycles in the remote troposphere. Thus, the transport and presence of NMHC can influence the formation or destruction of ozone—depending on the atmospheric concentration of NO—and may affect the tropospheric concentration of OH radical (Kasting and Singh 1986). Furthermore, the products of the photochemical degradation of NMHC (aldehydes, carboxylic acids, ketones, etc.) may enhance the impact of NMHC on the chemistry of the remote atmosphere. In particular, the formation of organic nitrogen species (peroxyacetyl nitrate [PAN] and other peroxyacyl nitrates, alkyl nitrates, etc.) from the oxidation of NMHC in the presence of nitrogen oxides can be important to the formation of ozone in the clean troposphere (Kasting and Singh 1986).

Although high-molecular-weight NMHC are generally present in the atmosphere at concentrations well below those of the low-molecular-weight species, their reaction-rate constants increase with carbon number (Atkinson 1985, Carter and Atkinson 1985, Greiner 1970). Thus the heavy compounds may be more important in the chemistry of the remote atmosphere than predicted by current models because of their higher reaction rates and the possibility of multiple reaction steps in their atmospheric degradation. In addition to higher molecular-weight compounds of anthropogenic origin, naturally occurring hydrocarbons and oxygenated compounds from plants and soils are used to characterize the long-range transport and deposition of organic material to the ocean (see Section 12.3, p. 262).

The C_1 and C_2 carboxylic acids, particularly formic (HCOOH) and acetic acid (CH_3COOH), are constituents of the troposphere (see Chapter 5, p. 105). These acids contribute substantial fractions to the free acidity in precipitation and cloud water (Keene and Galloway 1988) and are probably a major sink for OH radicals within clouds (Jacob 1986). As such, carboxylic acids interact directly and indirectly in the chemical cycling of numerous other atmospheric species. In addition, the production of low-molecular-weight acids and aldehydes by the photodecomposition of aliphatic materials in the marine boundary layer may influence the organic content of ocean surface water. The major processes involved in the cycling of these compounds must, therefore, be understood if we are to study other biogeochemical cycles in the troposphere.

Polynuclear aromatic hydrocarbons (PAH) have large emission sources and are potentially hazardous to aquatic ecosystems and remote environments. This class of compounds contains some well-known carcinogens, e.g., benzo(a)pyrene (Perera 1981). Although all the chemical species responsible have not been clearly identified, Tokiwa and Ohnishi's (1986) studies showed that some PAH and their atmospheric transformation products can be mutagenic. In urban areas, for example, nitroarene transformation products account for up to 10% of the direct-acting mutagenicity of ambient particulate organic matter (Arey et al. 1988, Strandell et al. 1987).

The studies of chlorinated pesticides and PCBs are motivated by a need to understand their transport to and toxic effects in remote areas. Organochlorine compounds, such as polychlorinated biphenyls (PCBs), DDT, and hexachlorocyclohexanes (HCH), are produced in large quantities and widely used throughout the world. Although most developed countries have prohibited or restricted the production and the use of some chemicals since the 1970s and early 1980s because of growing concerns about their persistent nature and biological effects, organochlorine insecticides (including DDT, HCH, cyclodienes, and toxaphene) are still heavily used in some developing countries. All of the PCBs used in older electrical equipment, such as transformers and capacitors, cannot be recovered. More than half of all the PCBs ever produced are still in use and most PCBs in current equipment will be disposed of during the next decade (Bletchley 1984). Therefore, although concentrations of organochlorine pesticides and PCBs are declining in the atmosphere and in other

environmental compartments of countries where they have been banned or
restricted, these hazardous chemicals are being emitted from countries
where they are still produced and used. Thus, contamination in developing countries and inputs to remote ecosystems are unlikely to decline, at
least in the near future (Tanabe 1988).

The open ocean is a vast reservoir and the final sink of persistent
man-made chemicals (National Academy of Science 1978b, 1979; Tanabe and
Tatsukawa 1986). Tanabe et al. (1988), in a study of PCB metabolism in
small cetaceans, have reported that, because marine mammals are slow to
metabolize persistent organochlorine compounds, they are vulnerable to
long-term toxicity of these chemicals. Through nursing their young,
marine mammals tend to transfer these chemicals to their offspring.
Addison (1986) has suggested that PCBs (possibly aggravated by the DDT
group) may also interfere with the reproduction of marine mammals. The
atmospheric transport rate of persistent organochlorine compounds from
continents to oceans must be found if we are to understand the future
trend of contamination and the possible biological effects of these chemicals on the marine environment.

Carbonaceous aerosols of natural and anthropogenic origins can be
transported far from their sources. We still do not completely understand the role of such long-range transport in the global atmospheric
carbon budget, and only sparse data were available at the time of our
workshop in Bermuda to assess the significance of carbonaceous particles
to the clean troposphere. However, ''soot'' carbon, i.e., the dark component of these carbonaceous aerosols, may affect the environment by
absorption of solar radiation (Chylek et al. 1984, Clarke et al. 1984,
Coakley et al. 1983, Wang et al. 1986); by possible participation in
heterogeneous reactions in the atmosphere (Brodzinsky et al. 1980); by
adsorption of other atmospheric pollutants (Barbaray et al. 1977, Brodzinsky et al. 1980); and by acting as condensation nuclei thereby altering the size distribution and optical properties of clouds (Penner and
Edwards 1986, Pueschell et al. 1981, Twomey 1977). Surface effects of
these carbonaceous particles are enhanced because most carbon is associated with submicrometer particles.

12.3. SOURCES

Emissions of organic compounds to the atmosphere are complex and result
from human activity and natural processes. The distinction between natural and anthropogenic contributions is simple for some classes of compounds, such as chlorinated pesticides and PCBs (which have no known
natural source), but can be difficult for others. For the organic compounds considered, we discussed their sources in terms of how and where
emissions occur, what was known about the magnitude of source strengths,
and the proportion of anthropogenic versus natural emissions.

12.3.1. Nonmethane Hydrocarbons (NMHC)

The main sources of light- and medium-molecular-weight NMHC are engine
exhaust, industrial emissions, solvent usage, refinery losses, fuel

evaporation, natural-gas leakage from oil and gas fields, distribution losses, biomass burning (including agricultural waste burning, forest fires, wood-fuel combustion, and slash-and-burn agriculture), soil production, and oceanic and vegetational emissions (Bonsang et al. 1988, Ehhalt and Rudolph 1984, Greenberg et al. 1984). The major hydrocarbons emitted from forests are isoprene and terpenoid compounds (Altshuller 1983, Graedel 1979, Greenberg and Zimmerman 1984, Greenberg et al. 1984, Zimmerman et al. 1988). These species are short-lived (typically with lifetimes of a day or less) and, therefore, have little relevance to long-range transport. However, the products of their photooxidations are longer lived and may be transported over greater distances. All known NMHC emanate from the earth's surface; and, except for oceanic emissions and biomass burning, most sources are in the midnorthern latitudes.

The qualitative NMHC composition for several sources is relatively well known. Nonvegetative biogenic sources (microbes in soils, oceans, etc.) produce mainly light alkenes, especially ethene and propene. Natural-gas and fuel losses consist almost entirely of saturated hydrocarbons whereas the hydrocarbon spectrum emitted by combustion sources (engine exhaust, biomass burning) is very complex and include alkenes, straight-chain and branched alkanes, and acetylene and aromatic hydrocarbons.

Naturally produced high-molecular-weight NMHC are distinguished from petrogenic residues by certain features: Terrestrial plants synthesize alkanes with a strong odd:even ratio (expressed by the carbon-preference index) dominated by $n-C_{27}$, $n-C_{29}$, and $n-C_{31}$; marine plankton synthesize mainly $n-C_{15}$, $n-C_{17}$, and $n-C_{19}$, and a few other specific markers. (The high concentrations of pristane in the North Pacific may be produced by zooplankton species [see Chapter 5, p. 105]). These biogenic signatures differ from those of petrogenic residues, which show a smooth distribution of n-alkanes over a wide carbon range. Chromatograms of hydrocarbons from fossil-fuel sources also show the presence of an unresolved complex mixture, which is lacking in hydrocarbon chromatograms from biogenic sources. Simoneit (1984, 1986) and Simoneit and Mazurek (1981) have discussed these and other features used to distinguish biogenic and fossil-fuel hydrocarbons.

Various researchers have published estimates of the global source strength of light NMHC (Blake and Rowland 1986, Bonsang and Lambert 1985, Bonsang et al. 1988, Ehhalt and Rudolph 1984, Greenberg and Zimmerman 1984, Greenberg et al. 1984, Rudolph and Ehhalt 1981). The total emission of light NMHC—not including isoprene and terpenes—is approximately $100-150 \times 10^9$ kg/yr, with an uncertainty of a factor of at least two. Insufficient data were available at the time of our workshop from which to estimate the source strength of heavier NMHC (C_{10} and above) emitted to the atmosphere.

12.3.2. Low-Molecular-Weight Halocarbons

The production of some volatile halocarbons in the Northern Hemisphere is fairly well known; the data are summarized by the National Academy of Science (1978a). The total chloromethane (methyl chloride +

dichloromethane + chloroform + carbon tetrachloride) production in the United States from 1908-1976 is estimated to be 15×10^6 tonnes, with carbon tetrachloride accounting for 53% of this total. In 1976, the United States was responsible for about 54% of the chloromethanes produced in the free world (i.e., the U.S. + Europe + Japan) that year.

In 1978, NAS (National Academy of Science 1978a) published estimates of the cumulative free-world emissions of chloromethanes. The losses from product manufacture and use between 1908-1976 are 7.7×10^6 tonnes, the bulk of which comes from the U.S., western Europe, and Japan. Cumulative emissions from eastern Europe, Australia, South America, Africa, and Asia are 10% of those in the U.S., western Europe, and Japan. Regulatory changes could substantially decrease industrial emissions of certain chloromethanes; for example, a ban on chlorofluorocarbon production could reduce emissions of the precursor carbon tetrachloride by 30%.

The production data may underestimate total emissions for some compounds since nonindustrial sources are not included. For example, additional sources of halomethanes are the chlorination of natural water and drinking water and pulp- and paper-industry effluents (Christman et al. 1983, Oliver and Lawrence 1979, Rook 1974). Combustion processes also release certain halocarbons, primarily methyl chloride and dichloromethane (NAS 1978a).

Some light halocarbons also have substantial natural sources. For example, Khalil et al. (1983) have estimated that about 57% of total $CHCl_3$ emissions--3.5×10^8 kg/yr--is from the world's oceans (largely in the tropics). Singh et al. (1983) find that ocean surface water is supersaturated with CH_2Cl_2 and C_2Cl_4 (as well as $CHCl_3$), inferring an oceanic source for these compounds. The same seawater samples are in approximate equilibrium with the chlorofluorocarbons (freons) in the air. Other halocarbons with predominantly marine sources are CH_3Cl, CH_3Br, and CH_3I. Total emissions of these compounds, as well as $CHCl_3$, to the atmosphere range from 7.7×10^7 kg/yr to 5.2×10^9 kg/yr (NAS 1978a).

12.3.3. C_1 and C_2 Carboxylic Acids

Experimental evidence, chemical-modeling investigations, and theoretical considerations support various hypotheses concerning natural sources for carboxylic acids in the atmosphere. Vertical profiles measured through the forest canopy in the Brazilian Amazon suggest that, although the photooxidation of isoprene emitted by vegetation may generate substantial HCOOH (approximately 1 ppbv), CH_3COOH is directly emitted by vegetation (Andreae et al. 1988, Jacob and Wofsy 1988). The direct emissions of HCOOH from plants, soils, or possibly formicine ants (Graedel and Eisner 1988) have been suggested to balance model predictions with the measured concentrations (Jacob and Wofsy 1988) even though other atmospheric sources may also be involved. Because HCOOH is found in continental regions where isoprene-emitting vegetation is rare, the relative importance of other precursor hydrocarbons or of direct emissions probably varies spatially (Keene and Galloway 1988). Heterogeneous processes involving formaldehyde (HCHO), OH radical, and transition metals may also generate substantial amounts of HCOOH within clouds (Chameides and Davis 1983, Graedel et al. 1986, Jacob 1986). Harvey and Lang (1985) have suggested that alkenes emitted by marine microorganisms (volatilized and subsequently photooxidized) are possible precursors for HCOOH and CH_3COOH in marine regions.

Carboxylic acids, especially CH_3COOH, are emitted directly to the atmosphere by the combustion of fossil fuels and biogenic materials (Talbot et al. 1988). Anthropogenic hydrocarbons may be additional precursors for these compounds (Norton 1985). The similarities between HCOOH and CH_3COOH concentrations in precipitation and in cloud water from impacted and remote regions suggest, however, that anthropogenic activities are probably not the main source of these acids over broad geographic regions (Keene and Galloway 1988).

The distributions of aqueous-phase concentrations and per-event depositions (see Chapter 5, p. 105) suggest that the sources of continental carboxylic acids vary in time and space. The strongest sources seem to be in tropical and subtropical regions and in temperate regions during the growing seasons. In marine regions, the deposition data suggest that sources are strongest during the spring and early summer, the highest periods of biological productivity.

12.3.4. Polynuclear Aromatic Hydrocarbons (PAHs)

PAHs are emitted to the atmosphere by many anthropogenic processes, including industrial activities, incineration of municipal wastes, automobile emissions, power generation, and residential wood combustion. PAHs are also produced naturally from volcanic eruptions, forest fires, and other biomass burning. The U. S. Environmental Protection Agency (Baum 1978) and NAS (National Academy of Sciences 1972) have published estimates of the PAH emission rates. The EPA figures for the anthropogenic benzo(a)pyrene (BaP) released in the United States from 1971 to 1973 are given in Table 12-1.

Because of their wide range of vapor pressures, individual PAH compounds enter the ambient atmosphere as gases or associated with particles. Thus, naphthalene ($p^o = 11.5$ pa) is gaseous although benzo(ghi)perylene ($p^o = 1.3 \times 10^{-8}$ pa) is found exclusively on particles. Five-ring and heavier PAHs emitted from combustion processes are typically associated with small particles (≤ 1 μm). As noted in other sections, the different physical state of PAHs in the atmosphere can dramatically influence their transport and deposition.

Various PAH reaction products are formed in urban atmospheres or during transport from source to receptor (Atkinson et al. 1989, Gibson et al. 1986). For instance, several species of nitroarenes, such as 1-nitropyrene, are emitted from combustion sources, e.g., diesel exhaust (Paputa-Peck et al. 1983, Tokiwa and Onishi 1986).

At the time of our meeting at the Bermuda Biological Station, substantial work had already been done to characterize ratios among various PAHs from different combustion sources. In some cases there are almost unique tracers, such as retene for the combustion of spruce and resinous soft woods and cyclopenta(cd)pyrene from spark-ignition engines (Daisey et al. 1986). In a review, Daisey et al. (1986) conclude that PAH ratios may be useful in distinguishing sources but that adequate data to be used in source-apportionment models do not exist. However, these organic

Table 12-1. Anthropogenic emissions of benzo(a)pyrene in the United
            States from 1971 to 1973.

| | BaP Emissions | |
Sources	Metric Tons/yr	Percentage of Total Tonnage
Coal-refuse fires	281	35.0
Residential coal furnaces	272	33.8
Coke production	154	19.1
Vehicle disposal (open burning)	23	2.9
Woodburning fireplaces	23	2.9
Mobile sources (gasoline)	10	1.2
Tire degradation	10	1.2
Forest- and agricultural-refuse burning	10	1.2
Open municipal refuse burning (domestic)	9	1.1
Intermediate coal furnaces	6	0.8
Enclosed municipal incineration (apartments)	3	0.3
Other	4	0.5
Totals	805	100.0

Source: Baum (1978).

tracers when combined with trace elements may be used to help identify
the nature and geographical distribution of sources.

12.3.5. Chlorinated Pesticides and PCBs

Of the persistent organochlorine compounds, the magnitude and distribution of the world production of PCBs and DDT have been the most well
established. The commercial production of PCBs started in 1929 and
Bletchley (1984) estimates that the cumulative world output is 1.2×10^6
tonnes. Approximately 7×10^5 tonnes of PCBs were sold in the United
States between 1930-1975, the peak year being 1970 (Addison 1986).
Because PCBs are used primarily in various industries, most sources are
in developed countries.
    Since DDT was banned in most industrialized countries in the early
1970s, its use has shifted increasingly toward tropical countries
(Chapin and Wasserstrom 1981, Goldberg 1975). Goldberg (1975) estimates
that nearly 39,000 tonnes/yr of DDT was used for agricultural purposes
and malaria control in Asia, Africa, and Latin America between 1971-1981.
Other chlorinated pesticides still in use throughout the world include
hexachlorocyclohexane (HCH), polychlorocamphenes (PCC, e.g., toxaphene),
and the cyclodienes (aldrin, dieldrin, chlordane, endrin) (Table 12-2).
    Both the location of sources and the rates of emissions would be
profoundly affected by any restrictions or bans placed on the production
or use of PCBs or DDT. The ''southward tilt'' of DDT referred to by
Goldberg (1975), the areas of consumption shown in Table 12-2, and the
environmental data available from Kaushik et al. (1987) and Tanabe et

Table 12-2. Yearly tonnages and locations of worldwide chlorinated-hydrocarbon-insecticide usage (metric tons).

Locations/ Years	DDT	Technical HCH	Lindane	Aldrin and Similar Insecticides	PCC*	Other
USA						
1985				2,500**		
1981-82					17,500	
1975-76					39,864	
Mexico						
1982-84	1,000	750	100		4,000	1,705
1974-76	1,382	975			1,950	1,669
Kenya						
1983-84	224	14	17	339		
India						
1982-83	1,532	48,521		134		4,713
1974-76	4,970	23,716		1,045		
Turkey						
1982	379	2,552	77			63
1974-76	2,772	1,490	36			
Jordan						
1982						3,000
Italy						
1982-83			2,732			926
1974-76	1,570	2,550	1,884	913		475
Hungary						
1982-84					165	368
1974-76			2,207		403	113
Poland						
1984			166		41	1,280
1974-76	17	195	294		134	225
Argentina						
1982-84		18	530	1,557		277

Sources: For the 1985 USA figure, U.S. Environmental Protection Agency (1987); for the other USA figures, Bidleman et al. (1988); all other figures are from Food and Agricultural Organization (1986).

*Polychlorinated camphenes (toxaphene and similar products).

**Aldrin + chlordane + heptachlor.

al. (Tanabe and Tatsukawa 1980; Tanabe et al. 1982a, 1982b) indicate that tropical areas are the major sources of atmospheric DDT both now and in the future. Toxaphene, a PCC mixture, has been banned in the United States since 1982 after a total of 178,000 tonnes were used between 1972-1982 (Bidleman et al. 1988). The use of chlordane and heptachlor, two termiticides of which a total of about 1900 tonnes/yr are applied in the

United States, has recently been reviewed by the U.S. Environmental
Protection Agency (1987).

The relative importance of the various sources of these contaminant
chemicals is difficult to judge. For example, How much PCC enters the
North American troposphere from countries where PCC products are still
being used (e.g. Mexico)? How much is being volatilized from previously
treated soils in the United States? Although the evaporation of pesticides from treated croplands had undergone considerable field study and
modeling (Glotfelty 1981; Jury et al. 1983, 1984), regional models of
evaporative input to the atmosphere were lacking.

12.3.5. High-Molecular-Weight Oxygenated Compounds

In their reviews, Gagosian (1983) and Simoneit (1986) point out the
importance, along with hydrocarbon markers, of many oxygenated compounds
as tracers of source material for organic matter in remote atmospheres.
Specific compounds in the series of fatty acids, fatty alcohols, and
sterols are the most useful of these markers. These oxygenated compounds
are emitted into the atmosphere from near the earth's surface. The
available data base was insufficient to compute source strengths for
these materials.

Gagosian et al. (1987) have shown that many compounds in these
classes can be used to distinguish terrestrial from marine sources.
Marine air samples from New Zealand and Enewetak Atoll show a strong
even:odd preference for the $C_{21}-C_{36}$ fatty alcohols and the $C_{19}-C_{32}$ fatty
acids, indicative of terrestrial plant wax input from distant sources.
Greaves et al. (1987) report on how they incorporate marker compounds
into a source-identification model using factor analysis. Oxygenated
compounds are also generated in the environment by the oxidation of precursors, e.g., long-chain hydrocarbons (see Section 12.6, p. 280).

12.3.6. Particulate Carbon and Soot Carbon

Industrial combustion processes and biomass burning are the main sources
of particulate organic carbon and elemental (''soot'') carbon in the
troposphere. The budget estimates of Cachier et al. (1985) suggest that
the natural emissions of particulate carbon from tropical regions are at
least as important as those from industry. They estimate that approximately 2×10^{13} g C/yr of particulate organic carbon is produced globally, most in tropical regions. Of this total, approximately two-thirds
is from biomass burning and one-third is from plant and soil emissions
(terpenes and their oxidation products, waxes, etc.). Duce (1978) has
estimated that the industrial input of particulate carbon is 1.6×10^{13} g
C/yr. Consequently, the budget of particulate carbon in remote tropical
regions (30°N to 30°S) may be dominated by input from vegetation. The
particulate carbon concentrations found in the higher latitudes of the
Northern Hemisphere may be the most influenced by industrial sources.

12.4. TRANSPORT

In this section we present the evidence that long-range transport exists.
Such evidence consists of actual measurements of gaseous-, particulate-,
and aqueous-phase species in the troposphere over continental and marine
areas; measurements of these species in precipitation; and the inferences
drawn from the accumulation of these species found in sediments.

12.4.1. Nonmethane Hydrocarbons

The average atmospheric residence times for NMHC range from a few hours
to several months. Consequently the NMHC composition of any given air
mass changes considerably depending on the transport time between the
source and receptor. Figure 12-1 shows the mixing ratio of air masses
passing over Amsterdam Island that reflect continental and marine influences (see the concurrent ^{222}Rn measurements). In the air mass with the
lower ^{222}Rn concentration, the short-lived NMHC of predominantly continental origin (toluene, xylenes, and medium-molecular-weight alkanes) are
depleted whereas the longer-lived NMHC (C_2H_6, C_2H_2, C_6H_6) and the oceanic
NMHC (C_2H_4, C_3H_6) change only slightly.
    In spite of the number of studies of the large-scale distribution of
light NMHC (see Chapter 5, p. 105), when we met the temporal and spatial
distributions and the related variability of light NMHC were not yet
fully understood. Only for the longer-lived NMHC, such as C_2H_6, C_2H_2,
C_3H_8, and C_6H_6, could the most prominent features in their latitudinal
distributions be explained. These appeared to reflect their source distributions. The concentrations of these species are highest at midnorthern latitudes and decrease towards the equator; comparably low values are
found in the Southern Hemisphere. However, there were indications that
even these relatively long-lived species with atmospheric lifetimes of a
few weeks to a few months show significant longitudinal gradients
(Rudolph 1988). This might have indicated that the longitudinal mixing
within a latitudinal band was slower than generally assumed.
    The only NMHC whose seasonal variability was reasonably well known
was ethane (Blake and Rowland 1986). The seasonal cycle of ethane
reflects the seasonal variation of the atmospheric removal process (reaction with OH) and agrees reasonably well with model predictions (Isaksen
et al. 1985).
    Ehhalt et al. (1985), Rudolph (1988), Singh and Salas (1982), and
Tille et al. (1985) have published their findings on the vertical distribution of light NMHC in the troposphere. On the average, the vertical
gradient of light NMHC between the lower and upper troposphere is not
pronounced. This is expected for the long-lived alkanes (C_2H_6, C_3H_8) but
is rather unexpected for the short-lived alkanes (C_4H_{10} and heavier) and
the reactive light alkenes (C_2H_4 and C_3H_6). Since the conventional one-
dimensional models predict vertical gradients more than an order of magnitude greater than those actually measured, an unrecognized mechanism
might be rapidly transporting NMHC upward from the tropospheric boundary
layer into the upper troposphere.
    The few measurements of long-chain n-alkanes (C_{10} - C_{36}) in the
remote oceanic atmosphere (see Chapter 5, p. 105; Duce et al. 1983,

270

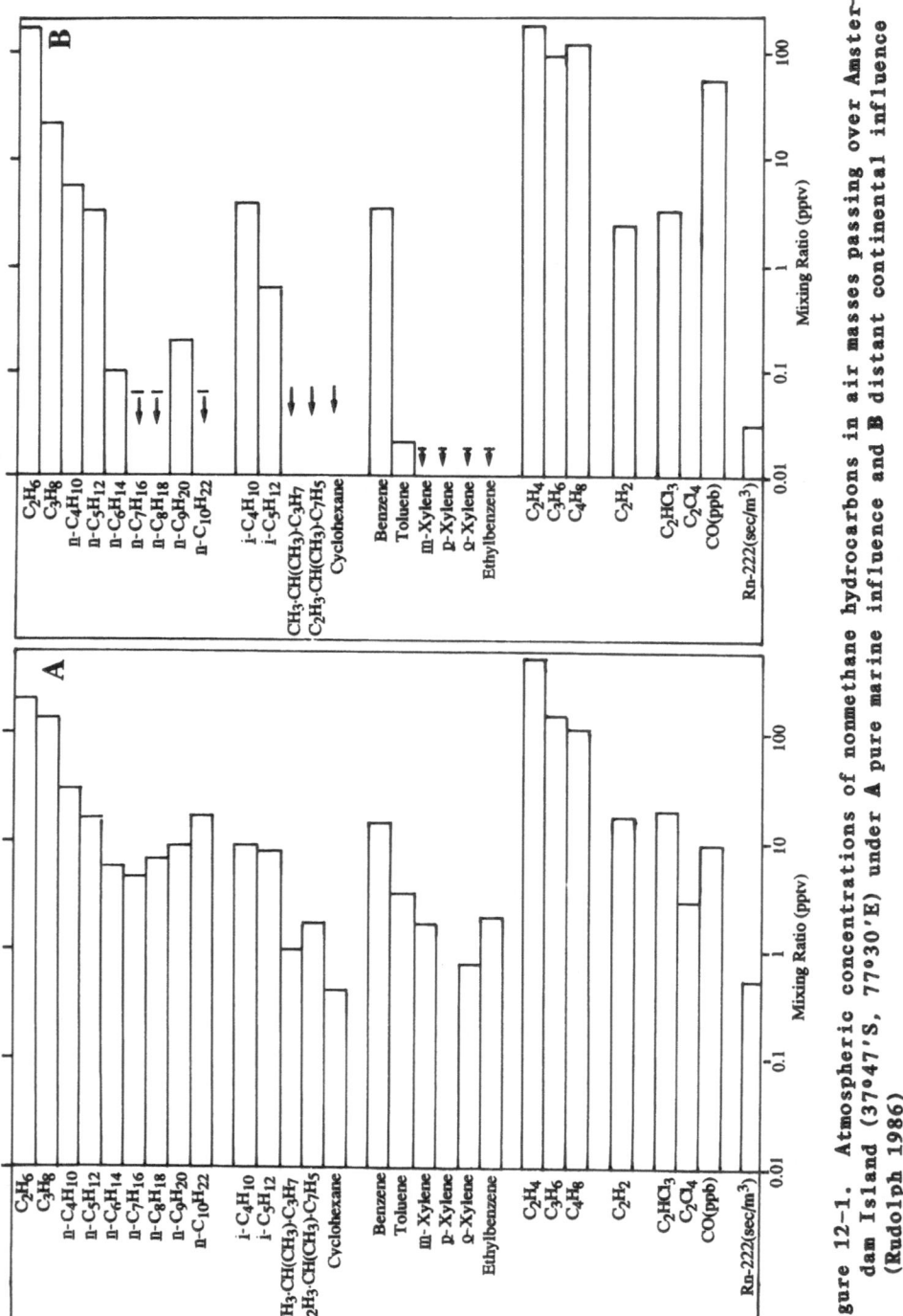

Figure 12-1. Atmospheric concentrations of nonmethane hydrocarbons in air masses passing over Amsterdam Island (37°47'S, 77°30'E) under A pure marine influence and B distant continental influence
(Rudolph 1986)

Gagosian et al. 1987, Zafiriou et al. 1985)) indicated that these compounds were present at levels ranging from ng/m^3 to pg/m^3. Heavy alkanes
are in the gas phase and associated with particles, and the gas:particle
ratio for heavy alkanes decreases systematically as the carbon number
increases. Because only a few studies had been conducted on heavy
alkanes in the marine troposphere, representative concentration distributions could not be inferred.

12.4.2. Low-Molecular-Weight Halocarbons

Light chlorinated hydrocarbons are introduced into the atmosphere within
the boundary layer, and their relatively long lifetimes allow their mixing into higher levels of the atmosphere. The tropospheric burden of
chloromethanes is estimated to be 9.8×10^6 metric tons (National Academy
of Science 1978a). These compounds have been observed in association
with Arctic haze (Rasmussen et al. 1983) and during experiments in the
Atlantic and Pacific Oceans (Prinn et al. 1987). The latitudinal dependence of tetrachloroethene shows a strong north-south gradient, with concentrations at midnorthern latitudes of roughly 30-35 parts-per-trillion
(pptr [10^{12}]) and about 1 pptr in the Southern Hemisphere (Rudolph et al.
1984). This indicate that tetrachloroethene is predominantly man-made.

In some cases, high concentrations of the chloroethenes mark polluted air in remote regions. Unfortunately these data were not being
reported but being used as a basis to discard ''contaminated'' samples
from ''background'' measurements. Prinn et al. (1987) have commented
that such pollution episodes are still rare at most background sites but
are becoming more common. The data being obtained at these background
monitoring stations could be valuable in establishing a long-term data
base for these compounds in remote areas.

12.4.3. C_1 and C_2 Carboxylic Acids

Most processes discussed in Section 12.3 (p. 262) inject or generate
carboxylic acids near the surface; some type of convective activity is
needed to facilitate the long-distance transport of substantial quantities of these acids.

Talbot et al. (1986) have reported enriched concentrations of particulate formate and acetate ions in haze layers of Saharan dust over the
western equatorial North Atlantic Ocean. However, Galloway et al. (1989)
have compared the distributions of formic acid (HCOOH) and sulfate in
precipitation associated with westerly and southwesterly storms on Bermuda and find little evidence for the transport of formic acid from eastern North America to the island. Therefore, carboxylic acids might be
rapidly scavenged in the marine boundary layer and not transported over
long distances. However, the maximum westerly flow across Bermuda is in
the winter when continental source strengths are low; in the summer when
the continental sources for carboxylic acids are stronger, the offshore
flow is weaker. Consequently the lack of significant continental signal
for formic acid in the wet deposition sampled on Bermuda might simply
reflect the temporal variability in source strengths and meteorological
conditions.

Pszenny (unpublished data available from Dr. Pszenny, NOAA/AOML/OCD, 4301 Rickenbacker Causeway, Miami, FL 33149) has investigated the transport of continentally derived materials to Amsterdam Island in the central Indian Ocean, a distance of approximately 5,000 km with a transport time of 3-6 days. He found no sign of significant concentrations of carboxylic acids above the marine background values.

12.4.4. Polynuclear Aromatic Hydrocarbons

Few measurements of particulate PAHs--and even fewer of gaseous PAHs--had been made in remote continental and marine areas (see Chapter 5, p. 105) and no data were available for naphthalene. PAH concentrations show strong gradients away from sources. Total atmospheric PAH concentrations measured at Barrow, Alaska, in 1979 are nine times higher during the Arctic haze period in March than they are in August when there is no haze (Daisey et al. 1981). Other researchers have corroborated the evidence of the long-range transport of PAHs in their findings of PAHs in sediment cores from lakes, regional seas, and oceans (Charles and Hites 1987, Gschwend et al. 1981, LaFlamme and Hites 1978, McVeety and Hites 1988, Tissier and Saliot 1981, Wickstrom and Tolonen 1987). Daisey et al. (1986) have stated that, because of their low background concentrations, PAHs are useful as tracers of combustion inputs to the atmosphere and specific PAH-related markers should be able to to characterize different combustion sources.

Gibson et al. (1986) have found PAH-transformation products, e.g., hydroxy- and nitro-PAHs, in remote locations thereby providing further evidence of long-range transport. Specific measurements of these transformation products, as well as of their parent PAHs, would have provided important information on their chemical reactivity.

12.4.5. Chlorinated Pesticides and PCB

Organochlorine pesticides have been found in the atmosphere over all the world's polar and ocean regions except the South Atlantic for which no data had yet been reported. These pesticides had been more extensively studied in oceanic areas than in remote continental areas. The most data had been amassed for DDT and HCH with much less for the other pesticides. The concentrations of tropospheric DDT measured since 1975 are in the 100-60,000 pg/m^3 range in the tropical Eastern Hemisphere, particularly near India, a heavy user of DDT (Bidleman et al. 1987, Kaushik et al. 1987, Tanabe et al. 1982b). DDT levels drop off rapidly away from regions where DDT is being used. When we compared the concentration figures for the North Atlantic--which is surrounded by countries where DDT had been banned for over a decade--to those for the eastern Indian/ western Pacific Ocean--which is adjacent to regions where DDT was still being used--the effect of the usage ban was apparent (Fig. 12-2.)

The highest HCH concentrations are in tropical areas of the eastern hemisphere; away from source regions, their levels are more uniform than those of DDT (Bidleman et al. 1987, Kaushik et al. 1987, Tanabe et al. 1982b). This could have reflected a longer residence time for HCH in the troposphere. The concentrations of DDT and HCH were probably also high

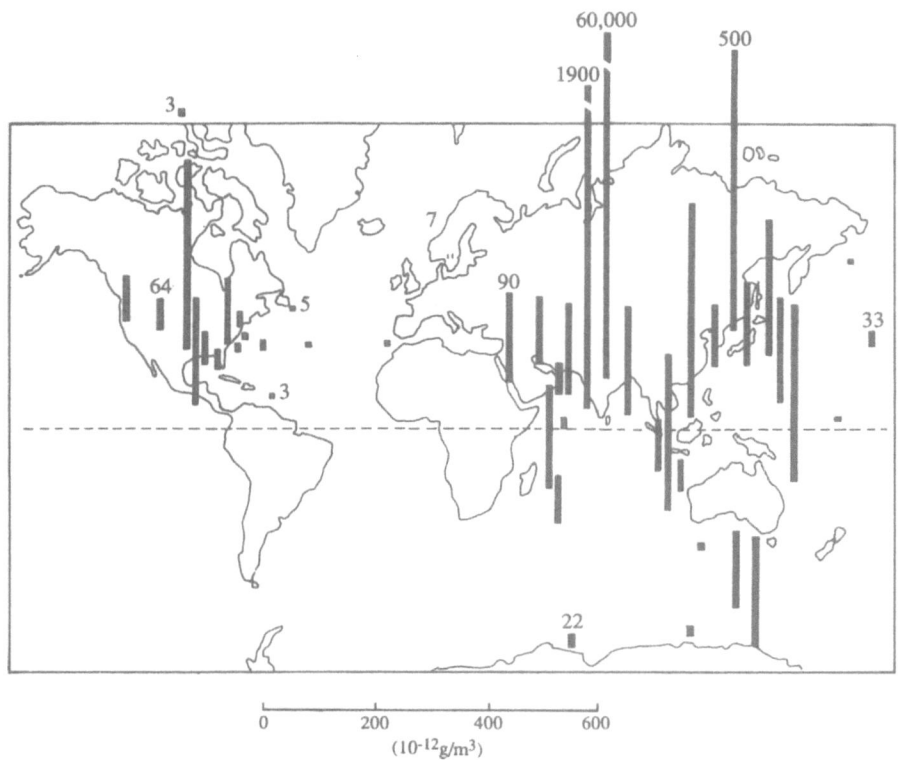

Figure 12-2. Tropospheric DDT measured since 1975. The base of each bar
locates the sample; the height of the bar indicates the concentration amount (pg/m³).

in the tropical areas of Africa and South America but atmospheric measurements had not yet been made for these continents.

Atlas et al. (1986) have reviewed the PCB concentrations reported
for continental and marine air as well as for precipitation and sediment
cores. The aerial concentrations in North American cities and over the
Great Lakes are in the ng/m³ range, followed by concentrations in the
hundred pg/m³ range over North American coastal areas away from cities.
Tanabe et al. (1982b) have also reported PCB levels of 56-1200 pg/m³ over
the eastern Indian and western Pacific Oceans; they suggest that the
uncontrolled use of PCBs in developing countries is responsible. The
lowest atmospheric concentrations of PCBs are in open-ocean areas of the
North and South Pacific, the North Atlantic, and the polar regions.

Of all the heavy organochlorines in the atmosphere, the most extensive effort to link atmospheric transport to loadings had been carried
out for PCBs. Strachan and Eisenreich (1987), in a recent modeling

effort, estimated that 7×10^3 kg/yr of atmospheric PCBs are being deposited to the five Great Lakes. Approximately 60-90% of total PCB loadings to the upper Great Lakes (Huron, Superior, Michigan) is atmospheric as opposed to only 6-7% for the lower lakes (Erie, Ontario). Atlas et al. (1986) estimated that 1.7×10^6 kg/yr of atmospheric PCBs is deposited to the world's oceans compared to a riverine PCB input of only 2% of this value.

Data for other heavy organochlorines in the troposphere were much less extensive. The compounds other than PCBs, DDT, and HCH that were generally being identified in atmospheric samples from remote regions were hexachlorobenzene (HCB), chlordanes, and polychlorocamphenes (PCCs). Higher values for concentrations of chlordanes and PCCs were being found over the western North Atlantic than over other oceans, reflecting a North American source. Although the number of samples was small, chlordanes had also been found in the North Pacific, the southern Indian Ocean, and the Canadian Arctic (Chapter 5, p. 105). The only atmospheric PCC data that existed covered the North Atlantic, Sweden, and the Canadian Arctic. Atlas and Bidleman (unpublished data available from Dr. E. L. Atlas, Department of Oceanography, College of Geosciences, Texas A & M University, College Station, TX 77843-33146) had recently identified PCC components in the North Pacific atmosphere but no quantitative data were yet available. Bidleman and Leonard (1982) reported what appeared to be PCC components in a few of their samples from the Arabian Sea, but in most samples the PCCs are undetectable.

When we met in Bermuda, polychlorinated dibenzodioxins (PCDDs) and polychlorinated dibenzofurans (PCDFs) had not yet been measured in remote atmospheres. However, several researchers had recently determined levels of these organochlorines, probably from the combustion of chemical and municipal wastes (Czuczwa and Hites 1986), in urban air (Eitzer and Hites 1986, Nakano et al. 1987, Rappe and Kjeller 1987), and in sediment cores from the five Great Lakes (Czuczwa and Hites 1986). Ono et al. (1987) also found PCDFs, but not PCDDs, in three specimens of the open-ocean killer whale (Orcinus orca), suggesting that at least PCDFs were being transported to the marine environment.

12.4.6. High-Molecular-Weight Oxygenated Compounds

Evidence that high-molecular-weight oxygenated compounds were being transported far from their sources came from the continentally derived lipids measured in samples from remote areas. Gagosian and Peltzer (1986) have reported finding these compounds in deep-sea sediments far from their point of release into the atmosphere. Zafiriou et al. (1985) have estimated the deposition of fatty acids at Enewetak Atoll. Fluxes at this oceanic site are rain-dominated. Their estimates, extrapolated from a limited data set, are 1.2×10^4 µg/m^2/yr for the C_{13-18} fatty acids and an order of magnitude lower for the heavier fatty acids. Dry-deposition fluxes range from 0.1 µg/m^2/yr to 34 µg/m^2/yr, with the n-C_{16} fatty acid being the most abundant. These authors hypothesize that the bulk of the flux to remote regions of the North Pacific is associated with Asian dust storms.

12.5. DEPOSITION

Organic compounds are found in the atmosphere in gaseous, particulate, and aqueous phases; the distribution among phases depends on such physio-chemical characteristics as vapor pressure and water solubility. Although low-molecular-weight hydrocarbons and halocarbons are entirely gaseous, no vapor-phase component had yet been identified for some heavier PAHs (e.g., benzo(ghi)perylene, coronene). Compounds with liquid-phase saturation vapor pressures (P_L°) roughly between 10^{-1} and 10^{-7} pa show some vapor-particle (V/P) partitioning. Common pollutants, such as PCBs, pesticides, phthalate esters, PCDDs, and PCDFs, fall within this wide window, as do compounds with both natural and anthropogenic origins, such as PAHs, high-molecular-weight alkanes, alcohols, and carboxylic acids.

Organic compounds are deposited through the wet and dry removal of gases and particles from the atmosphere. The relative contribution of either process to total deposition depends on whether the compounds are removed over land or water and the absolute amount and frequency of the precipitation or fog, the vapor/particle distribution in air, and some-times the concentrations of other chemical constituents, such as sea-salt aerosol.

12.5.1. Exchange of Gases

To estimate the amount of gas transferred across a planar air/water interface, most researchers were using a two-film model in which the exchange was controlled by molecular diffusion across interfacial air and water films (Liss 1983, Liss and Merlivat 1986, Liss and Slater 1974, Mackay and Yuen 1983). To use this model, one has to know (1) the exchange constants for the air and water films, which are functions of wind speed and molecular size (through the Schmidt number, Sc); (2) the equilibrium air/water partition coefficient (Henry's law constant, H); and (3) the solute concentrations in the bulk air and water phases, C_A and C_W.

This two-film model is useful for ascertaining whether the sea is a net source or sink for various organic gases and for estimating the flux magnitude. For example, two research groups had used this model to dis-cover that surface seawater appears to be close to equilibrium with atmospheric CCl_4 and Freon-11, with no net flux in either direction (Gammon et al. 1982, Hunter-Smith et al. 1983). Khalil et al. (1983), in using the model, found that the ocean is a source for $CHCl_3$, with an estimated strength of 3.5×10^8 kg/yr, which was comparable to the values being reported for other anthropogenic sources.

The dry flux of organic vapors is inextricably linked to water-column phenomena, which often limits our ability to describe the transfer process. The values for Henry's law constants are available for many low- and high-molecular-weight organic compounds (Mackay and Shiu 1981). However, most data at the time of our workshop were for compounds at 20-25° C only and were for freshwater samples. The variations of H with temperature ranging from -2° C to 30° C and with different salinities are necessary before the exchange of gases from equatorial to polar regions can be described. Recent progress had been made in obtaining H as a function of temperature for PCBs (Burkhard et al. 1985, Tateya et al. 1989) and of temperature and salinity for light halocarbons (Hunter-Smith et al. 1983).

The exchange of high-molecular-weight organic compounds which are sparingly soluble presented special problems. To use the two-film model, the truly dissolved and gaseous fractions of C_W and C_A must be known. In the water column, hydrophobic organic solutes might associate with dissolved organic carbon (DOC) or might bind to colloids that pass through a filter and be counted with the dissolved fraction during analysis. Recent laboratory investigations of dissolved organic carbon interactions with hydrophobic compounds permits the estimation of the dissolved fraction of C_W, even if its direct measurement is unobtainable (Baker et al. 1986, Chiou et al. 1987).

Detailed estimates of the air/sea fluxes for high-molecular-weight organic substances had not yet been attempted because the concentration data for seawater were lacking and, as discussed above, because the analytical difficulties in determining the truly dissolved and gaseous components had not yet been resolved. Atlas and Giam (1986, see also Chapter 5, p. 105) have restricted their estimates of gas exchange for organochlorine pesticides and PCBs to limiting cases in which the ocean is in equilibrium with the air (zero exchange) or serves as a perfect sink (maximum exchange). The rather large uncertainties (an order of magnitude for the PCBs and DDTs) in the estimated total-deposition figures caused by either having included or omitted gas-exchange values illustrated how important this gas-exchange process is to the global deposition of high-molecular-weight organic substances.

12.5.2. Distribution of Organic Compounds in the Atmosphere

Organic compounds are distributed among several phases in the atmosphere: gaseous, particulate (adsorbed to or occluded within a particle), and aqueous (dissolved in rain or cloud droplets). For this multiphase partitioning and its importance to the deposition of organic compounds to be understood, atmospheric sampling must include each phase.

Because the concentrations of the less volatile organic compounds in ambient air are low—pg to ng/m³—hundreds to thousands of cubic meters of air must be sampled to obtain enough material to analyze. The high-volume collectors commonly used in cloud-free air use filters (F) of glass, quartz fibers, or Teflon to retain particles; these filters are followed by a vapor-adsorbent trap (A). The ratio of the organic compounds found in the vapor-adsorbent trap to those found on the filter (A:F) is used to estimate the vapor:particle (V:P) ratio in ambient air (Bidleman 1988, Pankow 1987). However, how closely A:F represented V:P was unknown because of the nonequilibrium effects caused by changes of temperature and concentration over the collection period. For example, compounds adsorbed to particles deposited on the filter at night might be blown off in the heat of the next day. Also, the transformation of labile compounds during sampling (e.g., the oxidation or nitration of PAH) might cause the A:F ratio to differ from the V:P ratio.

An alternative to the high-volume collector is the diffusion denuder. This collector consists of a series of parallel tubes or concentric cylinders whose walls are coated with a high-boiling liquid. Vapors diffuse to the denuder walls and are stripped from the airstream. Particles diffuse slowly compared to the residence time of air in the denuder and pass through to be collected by a filter behind the denuder section. Organic compounds are partially stripped from the particles on the filter by the vapor-free airstream, but the volatilized material is collected in an adsorbent trap behind the filter. Thus, the sum of material found on the filter and adsorbent trap represents the original particle-bound fraction in ambient air. Denuders designed specifically to collect organic compounds were still being developed. However, comparisons of the preliminary data from those experiments using denuders with those using conventional, high-volume samplers showed differences in the apparent fraction of particulate organic compounds (Coutant et al. 1988, Johnson et al. 1987).

The V:P ratio can also be estimated from a theoretical equation of Junge's (1977).

$$f = cS_T/(P_L^o + cS_T), \qquad (12\text{-}1)$$

where f is the fraction of a specific organic compound in the particulate phase, S_T is the total surface area of airborne particulate matter per unit volume of air (cm^2/cm^3), and c is a parameter dependent upon, among other factors, the heat of desorption from the particle surface and the heat of vaporization of the liquid-phase adsorbate. Pankow (1987) has reviewed Junge's equation and empirical relationships for estimating V:P partitioning.

Values for the particulate fraction of PAHs and the organochlorines in the atmosphere of several cities have been estimated using data from high-volume collectors (Bidleman 1988, Bidleman et al. 1986, Keller and Bidleman 1984, Ligocki and Pankow 1989, Yamasaki et al. 1982). The figures agree within a factor of about 2-3 with f calculated using Equation 12-1 (Bidleman 1988). However, considering the limitations of the high-volume sampler and the probable differences in the adsorptive characteristics of particles from diverse locations, when we met the more reliable estimates of the vapor:particle ratio were probably those obtained from using Equation 12-1.

12.5.3. Phase Distribution and Atmospheric Removal Processes

The V:P distribution of organic compounds and their physical properties dictate how the compounds are removed from the atmosphere (Bidleman 1988, Eisenreich et al. 1981, Schroeder and Lane 1988). Some species (e.g., carboxylic acids) may also be subjected to substantial production or destruction within the droplets (Chameides and Davis 1983, Jacob 1986). Both Ligocki et al. (1985b) and Mackay et al. (1986) have used the following equation to describe the relative importance of vapor and particle scavenging for the wet deposition of nonreactive substances.

$$W = W_p + (RT/H)(1-f), \qquad\qquad (12\text{-}2)$$

where W and W_p are the overall scavenging ratio and the particle-scavenging ratio (the concentration in rain divided by the concentration in air, both on a mass/volume basis). W_p is highly variable since it is controlled by the size distribution of the particles and by meteorological conditions. The distribution of W_p for individual events is log normal (Chan et al. 1986, Slinn 1983), with a long-term average of 1×10^5 to 5×10^5 derived using small-particle trace-element data (e.g., Pb, Zn, Cd, As) (Arimoto et al. 1985, Chan et al. 1986, Talbot and Andren 1983). McVeety and Hites (1988) had reported that W_p values for particulate PAH are $(1.0-2.5) \times 10^5$ at Isle Royale, Lake Superior. However, Buat-Ménard (1986) had reported that even mean values of W_p vary substantially from site to site.

Ligocki et al. (1985a, 1985b) investigated rain scavenging of a large group of neutral organic compounds and determined the relative contributions of vapor and particle removal to total deposition. The scavenging ratios for those substances removed by vapor dissolution in raindrops agree well with those calculated using Henry's law constant, especially if the H at the actual rain temperature is used. Thus, rain appears to be close to equilibrium with the organic vapors in the air.

The available data base that could be used to estimate scavenging ratios of organic compounds was very limited. Bidleman (1988) had compiled the available information for organochlorines, PAHs, alkanes $\geq C_{18}$, and phthalate esters and compared them to the overall ratios (W) calculated using Equation 12-2. The field and predicted results he reported for compounds removed by vapor scavenging agree fairly well. However, the overall scavenging ratios of those substances deposited by particulate removal vary much more, probably because of the uncertainties involved in determining the V:P distribution in air and the scatter in the particle-scavenging ratio itself. The vertical structure through which a droplet falls might also have been causing problems not considered by Equation 12-2.

The removal of organic compounds by fog had only recently received attention, but the preliminary indications were that fog behaves quite differently from rain. Glotfelty et al. (1987) found that pesticides in fogs from California's Central Valley are enriched by up to 3000 times the concentrations predicted from their Henry's law constants. Moreover, there are distinct enrichment differences between the relatively neutral fogs in the Central Valley and the acidic fogs in Beltsville, Maryland. At least for organic compounds, fogs, in contrast to rain, seem to behave very differently from ideal solutions. We did not know the reasons for this behavior but some hypotheses were that high concentrations of surfactants in fog droplets or the presence of surface films were responsible for the enhanced uptake of pesticides in fogs.

The direct deposition of fog and cloud water contributes major proportions of hydrologic and chemical species to many ecosystems, particularly to those at higher elevations (Waldman and Hoffmann 1987, Weathers et al. 1988). Fog and cloud-water samples to be used for the analysis of major carbonaceous species had been collected in both active (Jacob et al. 1986) and passive (Keene and Galloway 1988) devices and Lovett and Reiners (1986) modeled deposition fluxes of cloud water to a coniferous forest. Nevertheless, because such variables as the nature of the impaction surface, the liquid-water content of clouds, 3-dimensional wind-fields, and the duration and frequency of the precipitation must be considered, deposition fluxes are highly variable and were, as yet, poorly quantified for most environments. The members of our working group agreed that the simultaneous multiphase sampling of cloud water and interstitial vapor was mandatory if we were to understand the cycling of such highly soluble species as carboxylic acids, which might have aqueous-phase sources and sinks (Jacob et al. 1986).

Understandably, ascertaining the dry deposition of particulate organic compounds had been hampered by the problems involved in determining V:P and the particle-deposition velocities ($V_{d,p}$) themselves (Georgii 1986, Slinn 1983). A major problem in modeling dry-deposition fluxes was the selection of $V_{d,p}$. In 1986 Georgii presented world maps on which he shows estimates of $V_{d,p}$ for continental and oceanic areas of 0.003-0.036 cm/s for accumulation-mode aerosols and 0.5-2.5 cm/s for coarse particles. Strachan and Eisenreich (1987) used a $V_{d,p}$ of 0.1 cm/s in modeling dry deposition of particles to the Great Lakes. Nevertheless, because of the problems involved, the uncertainties in the predicted $V_{d,p}$ remained large--at least an order of magnitude (Sievering 1984a, 1984b).

As with the data needed to establish scavenging ratios, the field data necessary to determine the deposition velocities of organic compounds to artificial surfaces were sparse. If the deposition surface collects only particulate organic compounds and C_A is taken as the sum of the particulate and gaseous species, the apparent deposition velocity (V_d) will increase with f, the particulate fraction. The few available field studies showed this trend, although V_d for some relatively volatile substances was higher than predicted (Bidleman 1988).

There were also problems in estimating the dry deposition of gases that involved such processes as dissolution, adsorption, chemical reaction, and biological activity of vegetation or soil microbiota (e.g., Liss 1983, Prospero et al. 1985). The temporal or spatial variability in biological uptake or release might also contribute to the nonlinear relationship between net flux and concentration. Such complex interactions among both the physical and the biological processes raised large uncertainties in estimating the dry-deposition velocities of gases. For some classes of organic compounds (e.g., carboxylic acids), the dry deposition of gases might be the most important atmospheric sink (Talbot et al. 1988). We agreed that the uncertainties in estimating this flux were one of the major gaps in our knowledge of the biogeochemical cycling of these compounds.

In summary, our understanding of the wet and dry deposition of organic substances was limited by our inability to identify and to quantify the species that were distributed among the different atmospheric phases--gaseous, particulate, and aqueous--as well as our inability to resolve the uncertainties in the deposition parameters for particles and gases.

12.6. TRANSFORMATIONS

In this section, we have briefly reviewed the chemical removal processes and estimated the atmospheric lifetimes resulting from these loss processes. Gaseous organic compounds in the troposphere are chemically transformed by photolysis and by chemical reaction with OH radicals (during daytime), NO_3 radicals (during nighttime), O_3 (typically throughout a complete 24-hour day in nonurban areas) and, for fused-ring PAHs, N_2O_5 (during nighttime hours). Particulate- and aqueous-phase organic compounds are removed from the atmosphere by photolysis and chemical reaction with O_3, OH radical, and other reactive species.

For the organic compounds we dealt with, concentrations of the nighttime NO_3 and N_2O_5 radicals in a clean maritime atmosphere are low enough (Noxon 1983, Winer et al. 1984) that reactions with these species are negligible. Therefore, we only considered the reactions of the gaseous organic compounds with the OH radicals and with O_3. Although the O_3 concentrations in the clean troposphere were reasonably well established (Oltmans 1981), the ambient concentrations of the OH radical were not. Hewitt and Harrison (1985) believed the ambient tropospheric OH-radical concentration at any given time and place are uncertain by a factor of at least 5 or, more likely, 10. However, Prinn et al. (1987) believed that the globally and annually averaged OH-radical concentration is known to within a factor of 2. The uncertainties in the ambient tropospheric concentrations of O_3 and OH radicals had led to corresponding uncertainties in the calculated lifetimes of organic chemicals because of the reactions with these species.

In the rest of this section, we have summarized our current knowledge of the kinetics and mechanisms of the photolytic and chemical reactions of selected organic compounds.

12.6.1. Nonmethane Hydrocarbons

Under tropospheric conditions, alkanes and aromatic hydrocarbons essentially react with the OH radical. We listed the measured rate constants at room temperature for reactions of the OH radical with selected alkanes and aromatic hydrocarbons in Table 12-3 along with their estimated atmospheric lifetimes. Although no data were available for the $\geq C_{14}$ alkanes, the OH-radical reaction-rate constants could be calculated reasonably accurately (Atkinson 1985, 1987a).

The OH-radical reactions with alkanes proceed by H-atom abstraction from the C-H bonds. If NO concentrations are high enough that the resulting alkyl peroxy (RO_2) radicals react predominantly with NO, the major products formed are alkyl nitrates (with yields ranging from ~ 3% for propane to ~ 30% for n-octane at 298°K and atmospheric pressure) and, from the smaller alkanes, aldehydes and ketones. Thus, ethane yields acetaldehyde; propane yields acetone and propionaldehyde; n-butane gives rise to acetaldehyde, methyl ethyl ketone, and n-butyraldehyde; and isobutane reacts to produce formaldehyde, acetone, and isobutyraldehyde. The reaction mechanisms are more complex and were much less understood for the $\geq C_5$ alkanes in the presence of NO and for all of the alkanes

Table 12-3. Room-temperature rate constants for gas-phase reactions of
the OH radical with organic compounds and the calculated atmospheric lifetimes.

Organic Compound	$k_{OH} \times 10^{12}$ (cm^3 molecule^{-1}s^{-1})	Lifetime Due to OH Radical Reaction[*]
DDT	~3.5[a]	~4 days
DDE	~8[a]	~2 days
Dieldrin	~13[a]	~1 day
Chlordane	~2[a]	~8 days
Hexachlorobenzene	0.02[a]	80 days
Hexachlorocylohexane	~1.0[a]	~15 days
Monochlorobiphenyls	2.8–5.2	3–6 days
Dichlorobiphenyls	1.4–2.9[a]	5–11 days
Trichlorobiphenyls	0.7–1.6[a]	10–22 days
Tetrachlorobiphenyls	0.4–0.9[a]	17–40 days
Pentachlorobiphenyls	0.2–0.4[a]	40–80 days
Malathion	64[a]	3 hr
Methyl chloride	0.044	1 yr
Dichloromethane	0.14	110 days
Trichloromethane	0.10	150 days
1,2-Dichloroethane	0.22	70 days
Trichloroethene	2.4	6 days
Tetrachloroethene	0.17	90 days
Ethane	0.27	55 days
Propane	1.2	13 days
n-Butane	2.5	6 days
n-Pentane	4.0	4 days
n-Hexane	5.6	3 days
n-Heptane	7.2	2.1 days
n-Octane	8.7	1.8 days
n-Nonane	10	1.5 days
n-Decane	11	1.4 days
n-Undecane	13	1.2 days
n-Dodecane	14	1.1 days
n-Tridecane	16	1.0 days
Ethene	8.5	1.8 days
Propene	26	7 hr
2-Methylpropene	51	4 hr
Benzene	1.2	13 days
Toluene	6.2	2.5 days
Acetylene	0.78[b]	20 days

Table 12-3. continued

Organic Compound	$10^{12} \times k^{OH}$ (cm^3 $molecule^{-1}s^{-1}$)	Lifetime Due to OH Radical Reaction[*]
Biphenyl	7.4	2.1 days
Naphthalene	22	8 hr
Fluorene	13[a]	1.2 days
Anthracene	130	1.4 hr
Fluoranthene	~50[a]	~4 hr
Pyrene	~50[a]	~4 hr
Dibenzo-p-dioxin	40[a]	5 hr
2,3,7,8-Tetrachlorodibenzo-p-dioxin	8.0[a]	2 days
Dibenzofuran	33[a]	6 hr
2,3,7,8-Tetrachlorodibenzo-furan	2.3[a]	7 days
Formic acid	0.46	35 days
Acetic acid	0.6	25 days
2-Propyl nitrate	0.18	85 days[**]
2-Butyl nitrate	0.67	23 days[**]
2-Pentyl nitrate	1.8	9 days[**]
3-Pentyl nitrate	1.1	14 days[**]
2-Hexyl nitrate	3.1	5 days[**]
3-Hexyl nitrate	2.7	6 days[**]

Sources: For room-temperature rate constants, Atkinson (1985, 1987a, 1987b, 1988).

[*]Assuming an average 12-hr daytime, OH-radical concentration of 1.5×10^6 molecule cm^{-3} (Prinn et al. 1987) and using the 298 K rate constant.

[**]Photolysis will also occur with a photolytic lifetime of approximately 10 days being estimated for CH_3ONO_2 (Taylor et al. 1980).

[a]Estimated; (see Atkinson 1987a, 1987b, 1988); uncertainties ~ a factor of 2 to 5.

[b]Rate constant is also pressure-dependent throughout the pressure range in the troposphere.

under low-NO conditions where the RO_2 radicals react mainly with HO_2 and other peroxy radicals.

The reaction products and mechanisms of the OH-radical reactions with the aromatic hydrocarbons were still incompletely understood. These OH-radical reactions proceed mainly by initial addition to the aromatic ring. Various ring-cleavage products are formed in the presence of NO, together with relatively small yields of phenolic compounds and, for the alkyl-substituted aromatics, aromatic aldehydes and ketones. However, only a fraction of the total reaction pathways had been accounted for and, under low-NO conditions, the reaction mechanisms are even more complex.

The alkenes react with the OH and NO_3 radicals and with O_3. As we said earlier, the NO_3-radical reactions are a minor or even negligible atmospheric loss process in clean maritime air. For propene and the higher alkenes, the O_3 reaction is a significant loss process, with lifetimes resulting from this reaction of ~1.5 days for propene, 1-butene, and 2-methylpropene and a matter of hours for the 2-butenes (Atkinson and Carter 1984) (Table 12-3). Clearly, both the OH-radical and O_3 reactions must be considered when calculating the atmospheric lifetimes of these alkenes and in assessing the products formed from their degradation in the atmosphere.

The O_3 reactions lead to the initial formation of a carbonyl compound and an energy-rich biradical, which can be stabilized or can decompose. The major products expected from the reactions of ethene and propene with O_3 are formaldehyde and formic acid from ethene; and formaldehyde, acetaldehyde, formic acid, and acetic acid from propene; together with a variety of other products, such as CO, CO_2 and radical species (Atkinson and Carter 1984). In the presence of NO, the OH-radical reactions of the alkenes also produce aldehydes and ketones; formaldehyde and glycolaldehyde ($HOCH_2CHO$) from ethene; formaldehyde and acetaldehyde from propene; and formaldehyde and acetone from 2-methylpropene (Atkinson 1985). Under low-NO concentration conditions, the OH-radical-reaction mechanisms are more complex and no definitive product data were available.

Under atmospheric conditions, acetylene reacts almost entirely with the OH radical, with the rate constant for this reaction being pressure-dependent for the total pressures encountered in the atmosphere. This reaction proceeds by initial OH-radical addition, forming glyoxal $[(CHO)_2]$ and formic acid in both the presence and absence of NO (Hatakeyama et al. 1988).

12.6.2. Low-Molecular-Weight Halocarbons

The compounds in this class that were of interest to our group were the chloromethanes (except CCl_4, which has no known chemical loss process in the troposphere), certain chloroethanes, and tri- and tetrachloroethene. Rate constants were available for the OH-radical reactions with these compounds (Table 12-3) and for the O_3 and NO_3-radical reactions with tri- and tetrachloroethene. The only significant tropospheric removal process is by reaction with the OH radical; we have listed the calculated tropospheric lifetimes in Table 12-3.

The OH-radical reactions with the chloromethanes and chloroethanes proceed by H-atom abstraction from the C-H bonds. In the presence of NO, at concentrations of $>$ (2-7) $\times$ 10^8 molecule/cm^3 (so that the chloroalkyl peroxy radicals formed from these chloroalkanes react with NO and not with HO_2 or other peroxy radicals), the expected products from CH_3Cl, CH_2Cl_2, and $CHCl_3$ are formyl chloride (HC(O)Cl), and phosgene ($COCl_2$), respectively.

The OH-radical reactions with tri- and tetrachloroethene proceed by initial OH-radical addition to the $>C=C<$ bond. In the presence of NO, these OH-radical reactions with trichloroethene lead to the formation of HC(O)Cl and $COCl_2$ and with tetrachloroethene to $COCl_2$, together with major amounts of other, as yet unknown, products (Tuazon et al. 1989).

12.6.3. C_1 and C_2 Carboxylic Acids

Under atmospheric conditions, formic and acetic acids (and their homo-logues) in the gas phase react mainly with the OH radical. The measured OH-radical reaction-rate constants and calculated atmospheric lifetimes caused by the OH-radical reactions are given in Table 12-3. The major product of the gas-phase OH-radical reaction with HCOOH is CO_2; no data were available for the corresponding reaction with acetic acid.

Because of their high aqueous solubility, formic and acetic acid are readily incorporated into raindrops, fog, and cloud water. The aqueous-phase destruction of formate and acetate by reaction with OH radical may be a major atmospheric sink for these compounds, particularly when pH >5 (Jacob 1986). Cloud-free conditions may then be more favorable for the long-distance transport of carboxylic acids over marine regions.

12.6.4. Polynuclear Aromatic Hydrocarbons

As discussed in Section 12.5 (p. 275), PAHs in the atmosphere are distri-buted between the gas and particle phases: 2-ring PAHs are almost entirely gaseous; $\geq$ 5-ring PAHs are particulate; and 3- to 4-ring PAHs are distributed between the two phases.

Atkinson et al. (1989) have published a detailed discussion of the tropospheric removal processes for the gaseous 2- to 4-ring PAHs and the nitroarene products formed from their atmospheric reactions. For PAHs not containing cyclopenta-fused rings, the only significant atmospheric gas-phase reaction is with the OH radical (Table 12-3). Atkinson et al. (1989), in their investigation of the products of gas-phase OH-radical reactions with naphthalene, biphenyl, fluoranthene, pyrene, and acephe-nanthrylene, found that nitroarenes are formed in small ($\leq$5%) yield. Although nighttime N_2O_5 reactions are a minor atmospheric loss process for many PAHs, these N_2O_5 reactions can, together with OH-radical reac-tions, be important to the formation of the urban nitroarenes (Atkinson et al. 1989).

Only limited data were available on the photolytic or reactive pro-cesses for particle-associated PAHs. Laboratory studies suggested that particle-associated PAHs are stabilized against photochemical reaction (Dunstan et al. 1989, Behymer and Hites 1988, Korfmacher et al. 1980). However, PAH containing cyclopenta-fused rings and certain other PAH,

such as benzo(a)pyrene, may react with O_3 in the particulate phase (Brör-
strom et al. 1983, VanVaeck and Van Cauwenberghe 1984).

Although no experimental data were available on the atmospheric
reactions of dibenzofurans, dibenzothiophenes, and dibenzo-p-dioxins,
reaction with the OH radical was probably the dominant removal process
for these compounds in the gas phase. (See Table 12-3 for the estimated
OH-radical reaction-rate constants for certain of these compounds.)

12.6.5. Chlorinated Pesticides and PCBs

The high-molecular-weight synthetic chlorinated compounds that had been
identified and quantified in areas remote from their sources included
DDT, DDE, dieldrin, chlordane, PCC, HCB, HCH, and PCB (summarized in
Bidleman et al. 1987). The only experimental kinetic data available were
for the gas-phase reactions of the OH radical with the monochlorobi-
phenyls (Table 12-3). However, Atkinson (1987a, 1987b) had published his
methods for estimating the OH-radical reaction-rate constants for other
organochlorines and we have included these estimates in Table 12-3.

For these chlorinated synthetic chemicals, the O_3 and NO_3-radical
reactions are negligible tropospheric removal processes. Photolysis may
be a significant removal process for the more highly chlorinated members
of this class, especially for heavy PCBs, which exhibit a relatively low
reactivity towards the OH radical. However, no data were available concerning the gas-phase photolysis rates of these compounds and we did not
know the products of the OH-radical-initiated reactions. The calculated
tropospheric lifetimes of the compounds caused by gas-phase reaction with
the OH radical are listed in Table 12-3. They were derived using a 12-
hr, average daytime, OH-radical concentration of 1.5×10^6 molecule cm^{-3}
(Prinn et al. 1987).

The lower volatility organochlorines are distributed between the
gaseous and particulate phases in the atmosphere and this partitioning
affects their atmospheric stability. Because these compounds may be
stabilized towards chemical reactivity when they are associated with
particles, wet and dry deposition may be a more important loss process.

12.6.6. Organophosphorus Compounds

Kinetic data concerning the potentially important atmospheric reactions
of organophosphorus compounds were available for only a few of the simplest members of this class; for example, $(CH_3O)_3PO$, $(CH_3O)_3PS$,
$(CH_3O)_2P(S)SCH_3$, and their isomers (Goodman et al. 1988). Based on
these data, the major gas-phase atmospheric removal process for organo-
phosphorous compounds was through their reaction with the OH radical.
These compounds were expected to be highly reactive, with correspondingly
short tropospheric lifetimes. Although no kinetic data were available
for the reactions of the OH radical with the commonly used organophosphorus compounds, Atkinson (1988) and Goodman et al. (1988) had published
methods for calculating these rate constants and an example for malathion
is given in Table 12-3. No product data were available for the gas-phase
reactions of these chemicals.

12.6.7. High-Molecular-Weight Oxygenated Compounds

As with other high-molecular-weight organic compounds in the atmosphere, fatty acids, alcohols, and other oxygenates are distributed between the vapor and aerosol phases. Evidence was beginning to accumulate suggesting that photo-oxidation might play a major role in the degradation of hydrocarbons and the aliphatic oxygenated compounds in the marine atmosphere and in surface water. Dicarboxylic acids, which were presumed to be ozonolysis products from common fatty acids, had been identified in the remote atmosphere (Kawamura and Gagosian 1987) and in urban air (Yokouchi and Ambe 1986). Knap et al. (1986) had reported a homologous series of aldehydes, methyl ketones, and aromatic ketones in air samples collected over the North Atlantic at least as high as 1,100 m. Atlas (see Chapter 5, p. 105) had measured a series of $\underline{n}$-aldehydes in air samples collected on a cruise track spanning the Atlantic and Pacific Oceans. In all these data sets, higher concentrations are recorded for the oxygenated compounds than for their precursor hydrocarbons.

The evidence that many oxygenated compounds are generated in the environment by oxidative decomposition of higher molecular-weight substances derives from the chemical structure of the oxygenate. Thus, $\underline{n}$-heptanal and $\underline{n}$-octanal, the predominant aliphatic aldehydes found in atmospheric samples collected between Bermuda and the east coast of the United States, are probably the products of an OH-radical-induced oxidative fragmentation at the double bonds of oleic and linoleic acids, both common in planktonic organisms.

Ehrhardt and Petrick (1984) have generated aromatic aldehydes in their laboratory identical to those in the water of Hamilton Harbour, Bermuda (Ehrhardt 1987), and the Arabian Gulf (Ehrhardt and DouAbul 1989), by sensitized photo-oxidation of alkylbenzenes under simulated environmental conditions. This sensitized photo-oxidation of long-chain aliphatic hydrocarbons in the condensed phase leads to the production of a homologous series of lower molecular-weight ketones and terminal alkenes (Ehrhardt and Petrick 1985). Therefore, biogenic and petrogenic aliphatic hydrocarbons may contribute to these oxygenated compounds in surface waters and in the marine boundary layer. Because of the similarity of photodecomposition products derived from atmospheric samples and seawater (Ehrhardt 1987), the mechanisms of oxidation in the atmosphere and seawater must, to some extent, follow similar pathways. The major difference is the initiation of the reactions with the abstraction of a hydrogen by hydroxyl radical in the atmosphere rather than by the intervention of a photosensitizer.

Several scientific teams had developed the chemical mechanisms necessary for the photodecomposition of aliphatic and aromatic hydrocarbons in the atmosphere, including the formation of aldehydes, ketones, and other carbonyl-containing organics (Aiken et al. 1982, Atkinson 1985, Atkinson and Lloyd 1984, Brewer et al. 1983, Calvert and Madronich 1987, Whitten et al. 1980). For these reactions to take place in the remote atmosphere, NO must be present. In air masses far from any continental sources of NO_x, peroxyacetyl nitrates may provide the mechanism of long-range transport of reactive nitrate species (Singh et al. 1986).

Alkyl nitrates are formed in the atmosphere from the parent alkanes, presumably mainly over continental areas. The major expected alkyl nitrates are 2-propyl nitrate, 2-butyl nitrate, 2- and 3-pentyl nitrates, and 2- and 3-hexyl nitrates. The atmospheric removal processes for these compounds are photolysis, with a lifetime of ~ 10 days (for methyl nitrate, Taylor et al. 1980) and reaction with the OH radical. Photolysis generates NO_2 and carbonyl compounds; the products of the OH-radical reactions were not known. We listed the rate constants at room temperature of the OH-radical reactions with 2-propyl, 2-butyl, 2- and 3-pentyl, and 2- and 3-hexyl nitrates in Table 12-3, together with the calculated atmospheric lifetimes caused by the OH-radical reaction. The overall lifetimes are, of course, shortened because of concurrent photolysis.

12.7. RECOMMENDATIONS

Having identified the gaps in our knowledge of the organic substances in the atmosphere, it was the consensus that future research should be designed to answer the following questions:

1. How does the long-range transport of organic substances impact tropospheric chemistry? (Section 12.7.1.)

2. How important is long-range transport in dispersing and depositing organic compounds that are toxic? (Section 12.7.2.)

3. How can atmospheric concentration data for organic species be used to check transport models, and vice-versa? (Section 12.7.3.)

12.7.1. Impact on Tropospheric Chemistry

In general, the long-range transport of organic compounds only affects the oxidative state of the troposphere if the product of the OH-radical reaction-rate constant and the concentration of the organic compounds is a significant fraction of or exceeds this same quantity for methane. Because of the many organic compounds emitted into the troposphere and subject to transport, we could not assess their cumulative impact but agreed that the classes of compounds that should be considered were the NMHC (the low-molecular-weight aromatic hydrocarbons, alkanes, alkenes, and acetylene), C_1-C_2 carboxylic acids, and the low-molecular-weight chloroalkanes and chloroalkenes. The importance of the higher molecular-weight compounds, such as the heavy alkanes, still needed to be addressed.

It was our consensus that the following steps would contribute much towards answering the first question, ''How does the long-range transport of organic substances impact tropospheric chemistry?''

1. The low-molecular-weight organic compounds mentioned above, their precursors (if applicable), and their photooxidation products should be simultaneously measured in multiphase

systems. The necessary ancillary data, especially for species affecting OH-radical concentrations (NO_x, O_3, other organic species, water vapor, liquid-water content of clouds, and radiation flux), should be collected concurrently (and preferably along the air-mass trajectory) so that a comprehensive model can be designed to study the transformations of these chemicals and their effect on tropospheric cycles.

2. Although the nature and kinetics of the major atmospheric loss processes were well understood for most simple organics in the gas phase, a larger data base must be built on the mechanisms and products of these processes, especially in remote oceanic atmospheres where NO_x concentrations are low.

3. The global distribution of organic nitrogen compounds in the troposphere must be determined to assess their potential for the long-range transport of NO_x. The data should include the global distribution of organic nitrogen compounds (e.g., alkyl nitrates), their photolysis lifetimes, and the products of their OH-radical reactions.

12.7.2. Dispersal and Deposition of Toxic Compounds

We considered the toxic compounds to be those with known physiological effects. Many are high-molecular-weight compounds, such as pesticides, PCBs, PAHs, PCDDs, and PCDFs. The following steps would contribute to answering the second question, ''How important is long-range transport in dispersing and depositing organic compounds that are toxic?''

1. The kinetics of atmospheric loss processes must be better understood for the heavy organic compounds, and products of such reactions need to be identified. Particularly important is the study of photolytic and reactive loss processes for particle-associated compounds. It is possible that the association of chemicals with particles may substantially extend their lifetimes over those expected for the same substances in the gas phase.

2. More research is needed on the deposition of heavy organic compounds. This research should include:
    a. Long-term measurements to determine the process and significance of the episodic nature of the deposition.
    b. Improved methods for distinguishing the physical states of organic chemicals in air and water (gaseous, dissolved, and particle-associated), especially for components of complex mixtures. Such data should be determined as functions of temperature and (for water-related properties) salinity.
    c. The collection of more data on the size distribution of organic compounds associated with particles and about the deposition characteristics of particles themselves.

3. The means to evaluate source strengths and their expected
    future trends need to be developed. For many high-molecular-
    weight compounds, production or usage figures could not be
    equated with figures for emissions into the atmosphere; these
    atmospheric emissions must be ascertained. One way of achiev-
    ing this would be to extend the already existing models of
    pesticide fluxes from agricultural fields to regional scales.

4. To assess future concentration trends and impacts on the marine
    environment, comprehensive compartment models of contaminant
    transfer, storage, and loss that include atmospheric deposition
    need to be designed. The importance of atmospheric deposition
    versus inputs from other sources (e.g., riverine) must be
    determined, especially on a regional scale.

5. Better estimates of the atmospheric burden of potentially toxic
    organic chemicals should be established. These estimates can
    then be used to assess the deposition potential of contami-
    nants. These data should include but not be limited to
    a. Vertical profiles of heavy organic compounds in the atmos-
        phere (practically nonexistent at the present).
    b. Expanded coverage of the spatial distribution, especially
        in the Southern Hemisphere and polar regions.
    c. An expanded list of compounds sought in the analyses of
        atmospheric samples, including toxic compounds and those
        that are produced in quantity; such as PCDDs and PCDFs,
        organophosphate and carbamate insecticides, and
        herbicides.

12.7.3. Organic Substances and Transport Models

The following suggestions should be implemented to answer the third question, ''How can atmospheric concentration data for organic species be
used to check transport models, and vice-versa?''

1. Acquire the following data to test transport models.
    a. Source data covering spatial distribution, composition,
        and rates of emission; data on 3-dimensional atmospheric
        concentrations; and kinetic data on loss processes (trans-
        formation and deposition).
    b. Data on the ratios of atmospheric concentrations for com-
        pounds having different atmospheric lifetimes (i.e., light
        aromatic hydrocarbons, alkanes versus alkenes, and certain
        isomeric PAHs, PCBs, and pesticides). This information can
        be exploited to follow transit times in air and to esti-
        mate the average OH-radical concentration.

2. Establish case studies covering the transport of organic com-
    pounds in well-defined meteorological regimes, such as trade-
    wind regions, monsoon regions, and the region off the coast of
    southern California.

3. Investigate the possibilities of using single back-trajectory
    analyses that were being used to study other atmospheric trace
    materials to provide an understanding of the transport of
    organic compounds. Because of the present uncertainties in the
    source strength of organic compounds, more complex transport
    models could not be applied. In conjunction with this investi-
    gation, flow climatology (covering 10 years or more) should be
    used to describe transport paths.

4. Encourage the ongoing research on the molecular and isotopic
    markers of individual and regional sources.

12.8. REFERENCES

Addison, R. F. 1986. PCBs in perspective. Canadian Chem. News 38:15–17.

Aikin, A. C., J. R. Herman, E. J., Maier, and C. J. McQuillan. 1982.
    Atmospheric chemistry of ethane and ethylene. J. Geophys. Res. 87:
    3105–3118.

Altshuller, A. P. 1983. Review: Natural volatile organic substances and
    their effect on air quality in the United States. Atmos. Environ.
    17:2131–2165.

Andreae, M. O., R. W. Talbot, T. W. Andreae, and R. C. Harriss. 1988.
    Formic and acetic acids over the central Amazon region, Brazil. 1.
    Dry season J. Geophys. Res. 93:1616–1624.

Arey, J., B. Zielinska, W. P. Harges, R. Atkinson, and A. M. Winer. 1988.
    The contribution of nitrofluoranthenes and nitropyrenes to the muta-
    genic activity of ambient particulate organic matter collected in
    Southern California. Mutation Res. 207:45–51.

Arimoto, R., R. A. Duce, B. J. Ray, and C. K. Unni. 1985. Atmospheric
    trace elements at Enewetak Atoll: 2. Transport to the ocean by wet
    and dry deposition. J. Geophys. Res. 90:2391–2408.

Atkinson, R. 1985. Kinetics and mechanisms of the gas-phase reactions of
    the hydroxyl radical with organic compounds under atmospheric condi-
    tions. Chem. Rev. 86:69–201.

Atkinson, R. 1987a. A structure-activity relationship for estimation of
    rate constants for the gas-phase reactions of OH radicals with
    organic compounds. Intern. Chem. Kinetics 19:799–828.

Atkinson, R. 1987b. Estimation of OH radical reaction rate constants and
    atmospheric lifetimes for polychlorobiphenyls, dibenzo-p-dioxins,
    and dibenzofurans. Environ. Sci. Technol. 21:305–307.

Atkinson, R. 1988. Estimation of gas-phase hydroxyl radical rate con-
    stants for organic chemicals. Environ. Toxicol. Chemistry 7:435–442.

Atkinson, R., and W. P. L. Carter. 1984. Kinetics and mechanisms of the
    gas-phase reactions of ozone with organic compounds under atmospher-
    ic conditions. Chem. Rev.84:437–470.

Atkinson, R., and A. C. Lloyd. 1984. Evaluation of kinetic and mechanistic data for modeling of photochemical smog. Phys. Chemistry 13: 315-444.

Atkinson, R., J. Arey, B. Zielinska, and A. M. Winer. 1989. Atmospheric loss process for PAH and formation of nitroarenes. Procs., 11th Int. Symp. on Polynuclear Aromatic Hydrocarbons, Chelsea, MI:Lewis, in press.

Atlas, E. L., and C. S. Giam. 1986. Sea-air exchange of high molecular weight synthetic organic compounds. In The Role of Air-Sea Exchange in Geochemical Cycling (P. Buat-Ménard, ed.) Series C, Vol. 185, Dordrecht:Reidel, 295-329.

Atlas, E.L., T. F. Bidleman, and C. S. Giam. 1986. Atmospheric transport of PCB to the oceans. In PCBs and the Environment (J. S. Waid, ed.) Boca Raton, FL:CRC Press, 79-100.

Baker, J. E., P. D. Capel, and S. J. Eisenreich. 1986. Influence of colloids on sediment-water partition coefficients of polychlorobiphenyl congeners in natural waters. Environ. Sci. Technol. 20:1136-1143.

Barbaray, B., J. P. Contour, and G. Mouvier. 1977. Sulfur dioxide oxidation over atmospheric aerosol--X-ray photoelectron spectra of sulfur dioxide adsorbed on V_2O_5 and carbon. Atmos. Environ. 11:351-356.

Baum, E. J. 1978. Occurrence and surveillance of polycyclic aromatic hydrocarbons. In Polycylic Carbons in Cancer, Vol 1. Environment, Chemistry and Metabolism (Harry V. Gelboin and Paul O. P. Ts'o, eds.) New York:Academic Press, 45-70.

Behymer, T. D., and R. A. Hites. 1988. Photolysis of polycyclic aromatic hydrocarbons adsorbed on fly ash. Environ. Sci. Technol. 22:1311-1319.

Bidleman, T. F. 1988. Atmospheric processes: Wet and dry deposition of organic compounds are controlled by their vapor-particle partitioning. Environ. Sci. Technol. 22:361-367.

Bidleman, T. F., and R. Leonard. 1982. Aerial transport of pesticides over the northern Indian Ocean and adjacent seas. Atmos. Environ. 16:1099-1107.

Bidleman, T. F., W. N. Billings, and W. T. Foreman. 1986. Vapor-particle partitioning of semivolatile organic compounds: Estimates from field collections. Environ. Sci. Technol. 20:1038-1043.

Bidleman, T. F., U. Wideqvist, B. Jansson, and R. Söderlund. 1987. Organochlorine pesticides and polychlorinated biphenyls in the atmosphere of southern Sweden. Atmos. Environ. 21:641-654.

Bidleman, T. F., M. T. Zaranski, and M. D. Walla. 1988. Toxaphene: Usage, aerial transport, and deposition. In Toxic Contamination in Large Lakes (N. Schmidtke, ed.) Chelsea, MI:Lewis Publishers, 257-284.

Blake, D. R. and F. S. Rowland. 1986. Global atmospheric concentrations and source strengths of ethane. Nature 321:231-233.

Bletchley, J. D. 1984. Polychlorinated biphenyls: Production, current use, and possible rate of future disposal in OECD member countries. Procs., PCB Seminar (M. C. Barros, H. Koemann, and R. Visser, eds.) Amsterdam:Ministry of Housing, Physical Planning, and Environment, 343-372.

292

Bonsang, B., and G. Lambert. 1985. Nonmethane hydrocarbons in an oceanic atmosphere. Atmos. Chemistry 3:257-271.

Bonsang, B., M. Kanakidou, G. Lambert, and P. Monfray. 1988. The marine source of C_2-C_6 aliphatic hydrocarbons. Atmos. Chemistry 6:3-20.

Brewer, D. A., Auguston, T. R., and Levine, J. S. 1983. The photochemistry of anthropogenic nonmethane hydrocarbons in the troposphere. J. Geophys. Res. 88:6683-6695.

Brodzinsky, R., S.-G. Chang, S. S. Markowitz, and T. Novakov. 1980. Kinetics and mechanism for the catalytic oxidation of sulfur dioxide on carbon in aqueous suspensions. Phys. Chemistry 384:3354-3359.

Brorström, E., P. Grennfelt, and A. Lindskog. 1983. The effect of nitrogen dioxide and ozone on the decomposition of particle-associated polycyclic aromatic hydrocarbons during sampling from the atmosphere. Atmos. Environ 17:601-605.

Buat-Ménard, P. 1986. The ocean as a sink for atmospheric particles. In The Role of Air-Sea Exchange in Geochemical Cycling (P. Buat-Ménard, ed.), Series C, Vol. 185, Doredrecth:Reidel, 165-183.

Burkhard, L. P., D. E. Armstrong, and A. W. Andren. 1985. Henry's Law constants for the polychlorinated biphenyls. Environ. Sci. Technol. 19:590-596.

Cachier, H. P. Buat-Ménard, M. Fontugne, and J. Rancher. 1985. Source terms and source strengths of carbonaceous aerosol in the tropics. Atmos. Chemistry 3:469-489.

Calvert, J. G., and S. Madronich. 1987. Theoretical study of the initial products of the atmospheric oxidation of hydrocarbons. J. Geophys. Res. 92, 221-2220.

Carter, W. P. L., and R. Atkinson. 1985. Atmospheric chemistry of alkanes. Atmos. Chemistry 3:377-405.

Chameides, W. L., and D. D. Davis. 1983. Aqueous phase source of formic acid in clouds. Nature 304:427-429.

Chan, W. H., A. J. S. Tang, D. H. S. Chung, and M. A. Lusis. 1986. Concentration and deposition of trace metals in Ontario--1982. Water Air Soil Pollut. 29:373-389.

Chapin, G., and R. Wasserstrom. 1981. Agricultural production and malaria resurgence in Central America and India. Nature 293:181-185.

Charles, M. J., and R. A. Hites. 1987. Sediments as archives of environmental pollution trends. In Sources and Fates of Aquatic Pollutants (R. A. Hites and S. J. Eisenreich, eds.) Advances in Chemistry Ser., Vol. 216, Washington:Amer. Chem. Soc., 365-389.

Chiou, C. T., D. E. Kile, T. I. Brinton, R. L. Malcolm, J. A. Leenheer, and P. MacCarthy. 1987. A comparison of water solubility enhancement of organic solutes by aquatic humic materials and commercial humic acid. Environ. Sci. Technol. 21:1231-1234.

Christman, R. F., D. L. Norwood, D. S. Millington, J. D. Johnson, and A. A. Stevens. 1983. Identy and yields of major halogenated products of aqueous fulvic acid chlorination. Environ. Sci. Technol. 17:625-628.

Chylek, P., V. Ramaswamy, and V. Srivastava. 1984. Graphitic carbon content of aerosols, clouds, and snow and its climatic implications. Sci. Total Environ. 36:117-120.

Clarke, A. D., R. E. Weiss, and R. J. Charlson. 1984. Elemental carbon
aerosols in the urban, rural, and remote marine troposphere and in
the stratosphere: Inferences from light absorption data and consequences regarding radiative transfer. Sci. Total Environ. 36:97-102.

Coakley, J. A., R. D. Cess, and F. B. Yurevich. 1983. The effect of
tropospheric aerosols on the earth's radiation budget: A parameterization of climate models. Atmos. Sci. 40:116-140.

Coutant, R. W., L. Brown, J. C. Chuang, R. M. Riggin, and R. G. Lewis.
1988. Phase distribution and artifact formation in ambient air
sampling for polynuclear aromatic hydrocarbons. Atmos. Environ.
22:403-409.

Czuczwa, J. M. and R. A. Hites. 1986. Airborne dioxins and dibenzofurans:
Sources and fates. Environ. Sci. Technol. 20:195-200.

Daisey, J. M., R. J. McCaffrey, and R. A. Gallagher. 1981. Polycyclic
aromatic hydrocarbons and total extractable particulate organic
matter in the Arctic aerosol. Atmos. Environ. 15:1353-1363.

Daisey, J. M., J. L. Cheney, and P. J. Lioy. 1986. Profiles of organic
particulate emissions from air pollution sources: Status and needs
for receptor source apportionment modeling. Air Pollut. Control
Assoc. 36:17-33.

Duce, R. A. 1978. Speculations on the budget of particulate and vapor-
phase non-methane hydrocarbons in the global troposphere. Pure Appl.
Geophys. 116:244-273.

Duce, R. A., V. A. Mohnen, P. R. Zimmerman, D. Grosjean, W. Cautreels, R.
Chatfield, R. Jaenicke, J. A. Ogren, E. D. Pellizzari, and G. T.
Wallace. 1983. Organic material in the global troposphere. Rev.
Geophys. Space Phys. 21:921-952.

Dunstan, T. D. J., R. F. Mauldin, Z. Jinxian, A. D. Hipps, E. L. Wehry,
and G. Mamantov. 1989. Adsorption and Photodegradation of pyrene on
magnetic, carbonaceous, and mineral subfractions of coal stack ash.
Environ. Sci. Technol. 23:303-308.

Ehhalt, D. H., and J. Rudolph. 1984. On the importance of light hydrocarbons in multiphase atmospheric systems. Berichte der Kernforschungsanlage Juelich JUEL-1942:1-43.

Ehhalt, D. H., J. Rudolph, F. Meixner, and U. Schmidt. 1985. Measurements of selected C_2-C_5 hydrocarbons in the background troposphere:
Vertical and latitudinal variations. Atmos. Chemistry 3:29-52.

Ehrhardt, M. 1987. Photo-oxidation products of fossil fuel components in
the water of Hamilton Harbour, Bermuda. Marine Chemistry 22:85-94.

Ehrhardt, M., and A. DouAbul. 1989. Dissolved petroleum residues and
alkylbenzene photooxidation products in the upper Arabian Gulf.
Marine Chemistry, in press.

Ehrhardt, M., and Petrick, G. 1984. On the sensitized photo-oxidation of
alkylbenzenes in seawater. Marine Chemistry 15:47-58.

Ehrhardt, M., and Petrick, G. 1985. The sensitized photo-oxidation of n-
pentadecane as a model for abiotic decomposition of aliphatic hydrocarbons in seawater. Marine Chemistry 16:227-238.

Eisenreich, S. J., B. B. Looney, and J. D. Thornton. 1981. Airborne
organic contaminants in the Great Lakes ecosystem. Environ. Sci.
Technol. 15:30-38.

Eitzer, B., and R. A. Hites. 1986. Concentrations of dioxins and dibenzo-furans in the atmosphere. Intern. Environ. Anal. Chemistry 27: 215-230.

Food and Agricultural Organization (FAO), United Nations. 1986. Production Yearbook, 1985, Vol. 39, pp. 300-301.

Gagosian, R. B. 1983. Review of marine organic geochemistry. Rev. Geophys. Space Phys. 21:1245-1258.

Gagosian, R. B., and E. T. Peltzer. 1986. The importance of atmospheric input of terrestrial organic material to deep sea sediments. Organic Geochemistry 10:661-669.

Gagosian, R. B., E. T. Peltzer, and J. T. Merrill. 1987. Long-range transport of terrestrially derived lipids in aerosols from the South Pacific. Nature 325:800-803.

Galloway, J. N., R. S. Artz, W. C. Keene, T. M. Church, and A. H. Knap. 1989. Processes controlling the concentrations of SO_4, NO_3^-, NH_4^+, H^+, $HCOO^-$, and CH_3COO^- in Bermuda precipitation. Tellus, in press.

Gammon, R. H., J. Cline, and D. Wisegarver. 1982. Chlorofluoromethanes in the northeast Pacific Ocean: Measured vertical distribution and application as transient tracers of upper ocean mixing. J. Geophys. Res. 87:9441-9454.

Georgii, F. 1986. A particle dry deposition parameterization for use in tracer transport models. J. Geophys. Res. 91: 9794-9806.

Gibson, T. L., P. E. Korsog, and G. T. Wolff. 1986. Evidence for the transformation of polycyclic organic matter in the atmosphere. Atmos. Environ. 20:1575-1578.

Glotfelty, D. E. 1981. Atmospheric dispersion of pesticides from treated fields. Ph.D. dissert., Univ. of Maryland, College Park, n.p.

Glotfelty, D. E., J. N. Seiber, and L. A. Liljedahl. 1987. Pesticides in fog. Nature 325:602-605.

Goldberg, E. D. 1975. Synthetic organohalides in the sea. Proc. Royal Soc. London 189:277-289.

Goodman, M.A., S. M. Aschmann, R. Atkinson and A. M. Winer. 1988. Kinetics of the atmospherically important gas-phase reactions of a series of trimethyl phosphorothioates. Arch. Environ. Contam. Toxicol. 17:281-288.

Graedel, T. E. 1979. Terpenoids in the atmosphere. Rev. Geophys. Space Phys. 17:937-947.

Graedel, T. E., and T. Eisner. 1988. Atmospheric formic acid from formacine ants: A preliminary assessment. Tellus 40B:335-339.

Graedel, T. E., M. L. Mandich, and C. J. Weschler. 1986. Kinetic model studies of atmospheric droplet chemistry: 2. Homogeneous transition metal chemistry in raindrops. J. Geophys. Res. 91:5205-5221.

Greaves, R. C., R. M. Barkley, and R. E. Sievers. 1987. Covariation in the concentrations of organic compounds associated with springtime atmospheric aerosols. Atmos. Environ. 21:2549-2561.

Greenberg, J. P., and P. R. Zimmerman. 1984. Non-methane hydrocarbons in remote tropical, continental, and marine atmospheres. J. Geophys. Res. 89:4767-4778.

Greenberg, J. P., P. R. Zimmerman, L. Heidt, and W. Pollock. 1984. Hydrocarbon and carbon monoxide emissions from biomass burning in Brazil. J. Geophys. Res. 89:1350-1354.

Greiner, N. R. 1970. Hydroxyl radical kinetics by kinetic spectroscopy. 6. Reactions with alkanes in the range of 300–500 K. Kinetics Chem. Phys. 53: 1070–1076.

Gschwend, P. M., P. H. Chen, and R. A. Hites. 1981. On the formation of perylene in recent sediments: Kinetic models. Geochim. Cosmochim. Acta 47:2115–2119.

Harvey, G. R., and R. F. Lang. 1985. Biogenic non-methane organics in and over the remote ocean. Paper presented at IAMAP/IAPSO Joint Assembly, Int. Union of Geodesy and Geophys., Honolulu, August (available from Dr. Harvey, NOAA, 4301 Rickenbacker Causeway, Miami, FL 33149).

Hatakeyama, S., N. Washida, and H. Akimoto. 1988. Rate constants and mechanisms for the reactions of hydroxyl (OD) radicals with acetylene, propyne, and 2-butyne in air at 297 ± 2 K. Phys. Chem. 90: 173–178.

Hewitt, C. N. and R. M. Harrison. 1985. Tropospheric concentrations of the hydroxyl radical—a review. Atmos. Environ. 19:545–554.

Hunter-Smith, R. J., P. W. Balls, and P. S. Liss. 1983. Henry's law constants and the air–sea exchange of various low molecular weight halocarbon gases. Tellus 35:170–176.

Isaksen, I. S. A., O. Hov, S. A. Penkett, and A. Semb. 1985. Model analysis of the measured concentrations of organic gases in the Norwegian Arctic. Atmos. Chemistry 3:3–27.

Jacob, D. J. 1986. The chemistry of OH in remote clouds and its role in the production of formic acid and peroxymonosulfate. J. Geophys. Res. 91:9807–9826.

Jacob, D. J., and S. C. Wofsy. 1988. Photochemistry of biogenic emissions over the Amazon forest. Geophys. Res. 93:1477–1486.

Jacob, D. J., J. W. Munger, J. M. Waldman, and M. R. Hoffmann. 1986. The H_2SO_4–HNO_3–NH_3 system at high humidities and in fogs: 1. Spatial and temporal patterns in the San Joaquin Valley of California. J. Geophys. Res. 91:1073–1088.

Johnson, N. D., S. C. Barton, D. A. Lane, and W. H. Schroeder. 1987. Further GAP sampler evaluations of semi-volatile chlorinated organic compounds in ambient air. In Procs., 1987 EPA/APCA Symp., Measurement of Toxic and Related Air Pollutants Pittsburgh:Air Pollut. Control Assoc.

Junge, C. E. 1977. Basic considerations about trace constituents in the atmosphere as related to the fate of global pollutants. In Fate of Pollutants in the Air and Water Environments (I. H. Suffet, ed.). Advances in Environ. Sci. Technol. Series, Vol 8, New York:Wiley, 1–25.

Jury, W. A., W. F. Spencer, and W. J. Farmer. 1983. Behavior assessment model for trace organics in soil: I. Model description. Environ. Quality 12:558–564.

Jury, W. A., W. J. Farmer, and W. F. Spencer. 1984. Behavior assessment model for trace organics in soil: II. Chemical classification and parameter sensitivity, III. Application of screening model, IV. Review of experimental evidence. Environ. Quality 13:567–586.

Kasting, J. F., and H. B. Singh. 1986. Nonmethane hydrocarbons in the troposphere: Impact on the odd hydrogen and odd nitrogen chemistry. J. Geophys. Res. 91:13,239-13,256.

Kaushik, C. P., M. K. K. Pillai, A. Raman, and H. C. Agarwal. 1987. Organochlorine insecticide residues in air in Delhi, India. Water Air Soil Pollut. 32:63-76.

Kawamura, K., and Gagosian, R. B. 1987. Implications of omega-oxycarboxylic acids in the remote marine atmosphere for photo-oxidation of unsaturated fatty acids. Nature 325:330-332.

Keene, W. C., and J. N. Galloway. 1988. The biogeochemical cycling of formic and acetic acids through the troposphere: An overview of current understanding. Tellus 40B:322-334.

Keller, C. D., and T. F. Bidleman. 1984. Collection of airborne polycyc-lic aromatic hydrocarbons and other organic compounds with a glass fiber filter-polyurethane foam system. Atmos. Environ 18:837-845.

Khalil, M. A. E., R. A. Rasmussen, and S. D. Hoyt. 1983. Atmospheric chloroform (CHCl$_3$): Ocean-air exchange and global mass balance. Tellus 35B:266-274.

Knap, A. H., Binkley, K., and Burns, K. 1986. Trace contaminants in the troposphere. EOS 67:885.

Korfmacher, W. A., E. L. Wehry, G. Mamantov, and D. F. S. Natusch. 1980. Resistance of photochemical decomposition to polycyclic aromatic hydrocarbons vapor-adsorbed on coal fly ash. Environ. Sci. Technol. 14:1094-1099.

LaFlamme, R. E., and R. A. Hites. 1978. The global distribution of poly-cyclic aromatic hydrocarbons in recent sediments. Geochim. Cosmo-chim. Acta. 42:1687-1691.

Ligocki, M. P., and J. P. Pankow. 1989. Measurements of the gas/particle distributions of polycyclic aromatic hydrocarbons and other selected compounds. Environ. Sci. Technol. 23:75-83.

Ligocki, M. P., C. Leuenberger, and J. P. Pankow. 1985a. Trace organic compounds in rain: II. Gas scavenging of neutral organic compounds. Atmos. Environ. 19:1609-1617.

Ligocki, M. P., C. Leuenberger, and J. P. Pankow. 1985b. Trace organic compounds in rain: III. Particle scavenging of neutral organic com-pounds. Atmos. Environ. 19:1619-1626.

Liss, P. S. 1983. Gas transfer: Experiments and geochmical implications. In Air-Sea Exchange of Gases and Particles (P. S. Liss and W. G. N. Slinn, eds.), Series C, Vol. 108, Dordrecht:Reidel, 241-298.

Liss, P. S., and L. Merlivat. 1986. Air-sea gas exchange rates: Introduc-tion and synthesis. In The Role of Air-Sea Exchange in Geochemical Cycling (P. Buat-Ménard, ed.), Series C, Vol. 185, Dordrecht:Reidel, 113-127.

Liss, P. S., and P. G. Slater. 1974. Fluxes of gases across the air-sea interface. Nature 247:181-184.

Lovett, G. M., and W. A. Reiners. 1986. Canopy structure and cloudwater deposition in sub-alpine forests. Tellus 38:319-327.

Mackay, D., and W. Y. Shiu. 1981. A critical review of Henry's law con-stants for chemicals of environmental interest. Physical Chemical Ref. Data 10:1175-1199.

Mackay, D., and A. T. K. Yuen. 1983. Mass transfer coefficient correlations for volatilization of organic solutes from water. Environ.
Sci. Technol. 17:211-217.

Mackay, D., S. Paterson, and W. H. Schroeder. 1986. Model describing the
rates of transfer processes of organic chemicals between atmosphere
and water. Environ. Sci. Technol. 20:810-816.

McVeety, B. D., and R. A. Hites. 1988. Atmospheric deposition of polycyclic aromatic hydrocarbons to water surfaces: A mass balance
approach. Atmos. Environ. 22:511-536.

Nakano, T., M. Tsuji, and T. Okuno. 1987. Levels of chlorinated organic
compounds in the atmosphere. Chemosphere 16:1781-1786.

National Academy of Sciences, (Committee on Biological Effects of Atmospheric Pollutants). 1972. Particulate Polycyclic Organic Matter.
Washington:National Academy Press.

National Academy of Science (Environmental Studies Board). 1978a. Chloroform, Carbon Tetrachloride, and other Halomethanes: An Environmental
Assessment. (Panel on Low-Molecular-Weight Halogenated Hydrocarbons,
Committee for Scientific and Technological Assessments of Environmental Pollutants, eds.) Washington:National Academy Press, 294 pp.

National Academy of Sciences (Ocean Sciences Board). 1978b. The Tropospheric Transport of Pollutants and Other Substances to the Oceans
(Workshop on Tropospheric Transport of Pollution to the Oceans,
eds.) Washington:National Academy Press, 243 pp.

National Academy of Science (Environmental Studies Board). 1979. PCB
transport through the environment. In Polychlorinated Biphenyls.
Washington:National Academy Press, 146-168.

Norton, R. B. 1985. Measurement of formate and acetate in precipitation
at Niwot Ridge and Boulder, Colorado. Geophys. Res. Ltrs. 12:769-
772.

Noxon, J. F. 1983. NO_3 and NO_2 in the mid-Pacific troposphere. J. Geophys. Res. 88:11,017-11,021.

Oliver, B. G., and J. Lawrence. 1979. Haloforms in drinking water: A
study of precursors and precursor removal. Amer. Water Works. Assoc.
71:161-163.

Oltmans, S.J. 1981. Surface ozone measurements in clean air. J. Geophys.
Res. 86:1174-1180.

Ono, M., N. Kannnan, T. Wakimoto, and R. Tatsukawa. 1987. Dibenzofurans,
a greater global pollutant than dioxins? Evidence from analysis of
open ocean killer whale. Marine Pollut. Bull. 18:640-643.

Pankow, J. F. 1987. Review and comparative analysis of the theories on
partitioning between the gas and aerosol particulate phases in the
atmosphere. Atmos. Environ. 21:2275-2283.

Paputa-Peck, M. C., R. S. Marano, D. Schuetzle, T. L. Rile, C. V.
Hampton, T. J. Prater, L. M. Skewes, T. E. Jensen, P. H. Ruehle, L.
C. Bosch, and W. P. Duncan. 1983. Determination of nitrated
polynuclear aromatic hydrocarbons in particulate extracts by
capillary column gas chromatography with nitrogen selective
detection. Anal. Chemistry 55:1946-1954.

Penner, J. E., and L. L. Edwards. 1986. Nucleation scavenging of smoke and aerosol particles in convective updrafts. In Procs., 23rd Conf., Radar Meteorol. and Cloud Phys., Boston:Amer. Meteorol. Soc., 2:83–86.

Perera, F. 1981. Carcinogenicity of airborne fine particulate benzo(a) pyrene: An appraisal of the evidence and the need for control. Environ. Health Perspective 42:163–185.

Prinn, R., D. Cunnold, R. Rasmussen, P. Simmonds, F. Alyea, A. Crawford, P. Fraser, and R. Rosen. 1987. Atmospheric trends in methylchloroform and the global average for the hydroxyl radical. Science 238:945–950.

Prospero, J. M., W. C. Keene, J. N. Galloway, R. J. Delmos, L. Granat, G. Gravenhorst, and G. E. Likens. 1985. The deposition of sulfur and nitrogen from the remote atmosphere: Working-group report. In The Biogeochmical Cycling of Sulfur and Nitrogen in the Rmote Atmosphere (J. N., Galloway, R. J. Charlson, M. O. Andreae, H. Rodhe, M. S. Marston, eds.), Series C, Vol. 159, Dordrecht:Reidel, 176–200.

Pueschell, R. F., E. W. Barrett, D. L. Wellman, and J. A. McGuire. 1981. Cloud modification by man-made pollutants. Effects of a coal-fired power plant on cloud drop spectra. Geophys. Res. Ltrs. 8:221–224.

Rappe, C., and L.-R. Kjeller. 1987. PCDDs and PCDFs in environmental samples: Air, particulates, sediments, and soil. Chemosphere 16:1775–1780.

Rasmussen, R. A., M. A. K. Khalil, and R. J. Fox. 1983. Altitudinal and temporal variations of hydrocarbons and other gaseous tracers of Acrtic haze. Geophys. Res. Ltrs. 10:144–147.

Rook, J. J. 1974. Formation of haloform during chlorination of natural waters. Water Treat. Exam. 23:234–243.

Rudolph, J. 1986. Comparison of biogenic and man-made light hydrocarbons. Large-scale distribution and impact on the chemistry of the atmosphere. Paper presented, 2nd Intern. Symp., Biosphere-Atmosphere Exchanges, Mainz, Germany.

Rudolph, J. 1988. The two-dimensional distribution of light hydrocarbons: Results from the STRATOZ III experiment. J. Geophys. Res. 93:8367–8377.

Rudolph, J., and D. H. Ehhalt. 1981. Measurement of C_2–C_5 hydrocarbons over the North Atlantic. J. Geophys. Res. 86:11,959–11,964.

Rudolph, J., C. Jebsen, A. Khedim, and F. J. Johnen. 1984. Measurements of the latitudinal distribution of light hydrocarbons and halocarbons. In Physico-Chemical Behaviour of Atmospheric Pollutants (B. Versine and G. Angeletti, eds.), Series C, Dordrecht:Reidel, 492–501.

Schroeder, W. H., and D. A. Lane. 1988. The fate of toxic airborne pollutants. Environ. Sci. Technol. 22:240–246.

Sievering, H., 1984a. Small particle dry deposition on natural waters: How large the uncertainty? Atmos. Environ. 18:2271–2272.

Sievering, H. 1984b. Small particle dry deposition to natural waters: Modeling uncertainty. J. Geophys. Res. 89:9679–9681.

Simoneit, B. R. T. 1984. Organic matter of the troposphere: III. Characterization and sources of petroleum and pyrogenic residues in aerosols over the western United States. Atmos. Environ. 18:51–67.

Simoneit, B. R. T. 1986. Characterization of organic constituents in aerosols in relation to their origin and transport: A review. Intern. Environ. Anal. Chemistry 23:207–237.

Simoneit, B. R. T., and M. A. Mazurek. 1981. Air pollution: The organic components. Critical Rev. Environ. Control 11:219–276.

Singh, H. B., and L. J. Salas. 1982. Measurement of selected light hydrocarbons over the Pacific Ocean: Latitudinal and seasonal variations. Geophys. Res. Ltrs. 9:842–845.

Singh, H. B., L. J. Salas, and R. E. Stiles. 1983. Selected man-made halogenated chemicals in the air and ocean environment. J. Geophys. Res. 88:3675–3683.

Singh, H. B., Salas, L. J., and W. Viezee. 1986. Global distribution of peroxyacetyl nitrate. Nature 321:588–591.

Slinn, W. G. N. 1983. Air-to-sea transfer of particles. In Air-Sea Exchange of Gases and Particles (P. S. Liss and W. G. N. Slinn, eds.), Series C, Vol. 108, Dordrecht:Reidel, 299–405.

Strachan, W. M. J., and S. J. Eisenreich. 1987. Mass Balancing of Toxic Chemicals in the Great Lakes: The Role of Atmospheric Deposition. Windsor, Ontario:Intern. Joint Commission, 113 pp.

Strandell, M., S. Hakansson, T. Alsberg, R. Westerholm, and L. Stillbron-Elfver. 1987. Chemical analysis and biological testing of moderately polar fractions of ambient air particulate extracts in relation to diesel and gasoline vehicle exhausts. Paper presented, 11th Intern. Symp., Polynuclear Aromatic Hydrocarbons, National Bureau of Standards, Gaithersburg, MD, Sept. 23–25.

Talbot, R. W., and A. W. Andren. 1983. Relationships between Pb and Pb-210 in aerosol and precipitation at a semiremote site in northern Wisconsin. J. Geophys. Res. 88:6752–6760.

Talbot, R. W., R. C. Harriss, E. V. Browell, G. L. Gregory, D. I. Sebacher, and S. M. Beck. 1986. Distribution and geochemistry of aerosols in the tropical North Atlantic tropsophere: Relationship to Saharan dust. J. Geophys. Res. 91:5173–5182.

Talbot, R. W., K. M. Beecher, R. C. Harriss, and W. R. Cofer, III. 1988. Atmospheric geochemistry of formic and acetic acids at a mid-latitude temperate site. J. Geophys. Res. 93:1638–1652.

Tanabe, S., and R. Tatsukawa. 1980. Chlorinated hydrocrabons in the North Pacific and Indian Oceans. Oceanog. Soc. Japan 36:217–226.

Tanabe, S., and R. Tatsukawa. 1986. Distribution, behavior, and load of PCBs in the oceans. In PCBs and the Environment (J. S. Waid, ed). Boca Raton, FL:CRC Press, 143–161.

Tanabe, S., M. Kawano, and R. Tatsukawa. 1982a. Chlorinated hydrocarbons in the Antarctic, western Pacific, and eastern Indian Oceans. Trans. Tokyo Univ. Fish. 5:97–109.

Tanabe, S., R. Tatsukawa, J. Kawano, and H. Hidaka. 1982b. Global distribution and atmospheric transport of chlorinated hydrocarbons: HCH (BHC) isomers and DDT compounds in the western Pacific, eastern Indian, and Antarctic Oceans. Oceanog. Soc. Japan 38:137–148.

Tanabe, S., Watanabe, S., Kan, H., and Tatsukawa, R. 1988. Capacity and mode of PCB metabolism in small cetaceans. Mar. Mammal Sci. 4:103–124.

Tateya, S., S. Tanabe, and R. Tatsukawa. 1989. PCBs in the globe: Possible trend of future levels in the open ocean environment. Great Lakes Res., in press.

Taylor, W. D., T. D. Allston, M. J. Moscato, G. B. Fazekas, R. Kozlowski and G. A. Takacs. 1980. Atmospheric photodissociation lifetimes for nitromethane, methyl nitrite, and methyl nitrate. Intern. Chem. Kinetics 12:231-240.

Tille, K. J. W., M. Savelsberg, and K. Bachmann. 1985. Airborne measurements of nonmethane hydrocarbons over western Europe: Vertical distribution, seasonal cycles of mixing ratios, and source strength. Atmos. Environ. 19:1751-1760.

Tissier, M. J., and A. Saliot. 1981. Pyrolytic and naturally occuring polyclic aromatic hydrocarbons in the marine environment. In Advances in Organic Geochemistry (M. Bjoroey, ed.) New York:Wiley, 268-278.

Tokiwa, S., and R. Ohnishi. 1986. Mutagenicity and carcinogenicity of nitroarenes and their sources in the environment. CRC Crit. Rev. Toxicol. 17:23-60.

Tuazon E. C., S. M. Aschmann, R. Atkinson, M. A. Goodman, and A. M. Winer. 1989. Atmospheric reactions of chloroethenes with the OH radical. Chem. Kinetics, in press.

Twomey, S. 1977. The influence of pollution on the shortwave albedo of clouds. Atmos. Sci. 34:1149-1152.

U. S. Environmental Protection Agency (Office of Pesticie Programs). 1987. Chlordane, Heptachlor, Aldrin, and Dieldrin: Technical Support Document. (draft) Washington:US Govern. Printing Office.

VanVaeck, L., and K. VanCauwenberghe. 1984. Conversion of polycyclic aromatic hydrocarbons on diesel particulate matter upon exposure to ppm levels of ozone. Atmos. Environ. 18:323-328.

Waldman, J. E., and M. Hoffman. 1987. Depositional aspects of pollutant behavior in fog and intercepted clouds. In Sources and Fates of Aquatic Pollutants (R. A. Hites and S. J. Eisenreich, eds.) Advance in Chemistry Series, Vol. 216, Washington:Am. Chem. Soc., 79-129.

Wang, W.-C., D. J. Wuebbles, W. M. Washington, R. G. Isaacs, and G. Molnar. 1986. Trace gasses and other potential perturbations to global climate. Rev. Geophys. 24:110-140.

Weathers, K. C., G. E. Likens, F. H. Bormann, S. H. Bicknell, B. T. Bormann, B. C. Daube Jr., J. S. Eaton, J. N. Galloway, W. C. Keene, K. D. Kimball, W. H. McDowell, T. G. Siccama, D. Smiley, and R. A. Tarrant. 1988. Cloud water chemistry from ten sites in North America. Environ. Sci. Technol. 22:1018-1026.

Whitten, G. Z., Hogo, H.., and Killus, J. P. 1980. The carbon-bond mechanism: A condensed kinetic mechanism for photochemical smog. Environ. Sci. Technol. 14:690-700.

Wickstrom, K., and K. Tolonen. 1987. The history of airborne polycyclic aromatic hydrocarbons (PAH) and perylene as recorded in dated lake sediments. Water Air Soil Pollut. 32:155-175.

Winer, A. M., R. Atkinson, and J. N. Pitts, Jr. 1984. Gaseous nitrate radical: possible nighttime atmospheric sink for biogenic organic compounds. Science 224:156-159.

Yamasaki, H., K. Kuwata, and H. Miyamoto. 1982. Effects of ambient temperature on aspects of polycyclic aromatic hydrocarbons. Environ. Sci. Technol. 16:189-194.

Yokouchi, Y., and Y. Ambe. 1986. Characterization of polar organics in airborne particulate matter. Atmos. Environ. 20:1727-1734.

Zafiriou, D. C., R. B. Gagosian, E. T. Peltzer, and J. B. Alford. 1985. Air-to-sea fluxes of lipids at Enewetok Atoll. J. Geophys. Res. 90: 2409-2423.

Zimmerman, P. R., J. P. Greenberg, and C. E. Westberg. 1988. Measurements of atmospheric hydrocarbons and biogenic emission fluxes in the Amazon boundary layer. J. Geophys. Res. 93:1407-1416.

13. CONCLUSIONS

Anthony H. Knap
Bermuda Biological Station for Research, Inc.
17 Biological Station Lane
Ferry Reach GEO1, BERMUDA

13.1. INTRODUCTION

The specific conclusions of each chapter can be found in the body of the book. A few general conclusions are mentioned here.

13.2. TRACE ELEMENTS

The working group on trace elements provides clear evidence in Chapter 9 (p. 177) that trace elements are being transported from continent to ocean and a strong possibility that they are also being transported from continent to continent. Two specific case studies undertaken by the group, on the Mediterranean and the Pacific, demonstrate how useful multiple tracers (metals, stable nuclides, or radionuclides) are in resolving emission sources. Other cases dealing with the Arctic and the North Atlantic provide budget calculations of trace elements using sulfur as an analogous element. A review of the trace-element data bases indicates that only fragmentary information was available and, therefore, it was only possible to make rough estimates. The highly variable character of atmospheric space and time makes it difficult to extrapolate the existing data bases to cover larger regions or times with any great confidence. However, this working group concluded that, at the time of the workshop, the problems of sampling and analyzing low concentrations encountered by past data-gathering efforts might be lessening which would enable better estimates to be made.

This group also found that they could not evaluate the impact of the atmospheric deposition of trace elements to the world's ocean because so little was known about the physico-chemical speciation of trace elements in aerosols and rain. Also, a poor understanding of the particle size distribution of aerosols presented much difficulty for designing models to evaluate the transport and deposition of trace elements.

13.3. MINERAL AEROSOLS

The sources of dust are the arid and semi-arid areas of the world--areas in the low midlatitudes of the Northern Hemisphere that extend across North Africa, the Middle East, and Asia. The generation and transport of dust rely on the production of fine-grained material through weathering,

A. H. Knap (ed.), The Long-Range Atmospheric Transport of Natural and Contaminant Substances, 303–305.

mobilization by wind, and transfer to the free troposphere by convective processes where the material then undergoes long-range transport.

The working group on mineral aerosols agreed that there are only a few places where long-range transport has been documented: the tropical North Atlantic and the Pacific Ocean (see Chapter 3, p. 59). In both cases, the evidence suggests that long-range continent-to-ocean transport exists and that the atmospheric source is an easily recognizable signal in the Pacific and Atlantic Oceans. Again in both studies, islands are used to investigate the transport as they do not have a continental dust source. There is also evidence that Saharan dust events undergo continent-to-continent transport, mainly in the summer. Measurements of mineral aerosol collected in Miami indicate transport of Saharan dust. It is not known whether the Asian dust does impact another continent but it certainly impacts the Pacific Ocean.

Although approximately 80-90% of aerosols in the atmosphere are removed by wet deposition, because of the high gravitational settling rate, dry deposition accounts for a significant fraction of the depositional flux. As much as 20-40% of the mineral-aerosol flux might be attributable to dry deposition. The deposition rates are extremely sporadic. In some cases day-to-day and year-to-year variability can be as high as two orders of magnitude. Therefore, the best estimates for evaluating deposition can only be achieved with long-term, continuous, time-series data. Once released from its source, mineral aerosol undergoes dramatic physical changes during long-range transport (Chapter 10, p. 197). The mean radius of the aerosol mass distribution rapidly shifts to the smaller particles. Large particles do not usually survive this very long transport (i.e., 1,000 km) but there are exceptions (see Chapter 3, p. 59). The best understood area is the Pacific because of the intensive SEAREX program as well as the work done by some European groups and the Japanese. There is little data on vertical distribution but it is assumed that the material is transported in the midtroposphere. It does appear that in the Northern Hemisphere, Europe and North America play very small roles as sources of mineral aerosols that undergo long-range transport.

This group also concluded that the effects of mineral aerosol must be taken into account when modeling the radiation balance of the earth. Aeolian inputs can be a significant source of nonbiogenic material found in oceanic sediment. The dissolution of these mineral aerosols can also provide a significant input into the surface ocean. Although this is a crossover into the trace-metal section, it does appear that there could be an effect of micronutrient input into the surface ocean.

13.4. SULFUR AND NITROGEN

The working group on sulfur and nitrogen (Chapter 11, p. 231) conclude that sulfur is transported over long distances and can undergo continent-to-continent transport. Episodic meteorological events (such as El Nino) as well as spatial and temporal variabilities of sources, deposition processes, and chemical transformations determine the most favorable conditions for long-range transport. Sulfur and nitrogen compounds in the middle troposphere are available for rapid long-range transport.

The studies outlined in Chapter 11 (p. 231) indicate that the use of fossil fuels in the Northern Hemisphere has had little impact on the nitrogen budget of the Southern Hemisphere. Although nitrogen compounds are transported to the western North Atlantic Ocean from North America and to the Pacific Ocean from Asia, there is little data to suggest that the downwind continents (Europe and North America, respectively) are significantly affected by these emissions. The transport and deposition of these nitrogen compounds to the ocean may be an important input of nutrient into the midoceanic oligotrophic gyres.

Volcanic emissions (Chapter 8, p. 163), biogenic emissions, sea-salt emissions, and industrial emissions are all major sources of atmospheric sulfur. Chapter 11 (p. 231) presents the evidence that sulfur is carried into the troposphere and then undergoes rapid long-range transport. Perhaps the best studied case covers the Atlantic Ocean where substances have been transported from North America as far as Europe. The Arctic haze phenomenon is an example of the transport of sulfur compounds from Europe to the Arctic (Chapters 6 and 11, pp. 137 and 231). These sulfur and nitrogen species are important both as tracers of atmospheric motion but they also impact the chemical reactivity of the troposphere and the overall biogeochemical cycles of sulfur and nitrogen.

13.5. ORGANIC COMPOUNDS

The organics working group had a difficult job because of the literally thousands of organic species to be considered. Their job was further complicated because so many of the species are highly reactive and are transformed into degradation products, which have significant impacts in their own right. For these reasons, this group investigated compounds that had (1) a major continental source, (2) a long enough atmospheric residence time to be transported hundreds to thousands of kilometers, and (3) a potential impact on remote areas. They agreed to assess the state of knowledge of emissions, transport, transformations, and deposition processes for seven groups of compounds: nonmethane hydrocarbons, low-molecular-weight halocarbons (excluding freon, CCl_4, and longer long-lived halocarbons), carboxylic acids, polynuclear aromatic hydrocarbons, chlorinated pesticides and PCBs, high-molecular-weight oxygenated compounds, and polymeric and soot carbon. Although in most cases there was a paucity of data, some compounds can act as tracers. For example, many of the chlorinated hydrocarbons and pesticides are found all over the atmospheres even in extremely remote locations. The accurate measurement of these compounds is still difficult on a worldwide basis. Because information on vertical distribution is so scarce, knowledge of continent-to-continent transport depends entirely on the type of compound involved. This group focused far more on the current state of knowledge and the gaps in this knowledge to determine the magnitude of continent-to-continent and continent-to-ocean transport.

14. GAPS AND FUTURE RESEARCH RECOMMENDATIONS

Anthony H. Knap
Bermuda Biological Station for Research, Inc.
17 Biological Station Lane
Ferry Reach GE01, Bermuda

14.1. GAPS

There were many overlapping gaps in our knowledge of the sources, transport, transformation, and deposition of the four groups of compounds
discussed during our workshop in Bermuda. In a plenary session on the
last day, we tried to synthesize these into common themes.

14.1.1. Sources

All groups keenly felt the striking lack of source data. The working
group on sulfur and nitrogen agreed that, although some information on
sulfur is available, data on organic nitrogen are lacking. The group on
trace elements knew more about sources in Europe than North America.
There is generally more information available on the Northern Hemisphere
than on the Southern Hemisphere.

14.1.2. Transport

For all compound classes, most data bases are from ground-based measurements and, therefore, data are needed on vertical distributions.

14.1.3. Transformations

Data for sulfur are better than for nitrogen. More information is needed
on high-molecular-weight organic sulfur compounds and on organic nitrogen
compounds. We know every little on the chemical interaction of mineral
aerosols with trace metals. There are large gaps in our knowledge of
multi-phase reactions and on photolysis reactions of all groups.

14.2. RECOMMENDATIONS FOR FUTURE RESEARCH

All workshop attendees were asked to identify what was hindering them the
most in trying to understand long-range transport, where the lack of data
was the most pronounced, and what could be done to remedy the situation.
Each group presented its responses in the form of the basic recommendations presented below (see Chapter 9-12, pp. 163-259, for details).

A. H. Knap (ed.), The Long-Range Atmospheric Transport of Natural and Contaminant Substances, 307-309.
© 1990 Kluwer Academic Publishers.

14.2.1. Trace Elements

More data are needed on the magnitude, location, and speciation of trace-element emissions. It is essential to determine trace-element particle size distributions in the remote atmosphere.

Data on the vertical distribution of trace elements are needed to validate a sulfur-trace-element analogy. Emission estimates could then be combined with atmospheric transport models to complement and corroborate field measurements.

Various physicochemical facts concerning trace elements need to be ascertained, such as those on solubility, the influence of acidity on solubility, the chemical forms of trace elements in aerosols and precipitation, and the particle size distributions of aerosols. To determine what impact trace-elements are having on the world's oceans, at-sea precipitation needs to be estimated with greater accuracy.

14.2.3. Mineral Aerosols

It was the consensus of this working group that an intensive 5- to 10-year field experiment should be established to study long-range transport over West Africa and the tropical Northern Atlantic. Extensive measurements should be taken at ground stations in source and receptor areas, from aircraft flying over source areas, from ships in receptor areas. This experiment should also include a remote-sensing program and the involvement of modelers would be mandatory.

This program would be designed to acquire data for vertical profiles of mineral dusts and related parameters and on mineral-dust deflation; an atlas of meteorological data would be established to develop a predictive ability for dust storms. A sun-photometer network would be included to monitor aerosol optical depth on a global scale. Specific studies would address the physicochemical properties of mineral aerosols, especially the surface sorptive property of aerosols and the solubility of aerosol components; the large, giant, and ultragiant atmospheric particles; and the relationship of transporting wind speed to particle size.

14.2.4. Sulfur and Nitrogen Compounds

This working group suggested the following directions for future research:

Shipboard measurements should be taken to cover regions where there are no data. A program that includes field measurements and laboratory studies is needed to quantify all aspects of the biological sources of sulfur so that the reaction path of DMS oxidation can be clarified. Measurement programs need to be established in major source regions and anthropogenic sources need to be inventoried. Regional chemical transport models should be developed to establish an effective data base on source strengths. Nitrate and its precursors must be measured in emissions in the boundary layer and free troposphere of coastal and remote regions; these measurements should then be used to develop chemical models of the NO_x conversion. There should be continued support for

global atmospheric chemistry models that treat fundamental transport
processes realistically.

14.2.5. Organics

The organic working group made their recommendations for future research
in the form of the understanding of three questions. The basic needs are
as follows:

Low-molecular-weight compounds, their precursors, and their
photooxidation products should be measured simultaneously in multiphase
systems so that a comprehensive model could be designed to study transformation and its effect on tropospheric cycles. An expanded data base
must be built on the mechanisms and products of major atmospheric loss
processes especially in remote oceanic atmospheres where NO_x concentrations are low. The global distribution of organic nitrogen compounds in
the troposphere must be determined.

A better understanding of long-range transport could be gained by
studing the kinetics of the atmospheric loss processes of the heavy organic compounds, particularly photolytic and reactive loss processes for
particle-associated compounds. More data needs to be collected on all
aspects of the deposition of heavy organic compounds, including the process and significance of episodic events, the physical state of organic
chemicals in air and water in the atmosphere, the size distribution of
the organic compounds, and the deposition characteristics of the compounds themselves. The scales of existing models of pesticide fluxes
should be extended to regions and compartment models of contaminant
transfer, storage and loss need to be designed. The importance of atmospheric deposition over the input from other sources needs to be resolved.
The atmospheric burden of potentially toxic organic chemicals needs to be
estimated more accurately by drawing vertical profiles, expanding spatial
distribution studies, and including toxic compounds, which are produced
in quantity, in the compounds that are already being analyzed in atmospheric samples.

Transport models should be tested with data covering spatial distribution, composition, and rates of emission; 3-dimensional atmospheric
concentrations; loss processes; and ratios of atmospheric concentrations
for compounds with different lifetimes. Case studies need to be established covering well-defined meteorological regimes. Single back-
trajectory analyses should be used to study the transport of these
compounds, including flow climatology to describe transport paths. Ongoing research should include more work on the molecular and isotopic
markers of individual and regional sources.

Atmospheric Concentration Unit Conversions
(To convert, multiply by the factors in this table)

FROM \ TO	$nmol\ m^{-3}$	$ng\ m^{-3}$	$molec\ cm^{-3}$	$ppbv$	$ppbx$	$ppbm$
$nmol\ m^{-3}$	-----	M_i	6.023×10^{8}	$\dfrac{RT}{P}$	RT	$\dfrac{M_i\cdot RT}{pM_{air}}$
$ng\ m^{-3}$	$\dfrac{1}{M_i}$	-----	$\dfrac{6.023\times10^{8}}{M_i}$	$\dfrac{RT}{pM_i}$	$\dfrac{RT}{M_i}$	$\dfrac{RT}{pM_{air}}$
$molec\ cm^{-3}$	1.66×10^{-9}	$1.66\times10^{-9}\,M_i$	-----	$\dfrac{1.66\times10^{-9}\,RT}{P}$	$1.66\times10^{-9}\,RT$	$\dfrac{1.66\times10^{-9}\,M_i\,RT}{pM_{air}}$
$ppbv$	$\dfrac{P}{RT}$	$\dfrac{pM_i}{RT}$	$\dfrac{6.023\times10^{8}\,P}{RT}$	-----	P	$\dfrac{M_i}{M_{air}}$
$ppbx$	$\dfrac{1}{RT}$	$\dfrac{M_i}{RT}$	$\dfrac{6.023\times10^{8}}{RT}$	$\dfrac{1}{P}$	-----	$\dfrac{M_i}{pM_{air}}$
$ppbm$	$\dfrac{M_{air}\cdot P}{M_i\,RT}$	$\dfrac{M_{air}\cdot P}{RT}$	$\dfrac{6.023\times10^{8}\,M_{air}\cdot P}{M_i\,RT}$	$\dfrac{M_i}{M_{air}}$	$\dfrac{pM_{air}}{M_i}$	-----

Source: Compiled by A. Pszenny, NOAA/AOML, 4301 Rickenbacker Causeway, Miami, FL 33149.

Notes:
P = pressure (atm)
T = temperature (K)
R = 8.2054×10^{-5} $m^{3}\ atm\ mol^{-1}\ K^{-1}$
M_i = molecular weight of i (g mol^{-1})
M_{air} = molecular weight of dry air (28.9 g mol^{-1})
1 atm = 1013.3 mb = 1013.3 hPa = 29.92" Hg = 760 mm Hg

T_m = mean T between h_1 and h_2
h_1, h_2 = altitudes (m), $h_2 > h_1$
P_1, P_2 = pressures, $P_2 < P_1$
C = nmol, ng, or molec
V = m^{-3} or cm^{-3}

Altitude correction:

$$h_2 - h_1 = 67.4\ T_m\ \log_{10}(P_1/P_2),\ \text{where}$$

Pressure and temperature correction:

$$(C/V)_{T_2,P_2} = (C/V)_{T_1,P_1}\cdot (T_1/T_2)\cdot (P_2/P_1),\ \text{where}$$